Electrical Energy Storage Technologies and Applications

Xisheng Tang • Zhiping Qi • Li Kong

Electrical Energy Storage Technologies and Applications

Springer

Xisheng Tang
Institution of Electrical Engineering
Chinese Academy of Sciences
Beijing, China

Zhiping Qi
Institution of Electrical Engineering
Chinese Academy of Sciences
Beijing, China

Li Kong
Institution of Electrical Engineering
Chinese Academy of Sciences
Beijing, China

ISBN 978-981-96-5624-0 ISBN 978-981-96-5625-7 (eBook)
https://doi.org/10.1007/978-981-96-5625-7

Jointly published with China Machine Press Co.,Ltd.

The print edition is not for sale in China (Mainland). Customers from China (Mainland) please order the print book from: China Machine Press Co.,Ltd..

Translation from the Chinese language edition: "电力储能技术及应用" by Xisheng Tang et al., © China Machine Press Co.,Ltd. 2020. Published by China Machine Press Co.,Ltd.. All Rights Reserved.

The original submitted manuscript has been translated into English. The translation was done using artificial intelligence. A subsequent revision was performed by the author(s) to further refine the work and to ensure that the translation is appropriate concerning content and scientific correctness. It may, however, read stylistically different from a conventional translation.

This Springer imprint is published by the registered company Springer Nature Singapore Pte Ltd.
The registered company address is: 152 Beach Road, #21-01/04 Gateway East, Singapore 189721, Singapore

Preface

The key to stable and efficient operation of the power system is to handle the instantaneous balance and spatiotemporal coordination of energy. Since the large-scale integration of renewable energy generation is particularly prominent, energy storage is an important means to maintain this balance and coordination. In recent years, various new energy storage technologies have gradually become practical, such as advanced pumped storage, new compressed air energy storage, lithium-ion batteries, lead-carbon batteries, flow batteries, sodium-sulfur batteries, as well as flywheel energy storage and supercapacitors. They have their own unique technical and economic characteristics, greatly enriching the connotation of electrical energy storage technologies and adding more choices for its application.

At present, there are more and more research, demonstration, and operation about energy storage applied to the power system, and various energy storage technologies and their systems are constantly improving themselves in their suitable fields. However, due to the involvement of multiple disciplines and specialties in the electrical energy storage system, how to choose suitable energy storage technology according to different application requirements, design a reasonable integration system, and achieve efficient control are important guarantees to improve its technical and economic performance.

This book starts with the application requirements of power systems and introduces typical electrical energy storage technologies, system composition, and control architecture, combined with the application of energy storage in microgrids, virtual power plants, renewable energy power generation fluctuation suppression, and frequency/voltage regulation. It focuses on system design, operation control, and other aspects and provides the big data analysis and application methods for energy storage system.

This book is a summary of the research conducted by the Power System Laboratory at the Institute of Electrical Engineering, Chinese Academy of Sciences, over the years. Teachers and students including Feng Zhiyue, Tong Jianzhong, Pei Wei, Wei Tongzhen, Huang Shengli, Zhou Long, Huo Qunhai, Deng Wei, Liu Wenjun, Miao Fufeng, Zhou Guopeng, Sun Yushu, Zhang Tianjiao, Li YuXuan, Li Ningning, Zhang Guowei, Shi Changli, Gao Chao, Hu Xiao, Zhang Guoju,

Yin Zhenggang, Wang Jianwei, Fan Menghan, and others have made great contributions to the research of this book. Thanks to editor Fu Chenggui of China Machine Press for her strong support for this book. Thanks to the strong support of many experts and enterprises in the field.

Given the author's limitations, and considering the rapid development of energy storage and power systems, some inadequacies are inevitable. We welcome any criticism and corrections.

Beijing, China Xisheng Tang
Beijing, China Zhiping Qi
Beijing, China Li Kong

Introduction

This book focuses on electrical energy storage systems and their application technologies, condensing the authors' theoretical knowledge and practical experience in the fields of electrical energy storage, distributed power generation, microgrids, and smart grids over recent years. This book is divided into three parts. The first part provides an overview, which first faces the transformative development of power systems, analyzes the role of energy storage in power system frequency regulation, peak-valley shaving, renewable energy consumption, transmission and distribution congestion management, distributed power generation and microgrids, and introduces the technical principles and development application situation of some typical energy storage technologies. The second part focuses on the system composition, grid connection topology, and control technology of battery and flywheel energy storage and analyzes the theoretical basis and control methods of composite energy storage, as well as the big data analysis methods and their applications in energy storage. The third part analyzes the dual-mode operation and peer to peer control methods of microgrids based on energy storage, the methods for smoothing and frequency/voltage regulation characteristics of renewable energy power generation based on energy storage, and the optimization scheduling methods of virtual power plants with energy storage and distributed power generation.

The purpose of this book is to envision the future development of the power system and build a bridge between energy storage technologies and the power system, with suitable energy storage and better technical and economic performance, to meet the application needs of the power system.

Overview

This section revolves around the development of future power systems, introduces the role and typical applications of energy storage in the main scenarios of the power system, and briefly introduces its technical principles and typical applications in combination with the current mainstream technologies of electrical energy storage.

Contents

Part I
Overview

Chapter 1
The Role of Energy Storage in Power Systems

The traditional power system is a continuous operation system that integrates power production, transmission, distribution, and consumption. The application of energy storage adds a link to store electrical energy to the traditional power system, transforming the power system from a "rigid" system to a "flexible" system, greatly improving the safety, flexibility, and reliability of the power system [1–3]. Especially, facing the inherent needs of the development of renewable energy scale integration and consumption, smart grid and energy internet, energy storage is given a "cornerstone" role positioning.

Energy storage, this book mainly refers to electrical energy storage, which is technically divided into mechanical energy storage, electromagnetic energy storage, and electrochemical energy storage. Mechanical energy storage converts electrical energy into mechanical energy for storage, and then reconverts it into electrical energy when needed, mainly including pumped storage, compressed air energy storage, and flywheel energy storage. Electromagnetic energy storage converts electrical energy into electromagnetic energy for storage, mainly including superconducting energy storage and supercapacitor energy storage. Electrochemical energy storage converts electrical energy into chemical energy for storage, and currently widely used ones include lead-acid batteries, lithium-ion batteries, flow batteries, and sodium-sulfur batteries.

Different energy storage technologies vary greatly in capacity level, charge and discharge ability, cycle life, efficiency, and cost, so their applicable application scenarios also vary greatly. Currently, the most widely used are pumped storage, lead-acid batteries, and lithium-ion batteries, and the research and application of other energy storage are also rapidly advancing.

Theoretically, energy storage can play an important role in all links of the power system's "generation, transmission, distribution, and use", can improve the stability, reliability, and quality of power system operation, can improve the utilization rate and operational economy of power assets, and can enhance the acceptance capacity of renewable energy. Of course, whether energy storage can be applied on a large

X. Tang et al., *Electrical Energy Storage Technologies and Applications*,
https://doi.org/10.1007/978-981-96-5625-7_1

scale in which link of the power system depends on its own technical and economic characteristics, the support of the power market, and the competition with other technical means.

1.1 Participate in Power System Auxiliary Services

Power system auxiliary services are for balancing small energy supply and demand differences in a very short period of time and responding to sudden events in the system, including frequency regulation reserve and operating reserve. Auxiliary services are generally completed by the centralized auxiliary service market, and resources that promise to provide service reserves, including generator sets, adjustable loads, energy storage devices, etc., are compensated according to the market clearing capacity and price [4] .

Since the purpose of configuring auxiliary service functions is to solve the problem of energy supply and demand balance in a short time, the total auxiliary service capacity is relatively small compared to the total load of the system, generally not exceeding 15% of the total load, of course, this depends on the actual power system's power source and grid structure.

Frequency regulation is divided into up-regulation and down-regulation, which is used to continuously and automatically balance very small energy supply and demand deviations, that is, imbalance energy, in a very short time. It is usually provided by generator sets equipped with AGC control devices within the market control area, and can also be provided by demand-side resources that can respond to AGC signals, such as demand response resources and energy storage. Operating reserve is to cope with load increases or sudden events in the system, which can be divided into: rotating or synchronous reserve, provided by online generators that have the ability to increase power; non-rotating reserve or asynchronous reserve, provided by generators that are not online but can start and provide power within a given time, usually within 10–30 min.

Frequency regulation mainly includes primary frequency regulation and secondary frequency regulation. Primary frequency regulation is a differential regulation spontaneously completed by the load in the system and generator sets with rotating reserve capacity; secondary frequency regulation mainly controls the frequency and tie-line power through real-time adjustment of the active power of the frequency-regulating power source in the power grid, balances the power of the regional power grid, and achieves non-differential regulation.

The general power grid frequency regulation demand is mainly provided by coal-fired thermal units, hydropower units, and gas turbines. Thermal power and hydropower respond to grid frequency changes by continuously adjusting the unit output power, achieving frequency regulation of the power system. However, whether it is a hydropower frequency regulation unit or a thermal power frequency regulation unit, they are all composed of rotating mechanical parts, and the effects of mechanical inertia and wear will affect the safety and quality of the grid frequency [5]. For instance, thermal power units have a relative long response delay, making them unsuitable for shorter frequency regulation cycles, and they face issues of insufficient frequency regulation due to thermal storage constraints. Meanwhile, the frequency regulation capacity of hydropower units is easily constrained by geographical and seasonal factors. At the same time, traditional power sources need to consider the unit's limitations on the frequency of changes in the amplitude and polarity of the response power during control, and even set a limit for the duration of the power signal in the same direction, blocking the reverse power signal during this period. All these restrictions can lead to delays, deviations, and reversals in regulation, resulting in inaccurate responses to frequency regulation signals.

Energy storage systems can reduce the imbalance of active power in the power system or regional control deviations to a certain extent through charging and discharging, thus participating in primary and secondary frequency regulation. Compared with the shortcomings of traditional power sources in power system frequency regulation, energy storage systems have certain technical advantages [6–8]:

1. Fast response speed, capable of full power output within hundreds of milliseconds, fully meeting the power change requirements within the frequency regulation time scale;
2. High control accuracy, energy storage can quickly and accurately track dispatch instructions, correspondingly reducing the power reserve margin of frequency regulation;
3. High operating efficiency, energy storage systems, especially various types of battery energy storage systems, have high charging and discharging efficiency, resulting in low losses during frequency regulation;
4. Capable of bidirectional regulation, energy storage systems can alternate between up-regulation and down-regulation without limitations, demonstrating high regulation capability.

Therefore, using energy storage systems for frequency regulation allows the frequency regulation curve to track the instruction curve well, avoiding issues of reverse regulation, regulation deviation and delay. Moreover, compared to thermal power units, the main value of energy storage in frequency regulation lies in its flexible control and efficient operation. For example, research has shown that the frequency regulation capability of flywheel energy storage is 1.7 times that of hydropower units, 2.7 times that of gas units, and nearly 20 times that of thermal power units and Gas-steam Combined Cycle Unit [9].

In terms of operating costs, since the process of AGC up and down regulation can be approximated as an energy balance process, the operating cost of energy storage systems is relatively low, mainly composed of the energy loss cost and maintenance cost of the energy storage system itself. Although it is still difficult to estimate the impact of energy storage on the total frequency regulation cost of the power system, its fast and accurate tracking characteristics can significantly reduce the rotating reserve capacity required by the power system. The saved rotating reserve capacity can be used for grid peak shaving, accident backup, etc., generating certain indirect benefits. Therefore, compared with traditional power sources, the technical and economic advantages of energy storage systems participating in frequency regulation are quite obvious. Therefore, deploying a certain amount of energy storage systems for frequency regulation in appropriate scenarios can effectively enhance the overall frequency regulation capability of the grid dominated by thermal power units, improve the pass rate of frequency and ACE control, and thus ensure the safety and stability of the grid.

Many demonstration verifications have been carried out in the industry for large-scale energy storage to participate in grid frequency regulation and other ancillary services. Since 2008, many demonstration projects have been completed, involving various types of energy storage such as lithium-ion batteries, with single station capacities gradually increasing from 1 MW to tens of MW [10–12], providing valuable exploration in system design, equipment performance, operation and maintenance, and standards formulation. In recent years, some commercial operation projects have gradually started construction and operation, and with the advancement of ancillary service market reforms in some areas, energy storage is playing an increasingly prominent role in regional grid ancillary services.

1.2 Energy Storage Participation in Power System Peak Regulation

Ensuring the reliability and quality of the power supply to meet the load requirement is a long-term task of the power system. Due to the uncontrollability and randomness of the load, the power system should have the ability to meet the load demand at any time. However, the demand for electricity by users varies greatly between day and night, and between different seasons, and the increasing proportion of renewable energy generation will further exacerbate this phenomenon. This requires the power system to reserve a large amount of spare capacity to meet the peak load, resulting in low operating efficiency of power equipment. Effectively regulating the peak-valley difference of the power system and improving the load rate is an important means to improve the asset utilization rate of the power system.

Peak-regulating power sources deliver electrical energy to the grid during peak electricity usage periods and obtain electrical energy from the grid during low electricity usage periods, achieving "peak shaving and valley filling" and regulating grid load. The role of peak-regulating power sources in modern power systems is becoming increasingly important and indispensable, and is a necessary means to achieve safe, reliable, economical, and efficient grid operation.

1.2.1 Conventional Peak Regulation Measures

Since the peak-valley load of the system can be accurately predicted, the peak regulation problem can be solved by the daily plan made by the day-ahead energy market or the power dispatching department. At present, the main methods used in the power system for peak shaving and valley filling are thermal power units, hydroelectric units, demand management, and pumped storage.

1. Using thermal power units for peak- valley regulation.
   The current power supply structure in China is mainly based on thermal power. Adjusting thermal power units to adapt to the changes in peak and valley loads is the main method of peak-valley regulation in the current power grid. However, there are the following problems with peak- valley regulation using thermal power units:
   First, the economy of using thermal power units for peak- valley regulation is low. Frequent start-stop and deep peak regulation of thermal power units will greatly increase the use of ignition and auxiliary fuel oil. At the same time, during peak and valley regulation, the operation of thermal power units will deviate from the economic operation point, causing the overall economy of thermal power units to decline.
   Second, using thermal power units for peak- valley regulation will increase the failure probability of thermal power units. Thermal power units that start and stop repeatedly for peak regulation are prone to various equipment problems, increasing maintenance workload and maintenance costs.
   Third, the adjustment speed of thermal power units is relatively slow, making it difficult to meet the requirements of quick changes in power system load.
   In addition, from the perspective of grid planning, a grid that relies solely on thermal power units for peak- valley regulation often needs to increase installed capacity to meet peak regulation requirements, which inevitably leads to excessive idle capacity in the system and low asset utilization.
2. Using hydropower units for peak- valley regulation.
   Compared with thermal power units, hydropower units have fast start-stop speeds, good economy, less pollution, and are suitable for use as peak-regulating power sources. However, hydropower has a disadvantage in that the power generation capacity varies greatly between abundant water periods and dry seasons, and the phenomenon of water abandonment for peak regulation often occurs, resulting in a lot of waste.
3. Using load management for peak- valley regulation.
   Through load management, the peak-valley difference of the power system can be regulated. The use of time-of-use electricity prices can motivate users to actively change their consumption behavior and electricity usage habits, while reducing electricity consumption and power demand.

4. Using pumped storage power stations for peak- valley regulation.
   Pumped storage power stations use the method of storing energy by pumping water during low electricity demand periods and releasing water to generate electricity during peak demand periods for peak-load regulation. The regulation capacity of pumped storage power station units is large, and they have the characteristic of rapid start-stop, making them a high-quality means of peak-load regulation in the power system. However, the construction of pumped storage power stations requires specific geographical conditions, and the construction period is long, and the project investment is large, which brings certain restrictions to the development of pumped storage power stations.

1.2.2 User-Side Energy Storage for Peak Regulation

By using energy storage systems located in the load side, under the guidance of time-of-use or real-time electricity prices, active peak shaving or peak shifting can be achieved by controlling user electricity consumption, which can save electricity costs for users and simultaneously achieve the effect of peak regulation for power system. The advantages of energy storage participating in peak regulation include:

1. Fast response. Energy storage devices have bidirectional power regulation functions, and their charging and discharging conversion speeds can reach below the millisecond level, much faster than traditional power sources.
2. High efficiency. The charging and discharging cycle efficiency of various battery energy storage systems is generally high, and the loss of electricity used for peak regulation is small.
3. Small losses. Energy storage can be distributed on the user side, directly matching with nearby loads temporally and spatially, which can avoid network losses from long-distance transmission.

Taking a processing enterprise operates round the clock on three shifts as an example, an analysis of the installation of an energy storage system is conducted. Figure 1.1 shows the time-of-use electricity prices for this enterprise, and Fig. 1.2 shows its typical weekday load curve.

In response to the electricity consumption characteristics of enterprise, a battery energy storage system is used for peak shaving and valley filling, and its operation sequence is shown in Fig. 1.3.

By adopting the charging and discharging strategy shown in Fig. 1.3, the peak shaving and valley filling of enterprise electricity consumption can be well achieved, thereby reducing electricity costs. With the maturation of battery energy storage system technologies and cost reduction, as well as the support of power market, user-side energy storage is showing more and more development prospects.

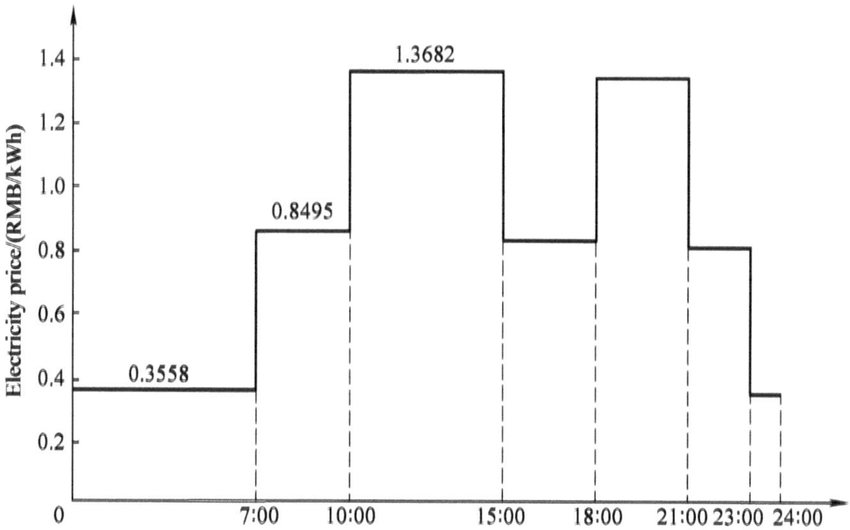

Fig. 1.1 Time-of-use electricity prices for a certain enterprise

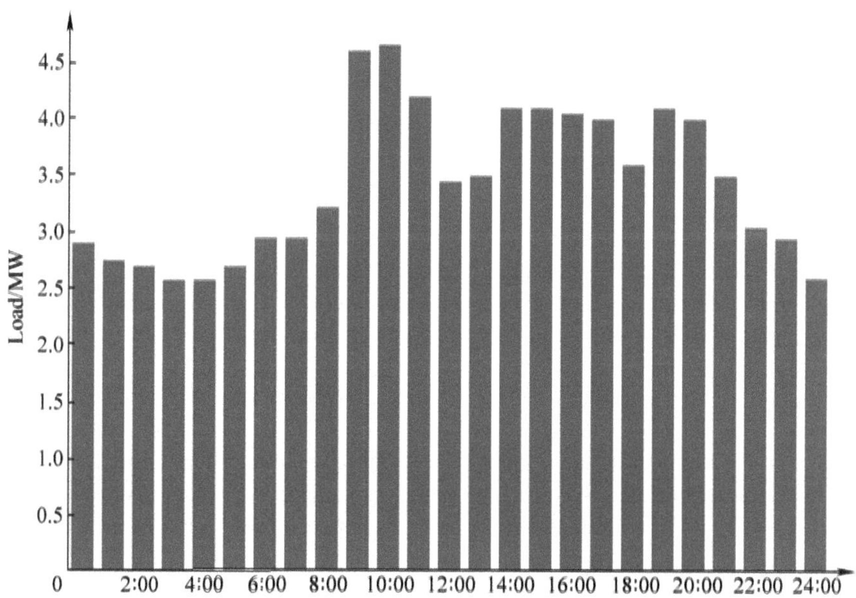

Fig. 1.2 Typical weekday load curve of a certain enterprise

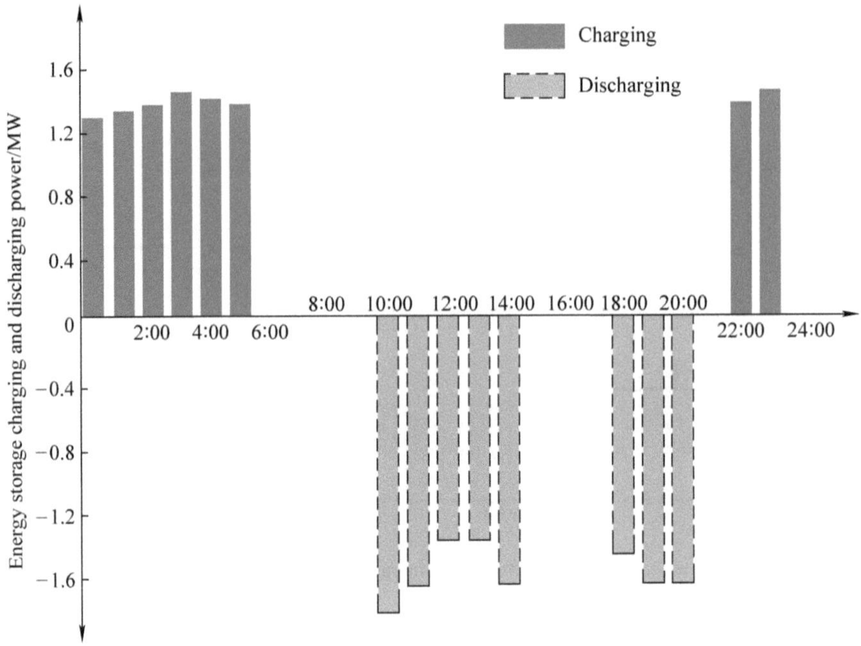

Fig. 1.3 Charging and discharging sequence of the battery energy storage system

1.3 Improving the Absorption Capability of Renewable Energy Generation

After many years of rapid development, the proportion of renewable energy generation in the power system is getting higher and higher. According to the power planning of various countries around the world, renewable energy will gradually change from supplementary energy to alternative energy. The large-scale development of renewable energy puts higher demands on the power grid, and improving the absorption capability of renewable energy is also a problem that needs to be solved urgently.

Due to the fluctuation and unpredictability of renewable energy generation such as wind power and photovoltaics, the power system needs to maintain regulation capacity, respond to sudden event capacity, and high power supply reliability and power quality level. It requires more accurate weather forecasts, stronger unit ramping capabilities, load tracking capabilities, and larger frequency regulation and operating reserve service capacities. The improvement of the absorption capability of renewable energy generation mainly revolves around improving power source characteristics, managing transmission blockages, and system safety and stability control.

1.3.1 *Improving the Characteristics of Renewable Energy Generation*

By utilizing the characteristics of energy storage, such as flexibility and quick response, the power characteristics of renewable energy generation can be improved, such as smoothing power output fluctuations, tracking prediction errors and dispatched output power, participating in power system frequency and voltage regulation, etc. This improves the power characteristics of renewable energy generation and enhances its grid-friendliness [13] .

Figure 1.4 shows the effect of energy storage in smoothing wind power output fluctuations. As can be seen, after smoothing, the output power of wind power has been significantly smoothed (red curve). Of course, the fluctuation rate of wind farm output power will vary depending on the installed capacity and the characteristics of the regional power grid, which is also the basis for the capacity configuration of energy storage.

In order to improve the characteristics of renewable energy generation, the energy storage system needs to meet control requirements in both power and energy aspects, reflected in the size of the power output of the energy storage system and its duration of action. For short-term and high-frequency energy storage needs,

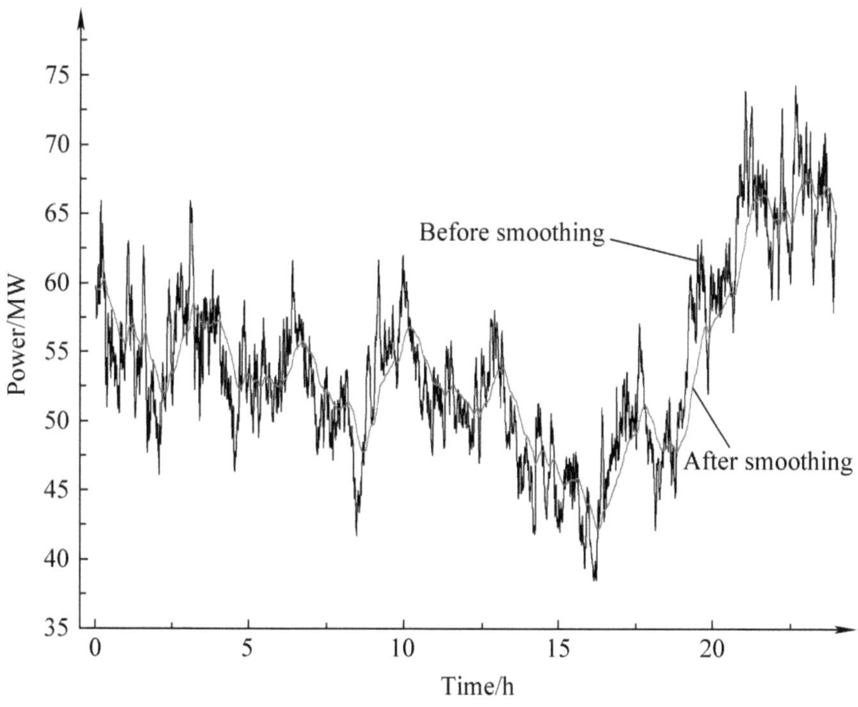

Fig. 1.4 Energy storage smoothing wind power fluctuations effect diagram [14, 15]

power-type energy storage can be competent; for long-term and low-frequency energy storage needs, energy-type energy storage can be competent.

In addition, since most energy storage systems are connected to the grid through power converters, they can fully utilize the characteristics of power electronics and their control, giving the energy storage system four-quadrant flexible operation capabilities, and when necessary, improve the voltage regulation capability of the grid coupling point through reactive power control.

1.3.2 Eliminating Wind and Solar Generations Abandonment through Time Shifting

Due to the limitations of transmission and distribution line capacity, renewable energy generation is often limited. However, whether it is wind power or photovoltaics, their high power output periods are often concentrated in a few hours of the day, and more time is in a state of no or little power generation. By using the time-shifting capability of energy storage, when wind or photovoltaic power generation is excessive, the abandoned wind or photovoltaic energy can be stored through energy storage, providing a peak-shifting-like function, alleviating the congestion of renewable energy power generation concentrated in a few hours of the day, improving the absorption capability of renewable energy power generation, and improving the utilization rate of transmission and distribution system.

In Fig. 1.5, when the illuminate is good and the photovoltaic power output is large (green curve), due to the limitation of the transmission line capacity, light needs to be discarded, at this time the discarded photovoltaic power can be used to charge the energy storage (blue curve), after the evening when the photovoltaic power generation weakens or stops, the energy storage will then discharge the stored power (purple curve), the overall power generation curve of photovoltaics and energy storage (yellow curve) can be controlled within the specified range.

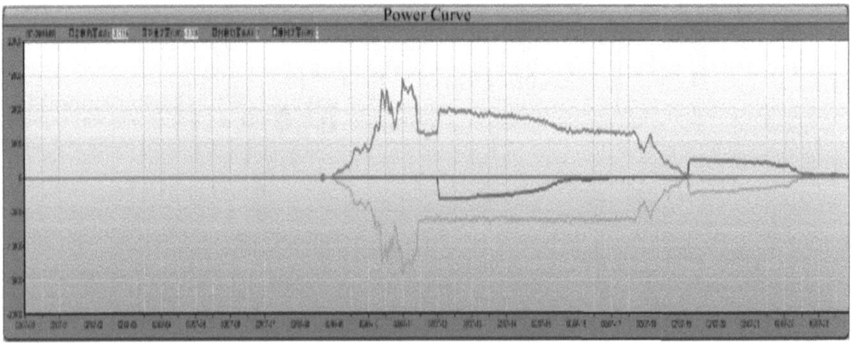

Fig. 1.5 Energy storage achieves photovoltaic time-shift absorption

1.3.3 Improving the Power Supply Adequacy of the Power System

As the load in the power grid is always fluctuating, and the start and stop of generator units also occur from time to time, the power balance in the system operation process is always relative, while imbalance is absolute. Therefore, in order to meet the power load demand in the system, it is necessary to ensure that the system has a certain power supply adequacy. In traditional power systems, power supply adequacy is guaranteed based on the static characteristics of the load. It aims at real-time balance of static power load, and is achieved through load forecasting, power source and power grid planning, and real-time dispatching of power source operation. Among them, the prediction of load static characteristics is the basis, and the planning of power sources and power grids is to reasonably plan the layout of power sources on the premise of "accurate" load prediction, so that in all possible operation modes, the system has enough spare power sources for dispatching.

The scheduling of backup power sources usually adopts three methods [16]: ① Start and stop power sources with online quick response characteristics as frequency regulation power sources (minute level); ② Start and stop power sources with quick response characteristics (such as hydropower, gas power generation, etc.) as load tracking power sources (hour level); ③ Start and stop power sources with slow response characteristics (such as thermal power, nuclear power, etc.) as base load power sources (day level).

In the future power system with high penetration of renewable energy, in addition to the power balance problems caused by load fluctuations and generator unit start-stop in traditional power systems, there are also fluctuations and uncertainties in the output of renewable energy, which pose new challenges to the power supply adequacy of the power system [17,18]. Taking wind power generation as an example, after the wind power is connected to the grid, the load in the system minus the wind power output constitutes the net load of the system, which must be provided by the output of conventional generator sets. In the power grid where the proportion of large-scale wind power output is large, the net load of the system shows static characteristics significantly different from the traditional power grid load. The wind power output causes the following two changes in the net load characteristics: ① The fluctuation rate and range of the net load increase; ② The uncertainty of the fluctuation rate and range of the net load increases.

Studies have shown that when the wind power penetration rate is relatively low, the change in net load characteristics will increase the system's demand for load tracking power sources; when the wind power penetration rate is high, the change in net load characteristics will further affect the system's demand for base load power sources, causing the load level of various types of conventional power sources to decrease and start-stop frequently, ultimately reducing the operating efficiency of power system. It can be seen that from the perspective of ensuring the power supply adequacy of grid operation, large-scale wind power grid connection will put higher requirements on the flexibility of the static output characteristics of conventional power generation in the system [3]. Energy storage has quick response capability and control flexibility, which can greatly improve the power supply adequacy of the power system.

1.4 Delaying Upgrade of Transmission and Distribution Lines

There are inevitably some lagging links in the transmission and distribution lines, facing the insufficient power supply capacity for special load demands, which may even cause failures in local system and extend to the upper and neighboring power grids. This problem can be solved by upgrading the transmission and distribution lines, but there may be problems such as large investment and low investment return, or on-site conditions restricting the upgrade.

After many years of upgrade, the power supply capacity and reliability of China's rural power grid have greatly improved. However, it still constantly faces some new problems, such as the surge in electricity consumption caused by migrant workers returning home during holidays; the tea frying and tobacco baking in some mountainous areas are also relatively concentrated in time and area. If the transmission and distribution investment is significantly increased to meet these electricity demands, the annual utilization rate of the increased transmission and distribution capacity is too low due to the short peak load time.

In some older operating areas in the city, such as commercial areas, financial areas, office buildings, etc., they are facing the problem of rapid growth of power load and heavy load of the original distribution lines. Due to the high load density in these areas and limited underground pipeline space, the possibility of system expansion is very small, or almost impossible. How to ensure the safe power supply during the peak electricity consumption period of the day is also a problem that needs to be solved urgently.

By configuring energy storage systems, we can precisely solve the bottleneck problem of power supply in the above-mentioned regions, delay or avoid the pressure of upgrading the existing power transmission and distribution system, and greatly improve the utilization rate of power equipment. The energy storage system can be installed and arranged compactly according to the needs of the load and the characteristics of the power transmission and distribution system, reducing the land and space required for the basic construction of the distribution system. This will effectively change the extensive construction mode of the existing power system and promote its transformation towards connotation enhancement.

Taking the typical daily load data of a 10 kV distribution station in a certain region as an example for analysis, Fig. 1.6 shows the active and reactive power curves of its typical daily load.

As can be seen from Fig. 1.6, the load is in the valley period from midnight to eight o'clock, the peak load period is from nine to twelve in the morning, and the

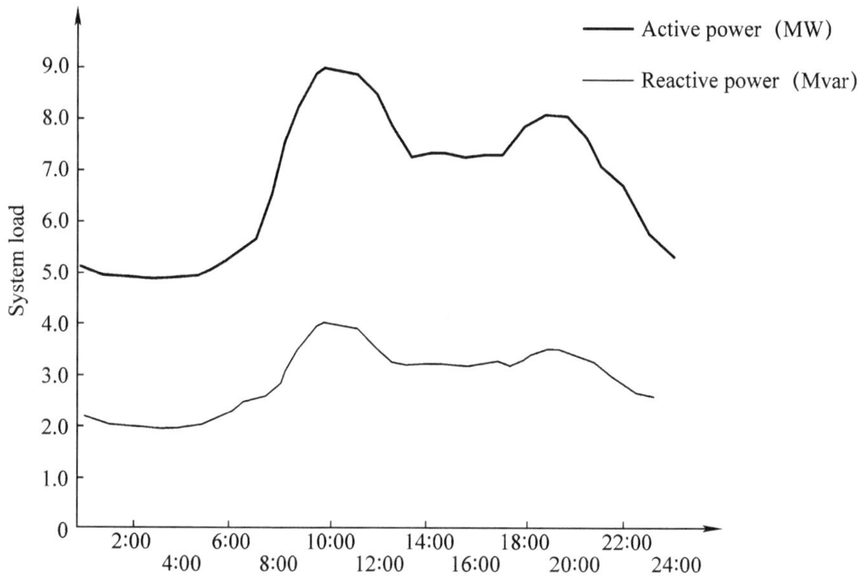

Fig. 1.6 Active and reactive power curves of typical daily load of distribution network

second peak load period is from six to eight in the evening. To reduce the pressure of the distribution system during the peak load period, it is possible to consider using battery energy storage to charge during the load valley period and discharge during the peak period. At the same time, to supplement reactive power, the energy storage should output as much reactive power as possible, as shown in Figs. 1.7 and 1.8.

After compensation with the battery energy storage system, the load curve of the distribution station is shown in Fig. 1.9. It can be seen that battery energy storage can effectively handle the power supply pressure of the local distribution system at specific times, avoiding overload operation of the distribution system. With the reform of the electricity market, various electricity sales entities start from the perspective of improving quality and efficiency, optimize the weak links of local power supply through energy storage, and optimize asset utilization efficiency, which is an effective method.

1.5 Distributed Generation and Microgrid

Distributed generation technologies, mainly distributed photovoltaics and dispersed wind power turbines, have been increasingly developed and applied due to their advantages such as proximity to users, high power generation efficiency, and well combination with various user electricity or energy needs. They have become one of the main application forms of renewable energy power generation and are also

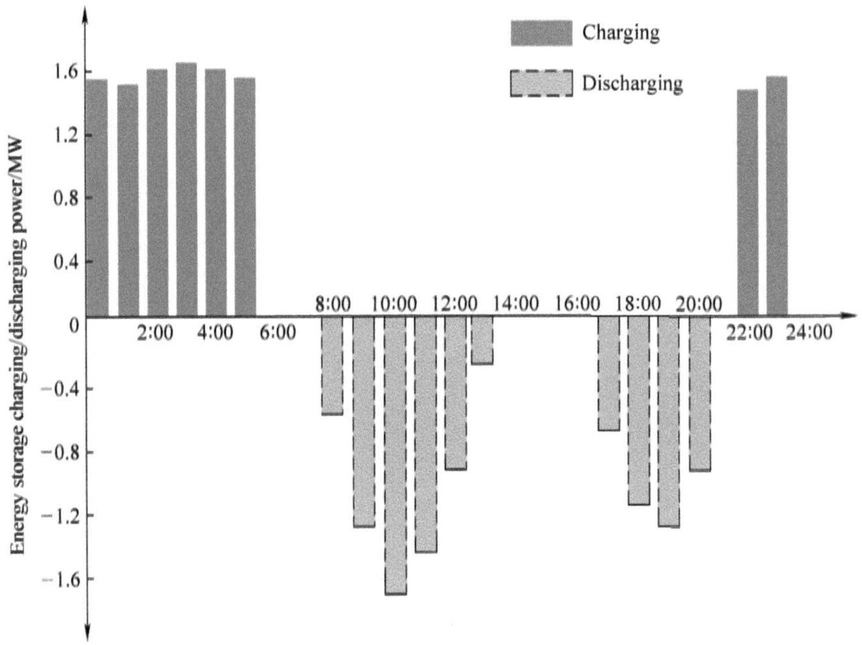

Fig. 1.7 Charging/discharging power of battery energy storage

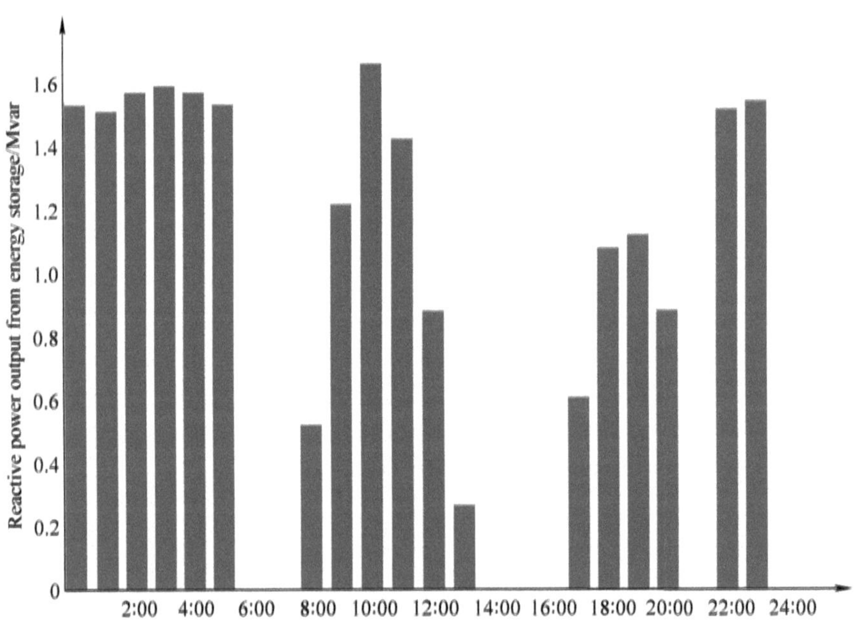

Fig. 1.8 Reactive power provided by battery energy storage

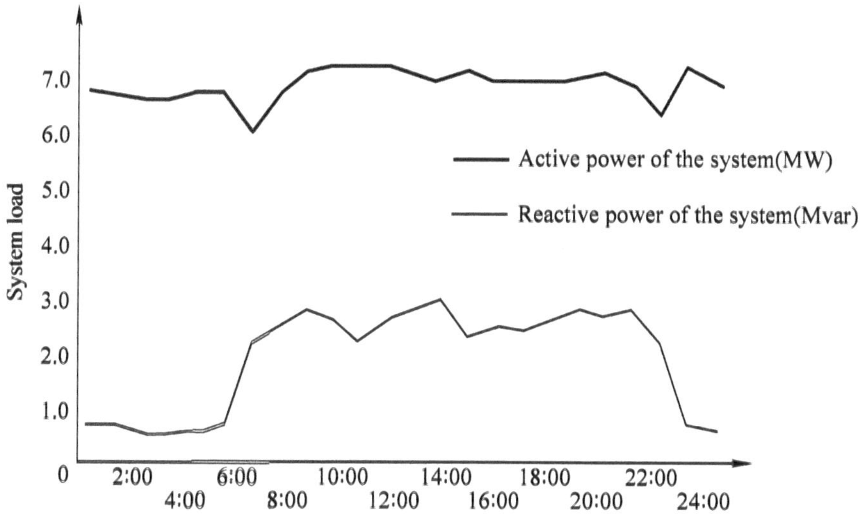

Fig. 1.9 Load curve of the system after using energy storage

important features and major challenges of future power systems. The large-scale development of distributed generation has changed the unidirectional power flow characteristics of traditional distribution networks, bringing changes to system operation control, relay protection, operation and maintenance, etc. Existing distribution networks need to adapt to the large-scale access of distributed generation, and new distribution networks need to better integrate distributed generation.

Energy storage has developed along with distributed generation. The distributed energy storage system, which is dispersedly located, can be well integrated with distributed electricity, solve the technical shortcomings of distributed generation itself, and play an important role in the safety and stability of the distribution network, power flow optimization, efficient operation, power trading, etc. Photovoltaic + energy storage will become one of the protagonists of the distribution network and electricity market.

The microgrid is an advanced organizational form for achieving efficient and highly reliable operation of distributed generation and load. The microgrid can operate on-grid and off-grid, and has its own energy management function, making it a "controllable unit" for the large grid. Especially for microgrids mainly powered by renewable energy, energy storage becomes an essential unit. Energy storage can regulate the imbalance of active and reactive power within the microgrid, improve the stability and power quality of the microgrid operation. Energy storage can optimize the power flow between the microgrid and the large grid, respond to dispatch instructions or maximize the benefits of power trading. When the large grid fails or is under maintenance, the microgrid can operate off-grid to ensure the reliability of power supply to important loads within it.

Virtual power plants emerged with the development of distributed power generation and the electricity market. They integrate distributed power sources, energy storage, and controllable loads through advanced communication technologies and software systems, forming an well-being aggregate and a main body of the electricity market to participate in grid dispatch or electricity market transactions. Virtual power plants integrate various distributed resources on a larger scale, allowing them to better integrate into the power system in terms of technical performance and operation mode, matching the operation control needs of the power system, and thus guiding the orderly development of renewable energy distributed power generation from a mechanism perspective. Energy storage, as an important unit and adjustment means of the virtual power plant, compensates for the shortcomings of distributed power generation and controllable load, and is the guarantee of the overall performance of the virtual power plant.

1.6 Combination of Main and Auxiliary, Showing Multiple Values of Energy Storage

With the large-scale development and utilization of renewable energy, the future power system is a complex system that includes thermal power, hydropower, wind power, photovoltaic, and energy storage. Energy storage will play an important supporting role in it, and its functions will run through all stages such as power generation, transmission, distribution, and demand side. The flexible controllability of energy storage in terms of energy and information enables it to participate more widely in applications such as frequency regulation, peak shaving, voltage regulation, emergency support, and smoothing of renewable energy fluctuations in the power system. The role of energy storage in the power system is shown in Table 1.1 [19–22] .

However, it should also be noted that energy storage systems often have high investment costs and limited construction sites. Therefore, whether from the perspective of investors or users, it should be treated as a scarce resource to fully realize its value. And all kinds of energy storage technologies, especially battery energy storage, also have the ability to achieve multiple functions at the same time. For example, energy storage configured on the demand side often mainly uses electricity price differences for peak shaving and valley filling to save electricity costs, but can also participate in the power system's demand response to obtain additional income, and can also provide emergency power supply for the user's important load, saving investment. Energy storage configured in wind farms and photovoltaic power stations can participate in power output fluctuation smoothing, tracking prediction errors and planned output, frequency and voltage regulation auxiliary functions, etc., to improve the grid-connected performance of renewable energy power generation. The combination of main and auxiliary functions better utilizes the technical and economic advantages of energy storage, improving the return on investment.

Table 1.1 The role of energy storage in the power system

Application field	Functional roles	Specific applications
Power generation side	Assist thermal power unit operation	Improve the efficiency of thermal power units participating in grid regulation; Can be used as a black start power source for thermal power units; Increase reserve capacity
	Improve the grid-connected absorption capability of renewable energy power generation	Smooth wind or photovoltaic output fluctuations, reduce prediction errors; Track wind or photovoltaic planned output; Time-shift absorption, reduce wind and solar power abandonment
	Substitute or delay the construction of new units	For areas with high peak load, energy storage can replace the construction of new power generation units, reducing investment
Grid side	Improve system operation stability	Increase the flexibility resources and system inertia of the power system, improve power supply quality, reliability, and dynamic stability
	Improve system operation economy	Optimize system power flow, reduce grid loss, improve system operation economy; Reduce the pressure on the grid during peak load times
	Delay grid upgrade and rebuilding	Energy storage can block the grid, delay the upgrade and rebuilding of the transmission and distribution system, and improve asset utilization
Demand side	Peak-valley shaving	According to the difference in peak and valley electricity prices, use energy storage for peak shaving and valley filling to reduce electricity bills
	Load tracking	Use energy storage to track user peak load, which can reduce electricity consumption and lower electricity costs
	Uninterruptible power supply	As a backup power supply to ensure uninterrupted power supply for important user loads, it can replace backup diesel generator sets
	Distributed generation and microgrids	Improve the operational stability of high-penetration distributed generation; Enhance power control and energy management capabilities in microgrids; Enhance the orderly grid-connection capability of distributed generation equipment
Auxiliary services	Frequency regulation	Can participate in the primary frequency regulation and AGC of the power system; Assist the frequency regulation operation of renewable energy generation, improving frequency regulation performance
	Voltage regulation	Can participate in AVC operation, improving system voltage stability and voltage quality Assist the voltage regulation operation of renewable energy generation, improving voltage regulation performance
	Reserve	Rotating and non-rotating reserve, enhancing the system's ability to deal with sudden disturbances and accidents
	Black start	Can be used as a black start power source

Chapter 2
Typical Power Storage Technologies

In a broad sense, energy storage includes the storage of basic fuels (coal, oil, natural gas, etc.), secondary fuels (gas, hydrogen, etc.), electrical energy storage, and heat storage. In a narrow sense, energy storage refers to a series of measures to store the energy produced by chemical or physical methods [24]. The scope of this book is electrical storage, i.e., electrical energy storage, which mainly includes pumped storage, compressed air energy storage, various electrochemical energy storage, flywheel energy storage, supercapacitors, etc.

2.1 Pumped Storage

Pumped storage is a special form of hydropower system that uses a certain amount of water as an energy carrier and provides electrical energy to the power system through the energy conversion between potential energy and electrical energy. Pumped storage power stations are equipped with upstream and downstream reservoirs. During low-load periods, the pumped storage power station operates in motor mode, pumping water from the downstream reservoir to the upstream reservoir for storage. During peak load periods, the pumped storage power station operates in generator mode, and the water stored in the upstream reservoir flows to the downstream reservoir through the turbine, driving the turbine to generate electricity. Since the water only circulates between the two reservoirs, after the first storage is completed, a large amount of water is not needed, only a small amount of water loss caused by evaporation and leakage needs to be replenished. The construction site of the pumped storage power station requires a high water head, large power generation capacity, small water leakage, short pressure water pipeline, and close to the load center [23,24] .

The characteristic of pumped storage is that it stores a very large amount of energy, and the energy storage release time is several hours to several days, with an

X. Tang et al., *Electrical Energy Storage Technologies and Applications*,
https://doi.org/10.1007/978-981-96-5625-7_2

overall efficiency of 70%–85%, which is very suitable for peak shaving in power systems and long-term occasions as reserve power sources [25]. From the development direction, pumped storage units are developing towards a better economy, high water head, and large capacity. Variable speed pumped storage units, seawater pumped storage units, and even pumped storage power stations built with underground reservoirs in plain areas are also the main directions.

As the only large-scale energy storage technology currently applied to power systems, pumped storage accounts for about 3% of the total installed capacity of power generation globally. As of 2018, the global installed capacity of pumped storage reached 170.7GW, accounting for 94.3% of the total installed capacity of energy storage [26], undertaking tasks such as peak shaving, frequency regulation, phase regulation, accident backup, and black start. Countries and regions such as Japan, the United States, and Western Europe entered the peak period of pumped storage power station construction in the 1960s and 1970s. The installed capacity of pumped storage power stations in some countries accounts for more than 10% of the total installed capacity of power generation, including France 18.7%, Austria 16.3%, Switzerland 12%, Italy 11%, and Japan 10%. The construction of pumped storage power stations in China started late, and it officially entered the development stage in the 1990s, and a batch of large-scale pumped storage power stations such as Guangzhou Pumped Storage Phase I, Beijing Thirteen Tombs, and Zhejiang Tianhuangping were built [3]. Since the "13th Five-Year Plan", the development of pumped storage in China has been rapid. As of the end of 2017, the scale of pumped storage power stations in operation in China was 28.49 million kilowatts, and the scale under construction was 38.71 million kilowatts. Both the installed capacity in operation and under construction rank first in the world [27]. At the same time, the construction technology of pumped storage power stations in China has reached the world's top level. Large-scale electromechanical equipment used to rely on imports, but after several years of technology introduction, digestion and absorption, it basically has production capacity of the equipment.

According to the 2014 notice from the National Development and Reform Commission on "Improving the Pricing Mechanism for Pumped Storage Power Stations", before the formation of the electricity market, pumped storage power stations implement a two-part electricity price. The capacity price mainly reflects the value of auxiliary services provided by the pumped storage power station, such as power reserve, frequency modulation, phase modulation, and black start, and is determined according to the principle of compensating the fixed costs and permitted profits of the pumped storage power station, gradually implementing the benchmark capacity price for newly commissioned pumped storage power stations. The electricity price mainly reflects the peak shaving and valley filling benefits achieved by the pumped storage power station through pumping power generation. It mainly compensates for the variable costs such as the power loss of the pumped storage power station. The price level is executed according to the local coal-fired unit on-grid benchmark electricity price. The pumping power provided by the power grid company to the pumped storage power station is priced at 75% of the coal-fired unit on-grid benchmark electricity price.

2.2 Compressed Air Energy Storage

Traditional compressed air energy storage systems are energy storage systems based on gas turbine technology. During charging, air is compressed and stored in the air storage chamber, converting electrical energy into the internal energy of the air; when electricity is used, high-pressure air is released from the storage chamber, enters the combustion chamber of the gas turbine for combustion, and drives the turbine to generate electricity. Traditional compressed air energy storage systems must be used in conjunction with gas turbine power stations and still rely on the combustion of fossil fuels to provide heat energy [28,29], and the energy storage efficiency is relatively low. In addition, similar to pumped storage power stations, compressed air energy storage systems also require special geographical conditions to build large-scale air storage chambers, such as rock caves, salt caves, abandoned mines, etc., and the station construction conditions are relatively severe.

Traditional compressed air energy storage systems are all large-scale, with a single machine capable of reaching 100 MW. At present, large-scale compressed air energy storage power stations have been put into commercial operation worldwide. The first is the Huntorf 290 MW*2 h power station in Germany, which was put into commercial operation in 1978 (later upgraded to 321 MW). This power station only needs a few minutes to start from a hot standby state to reach the maximum energy storage capacity, and it has been successfully used to smooth out power fluctuations in wind farms [30]. The second is the McIntosh 110 MW*26 h compressed air energy storage power station in Alabama, USA, which was put into commercial operation in 1991. The third is located in Sunagawa, Japan, built in 1997, with an installed capacity of 35 MW*6 h [24] .

In response to the shortcomings of traditional compressed air energy storage, various advanced compressed air energy storage technologies have been continuously developed in recent years, such as adiabatic compressed air energy storage, liquefied air energy storage, supercritical compressed air energy storage, and compressed air energy storage coupled with renewable energy.

Although there are currently no compressed air energy storage power stations in operation in China, efforts have been made to research and conduct small-scale experimental verification of new compressed air energy storage technologies. Since 2014, the Institute of Engineering Thermophysics, Chinese Academy of Sciences has successively built the world's first 1.5 MW supercritical compressed air energy storage system integration experiment and demonstration platform; completed the demonstration operation of a 10 MW advanced compressed air energy storage system, conducted related experiments, and optimized and upgraded the 10 MW turbine expander; completed the design of a 100 MW compressed air energy storage system and key components, and some components have begun to be processed [31] .

2.3 Electrochemical Energy Storage

Electrochemical energy storage is achieved through the mutual conversion of electrical energy and chemical energy. So far, electrochemical energy storage technology has been developed and applied for over a hundred years, from traditional lead-acid batteries to nickel-cadmium batteries, nickel-hydrogen batteries, lithium-ion batteries, etc., widely used in various medium and small power storage occasions. In recent years, with the continuous progress of materials and processes, some new energy storage technologies have gradually matured, including lithium-ion batteries, lead-carbon batteries, flow batteries, sodium-sulfur batteries, lithium-carbon batteries, etc.

2.3.1 Lead-Acid Battery

The lead-acid battery is a rechargeable battery that realizes charging and discharging through the solid-phase reaction between lead in different valence states. It has a history of more than 150 years and is the earliest secondary battery used on a large scale. Lead-acid batteries have abundant raw material sources, low prices, excellent performance, good safety, and a mature waste battery recycling system. They are currently the largest production and the most widely used secondary batteries in industry, communication, transportation, and power applications [26] .

In recent years, some new technologies of lead-acid batteries have been developed, such as lead-carbon batteries. By adding special carbon materials to the negative electrode, the short cycle life of traditional lead-acid batteries is compensated. Its cycle life can reach more than 4 times that of traditional lead-acid batteries, which is very suitable for energy storage.

Lead-acid batteries are also widely used in energy storage systems. In 1988, the city of Chino in California, USA, built the world's largest 10 MW/40MWh lead-acid battery storage power station at the time. At present, China's lead-acid battery production exceeds one-third of the world's battery production, becoming the main production place of the world's batteries, and the research and production technology has reached the international advanced level. Large-capacity energy storage systems based on lead-acid batteries have also been built one after another, especially in the applications of peak-valley shaving on the user side and microgrids, such as the 160MWh lead-carbon battery storage power station in Wuxi Singapore Industrial Park, Jiangsu Zhongneng Silicon Industry 12MWh lead-carbon energy storage station, the National Wind and Light Storage and Transmission Demonstration Station's 12MWh tubular colloidal lead-acid battery energy storage system, the 36MWh lead-carbon battery energy storage system of the Tibet Nima County Renewable Energy Local Area Network Project, the 19.2MWh lead-carbon energy storage system of the Tibet Yangyi Photovoltaic Power Station, etc.

The common disadvantage of lead-acid batteries is slow charging and discharging speed, generally requiring 6–8 h, and low energy density. Overcharging can easily lead to hydrogen evolution and reduce life. Therefore, lead-acid batteries have their suitable application scenarios. When the site requirements are not strict and there is a long charging and discharging time, lead-acid batteries are still a very competitive energy storage technology. For large-capacity lead-acid battery storage power stations, environmental temperature, especially temperature difference, has an important impact on battery consistency, so it is necessary to strengthen the dynamic environmental control of the energy storage system. In addition, regular equalization charging is very necessary for the full activation of active materials and the improvement of cycle life.

2.3.2 Lithium-ion Battery

Lithium-ion batteries use lithium ions as active ions. During charging, lithium atoms in the positive electrode material lose electrons to become lithium ions, which migrate to the negative electrode through the electrolyte. At the negative electrode, they combine with external electrons and are intercalated and stored, achieving energy storage. The process is reversible during discharge. The electrochemical performance of lithium-ion batteries mainly depends on the structure and performance of the electrode materials and electrolyte materials used. The negative electrode materials are mainly carbon or lithium titanate, and the positive electrode materials mainly include lithium manganese oxide, lithium cobalt oxide, lithium iron phosphate, ternary materials of nickel cobalt manganese, ternary materials of nickel cobalt aluminum, etc.

Lithium-ion batteries have the advantages of high energy density, low self-discharge rate, no memory effect, wide working temperature range, fast charging and discharging, and long service life [32]. The specific energy of currently commercialized lithium-ion batteries has exceeded 200 Wh/kg, and the specific power of EV batteries can reach more than 3000 W/kg. Compared with lead-acid batteries, lithium-ion batteries have strong large current discharge capability, the cycle life can reach 3000–5000 times, and the energy storage efficiency is over 90%. However, lithium-ion batteries have poor overcharge/discharge resistance, complex combination and protection circuits, and relatively high costs. Lithium-ion batteries have been widely used in consumer electronics, EVs, military equipment, aerospace and other fields [33]. In power systems, lithium-ion batteries have broad prospects in the fields of power system frequency regulation, peak shaving, renewable energy absorption, microgrids, etc.

In 1991, Japan's SONY was the first to commercialize lithium-ion batteries, which were first widely used in the field of electronic products. In 2008, the American company A123 was the first to build a 2 MW lithium-ion battery energy storage system for power quality regulation and backup power. In China, the Southern Power Grid Company built a 4 MW/16MWh lithium battery energy storage system in the first phase of the Shenzhen Baoqing Energy Storage Station, a 16 MW/60MWh lithium battery energy storage system in the first phase of the National Wind and Solar Storage and Transmission Demonstration Station, a 15 MW lithium battery energy storage system in the Qinghai Geermu Photovoltaic Energy Storage Station, and a 101 MW/202MWh lithium battery distributed energy storage station on the Zhenjiang power grid in Jiangsu; in South Australia, the Hornsdale 100 MW/129MWh lithium battery energy storage system (Hornsdale Power Reserve, HPR), etc. The scale of lithium-ion battery energy storage systems is getting larger and larger, and their role in power systems is gradually being reflected.

From the perspective of large-scale promotion and application, high performance and low cost are the development directions of lithium-ion batteries and their key materials, involving technical development and industrialization in many aspects such as key material/battery manufacturing, key equipment development, battery system integration, and battery cascade utilization and waste battery recycling. Due to the good application value of lithium-ion batteries in EVs and energy storage, it has now become the focus of the development of the energy storage industry, and large-capacity energy storage is increasingly favoring lithium-ion batteries. Continuously improving the safety of cells, modules, and systems, reducing costs, and forming a complete evaluation system are the keys to the application of lithium batteries in power systems.

2.3.3 Sodium-Sulfur Batteries

Sodium-sulfur batteries are molten salt secondary batteries with metallic sodium as the negative electrode, sulfur as the positive electrode, and a ceramic tube as the electrolyte membrane. The specific energy of sodium-sulfur batteries can reach 3 ~ 4 times that of lead-acid batteries, and they can discharge at high currents. Their discharge current density can generally reach 200-300 mA/cm^2., it has a high charging and discharging efficiency. Because it uses a solid electrolyte, it does not have the self-discharge and side reactions of liquid electrolyte batteries. The high energy density, good power characteristics, long cycle life, and no self-discharge of sodium-sulfur batteries make them the main force in early electrochemical energy storage. However, sodium-sulfur batteries require a temperature of 300–350 °C during operation. If the core reaction component, the ceramic electrode, is damaged, it will cause intense combustion. Moreover, the battery needs to be heated and insulated during charging, and it needs a good heat dissipation design during discharging. There are problems such as strict operating environment requirements and high heat dissipation requirements. Sodium-sulfur batteries are more suitable for high-power, large-capacity energy storage applications, and there are already more than 100 MW-level applications worldwide.

Japan's NGK company is currently the only company that has commercialized the mass production of sodium-sulfur batteries. It cooperated with Tokyo Electric

Power Company TEPCO to develop and use stationary applications as a breakthrough. It completed many sodium-sulfur battery energy storage system demonstration projects in the late 1990s and achieved commercialization in 2002. There are currently more than 200 power stations with a power greater than 500 kW and a total capacity of about 300 MW in operation. The Shanghai Institute of Ceramics,Chinese Academy of Sciences, has launched a 650 Ah sodium-sulfur battery single-cell prototype product, built a 2 MW pilot line, and a 100 kW/800kWh sodium-sulfur battery energy storage system [34,35]. At present, the bottleneck of sodium-sulfur batteries mainly lies in the high manufacturing cost and the reliability of long-term operation, and it needs to continue to improve in materials and structure. In addition, because the battery operates at high temperatures, the safety issues in the direct reaction process of molten sodium and sulfur have always been hidden dangers in its installation and usage.

2.3.4 Flow Battery

Flow batteries store and release electrical energy through the valence state changes of ions in the electrolyte, mainly including zinc-bromine system batteries, sodium polysulfide/bromine system batteries, iron-chromium batteries, and all-vanadium redox flow batteries [36]. The output power and capacity of flow batteries are independent of each other, the system design is flexible, the overload capacity and deep discharge capacity are good, and the cycle life is long; but pumps are needed to maintain the operation of the battery, so the maintenance requirements of the battery system are high, and the efficiency is low at low loads.

The all-vanadium flow battery is a flow battery technology that has begun to be applied on a large scale. It uses a vanadium ion solution as the active material for the battery reaction, and uses the redox reaction of different valence ion pairs to realize the conversion between chemical energy and electrical energy. The cycle life of the all-vanadium flow battery can be as long as 10,000 times. The main problems are environmental adaptability (10 °C–35 °C) and the precipitation of the positive electrode material V_2O_5 at low temperatures.

The research, development and manufacturing of all-vanadium flow batteries are mainly concentrated in a few countries such as Japan, Canada, China, and Australia [4]. In terms of engineering, the Canadian VRB company built a 2 MW*6 h in the Irish wind farm in 2008. After Beijing Puhua Company bought VRB Company, they provided a 2 MW*4 h all-vanadium flow battery energy storage system for the National Wind and Solar Storage and Transmission Demonstration Power Station. In July 2013, Sumitomo Electric Industries of Japan and Hokkaido Electric Power Company jointly constructed a 15 MW all-vanadium flow battery energy storage system at the Minamisoma substation to smooth the output fluctuations of wind and photovoltaic power generation. In 2008, the Dalian Institute of Chemical Physics, Chinese Academy of Sciences, successfully developed a 10 kW battery module and integrated the first domestic all-vanadium flow energy storage battery system with a capacity of 100 kW/200kWh; Dalian Rongke Energy Storage Company provided a 5 MW/10MWh flow battery energy storage system for the 50 MW wind farm in Woniu Stone, Faku County, Liaoning, and officially launched the Dalian 200 MW/800MWh flow battery energy storage peak-valley shaving power station national demonstration project in 2016.

Zinc-bromine flow battery is another commercially applied flow battery technology, which is an electrochemical energy storage system designed based on the principle of reciprocating movement of zinc bromide solution, with zinc bromide electrolyte as the reaction substrate. The energy density and power density of zinc-bromine flow batteries are higher than all-vanadium flow batteries, and the cost is lower, but there is also a problem of high self-discharge rate caused by the formation of complexes due to electrode reactions. At present, ZBB Energy Company in the United States has realized commercial mass production and has built many zinc-bromine battery energy storage systems with capacities ranging from several kWh to several MWh, including a 1 MW/3MWh mobile energy storage station.

2.4 Flywheel Energy Storage

Flywheel energy storage is a physical energy storage technology that uses a motor to drive a flywheel to rotate at high speed, converting electrical energy into mechanical energy for storage, and when needed, the flywheel drives the motor to generate electricity. The energy storage of flywheel energy storage is determined by the mass and speed of the flywheel rotor, and its power output is determined by the characteristics of the motor and converter. Theoretically, its energy storage and output power can be independently designed and controlled. Flywheel energy storage technology is mainly divided into two types, one is low-speed flywheels represented by contact mechanical bearings, which are characterized by large storage power but short support time, generally used for high-power occasions. The other is high-speed flywheels represented by magnetic levitation bearings, which are characterized by compact structure and high efficiency, but smaller single capacity, which can be used for longer power support.

A prominent advantage of flywheel energy storage is its high power density, which can reach 5–10 times or more of battery energy storage [37,38], and also has the advantages of fast response speed and long life. It has good application prospects in industrial UPS, rail transit braking energy recovery, power grid frequency regulation, power quality control and other occasions.

Flywheel energy storage systems require strong industrial support. Many countries in America and Europe have made significant progress in the research and development and industrialization of flywheel energy storage systems. The 100–2000 kW CleanSource series UPS of Active Power Company in the United States, the 65–1000kVA VSS series UPS of Pentadyne Company, the 20 MW Smart Energy Matrix of Beacon Power Company, the high-temperature superconducting magnetic levitation bearing 100 kW/5kWh flywheel energy storage of Boeing's

Phantom factory, and the 315–2200kVA series Rotary UPS of SatCon Company, have begun to be applied to multiple energy storage projects. China's research and development of flywheel energy storage is relatively lagging years ago, mainly focusing on key technology research and demonstration, but industrial development has accelerated in recent years [39]. In 2008, Beijing 306 Hospital installed a flywheel energy storage device with a capacity of 250 kW, which uses magnetic levitation bearings and can provide 15 seconds of power support. It can be used in conjunction with standby diesel generators. Although the core device is imported, this is the first time that flywheel energy storage has been installed and used in China's distribution power system.

Flywheel energy storage needs to continue to improve its energy density for large-capacity energy storage applications. By connecting standardized, modularized, and serialized flywheels in parallel to form a flywheel array, the economic efficiency of the system can be improved, which is the development direction of large-capacity flywheel energy storage [40]. Beacon Power's 20 MW frequency regulation power plant is an array of 200 individual flywheel devices of 25 kWh/100 kW, operated through intelligent control. The flywheel energy storage system of the New York subway consists of 10 flywheels of 100 kW/1.6kWh, used to absorb train braking energy and provide startup support. The US Department of Energy and the California Energy Commission have supported the construction of a flywheel energy storage system for grid support and frequency control, including 7 flywheels of 15 kW, with a total power of 105 kW, and will eventually build a 1 MW/250kWh flywheel array system.

There are more and more typical projects of flywheel energy storage applications. For example, in July 2011, Beacon Power's 20 MW flywheel energy storage system was built in Stephen Town, New York State. It uses a carbon fiber composite flywheel rotor, which can quickly absorb electricity from the grid and release it quickly. It only takes 15 min to absorb and release 1 MWh of electrical energy. The entire system can meet 10% of New York State's frequency regulation needs [24]. In 2013, Beacon Power built another 20 MW flywheel energy storage power station in Hazle, Pennsylvania, to participate in the frequency regulation service of the PJM regional grid. In July 2010, Vycon completed the construction of a 7 MW flywheel energy storage system for the Northeast Data Center. In early 2014, it provided an 8 MW flywheel energy storage system for a data center in Texas. In addition, it provided a flywheel for the Los Angeles Metro WESS substation for braking energy recovery [41]. China's Kinetic Traction System Company has introduced KTSI's flywheel energy storage technology, and its MW-level flywheel energy storage was commercially applied in the braking energy recovery of Beijing subway in 2019 [42] .

2.5 SuperCapacitors

Supercapacitors can be divided into three types according to the energy storage mechanism: double-layer capacitors with both positive and negative electrodes, Faraday pseudocapacitors with both positive and negative electrodes, and hybrid electrochemical capacitors with double-layer and Faraday pseudocapacitors on two electrodes. The charge in the double-layer capacitor based on porous carbon materials is stored in the double-layer interface between the electrode and the electrolyte in an electrostatic manner, only undergoing electrochemical polarization without electrochemical reactions. Therefore, it has the advantages of fast charging and discharging speed, long cycle life, high charging and discharging efficiency, good high and low temperature performance, safety and reliability, and environmental friendliness. It is the mainstream of commercial supercapacitor products and has been widely used in hybrid vehicles, rail transit, communications, aerospace, and instruments [43]. Due to the low energy density of super capacitors, they need to be combined with energy storage technologies in most other applications, in order to improve the overall technical and economic performance of the energy storage system.

Japan, the United States, Germany and other countries have obvious advantages in the research and development of supercapacitors. Companies such as Nippon Chemi-Con, NEC-Tokin, Matsushita, NEC, Elna, Tokin, Murata in Japan occupy more than half of the global supercapacitor production. Maxwell is the main supercapacitor manufacturer in the United States, producing large and medium-sized C/C organic electrolyte double-layer capacitors. Supercapacitors of the C/C organic electrolyte system produced by companies such as NessCap in South Korea, EPCOS AG in Germany, and Cap-XX in Australia, and large $C/Ni(OH)_2$ Hybrid supercapacitors also have a high level of performance. China has made many deployments in the research and development and industrialization of supercapacitors, with both double-layer capacitors and Faraday pseudocapacitors having product applications.

In terms of supercapacitor energy storage applications, in 2005, the state of California in the United States built a 450 kW supercapacitor energy storage device to reduce the power fluctuations of a 950 kW wind turbine generator delivering power to the grid. In 2011, Siemens successfully developed a supercapacitor energy storage system with a storage capacity of 21 MJ/1 MW, installed in the 750 V DC power supply network of the Cologne subway in Germany. In 2013, Malverne in New York built a 2 MW supercapacitor energy storage station, mainly to provide voltage support for the local power system. China has also carried out some applications of supercapacitor energy storage systems, such as the State Power Group's construction of a 1 MW supercapacitor energy storage system in the wind farm project in Jinzhou, Liaoning and Fengbei Town in 2012, which, together with lithium-ion batteries and all-vanadium flow batteries, improves the power characteristics of wind power; a 500 kW*15 s supercapacitor energy storage system is built in Luxi Island, Zhejiang, which is used together with lead-acid batteries for the operation management of grid-connected microgrids.

2.6 Superconducting Magnetic Energy Storage

Superconducting magenetic energy storage (SMES) uses a coil made of superconductors, which stores energy in the magnetic field generated by grid power magnetization. It is a way to directly store electrical energy without energy conversion. Currently, most superconducting coils are wound with wires made of conventional materials such as niobium titanium (NbTi) or niobium tin (Nb_3Sn), which operate in the low-temperature region of liquid helium (4.2 k°) with large storage capacity; there are also energy storage coils wound with Bi-series (first generation) high-temperature superconducting materials, but their magnetic field characteristics in the liquid nitrogen temperature region are poor, and it is difficult to increase the energy storage capacity. Superconducting energy storage has a fast response speed, about a few milliseconds to tens of milliseconds, and also has extremely high power density and energy conversion efficiency, long service life, etc., mainly used for power system stability control, especially for suppressing low-frequency oscillations and improving power quality.

At present, the research and development of SMES based on low-temperature superconducting materials and high-temperature superconducting materials are developing simultaneously, with capacities mostly at the MJ/MVA level. Since the 1990s, the function of superconducting energy storage in improving power quality has been highly valued and actively developed. The United States, Germany, Italy, South Korea and other countries have carried out the research and development of MJ-level SMES and put them into actual power system demonstrated operation. As a typical application of superconducting energy storage systems, the United States built a small superconducting energy storage system with six 3 MJ/8MVA unites based on low-temperature super materials in 2000, installed in the northern ring transmission network of Wisconsin, to achieve active and reactive power regulation, to improve the reliability and power quality of the region's power supply. From 2002 to 2004, eight 3 MJ/8MVA small superconducting energy storage systems were installed in the 500 kV transmission network in Tennessee to maintain the stability of the regional power grid. Germany's ACCEL Instrument GmbHh and EUS GmbH jointly developed a 2 MJ superconducting magnet for laboratory UPS systems, with an average power of 200 kW, up to 800 kW. In 2007, Japan's Chubu Electric Power Company and NEDO launched a research and development project for a 10 MW/20 MJ SMES device for grid support, and achieved demonstrated operation in 2008.

In China, major research units such as the Institute of Electrical Engineering Chinese Academy of Sciences, Tsinghua University, and Huazhong University of Science and Technology have carried out research and development work on SMES [44] The Institute of Electrical Engineering Chinese Academy of Sciences has made a breakthrough in the new type of multifunctional integrated superconducting power device, proposing an integrated superconducting current limiter-energy storage system that integrates current limiting and power quality regulation. They developed the world's first 100 kJ/25kVA superconducting current limiter-energy storage system prototype, and by the end of 2008, they achieved the demonstrated operation of China's first 1 MJ/0.5MVA high-temperature SMES. In 2005, Tsinghua University developed a 500 kJ/150kVA SMES, and Huazhong University of Science and Technology developed a 35 kJ/7kVA miniature high-temperature SMES for use in dynamic simulation of power systems [41].

2.7 Others

There are also some new energy storage materials and technologies under development. Although it is currently impossible to accurately judge the feasibility and prospects of these new materials and technologies for large-scale applications, they are worth continuous attention. For example, the emergence of nanotechnology may overturn the structure of some existing batteries, greatly improve the power density and energy density of the batteries, improve the stability of electrical/thermal performance, and help reduce costs, playing a greater role in energy and power systems. Here, special attention is paid to solid-state lithium batteries, metal-air batteries, liquid air energy storage, metal liquid batteries, phase change energy storage, hydrogen storage, etc. [45, 46]. Since energy storage technology itself is not the focus of this book, it will not be elaborated here.

2.7.1 Part Two: Electrical Energy Storage Systems

Energy storage is a systematic engineering. In addition to paying attention to various energy storage technologies, electrical energy storage systems also need to consider grid connection, operation control, equipment management and other links. As a whole, any shortcomings in any link will affect its application and even bring hidden dangers. This part starts from the energy storage system or complete equipment itself, mainly focusing on the typical battery energy storage system and flywheel energy storage system, and analyzes the composition, design, control and operation and maintenance management of the energy storage system.

Part II
Electrical Energy Storage Systems

Chapter 3
Battery Energy Storage System

Battery energy storage applied to power systems requires a large number of individual batteries to be connected in series and parallel, and connected to the grid through power electronic conversion circuits. Therefore, accurately grasping the characteristics of the battery and the needs of the grid, and optimizing the design, is the key to the battery energy storage system. This chapter mainly introduces the system composition, grid connection and operation control methods for lithium-ion batteries and lead-carbon batteries and other battery energy storage systems.

3.1 Battery Energy Storage System Composition

The battery energy storage system includes a battery pack, a Battery Management System (BMS), a Power Conversion System (PCS), a monitoring management system, and a power and environment supervision system. Large-scale energy storage power stations are often composed of multiple energy storage systems with basically independent configurations and functions. The standardization, modularization, and serialization of energy storage systems provide good support for shortening the construction period of energy storage power stations, operational reliability, and maintenance levels.

A typical battery energy storage system main structure is shown in Fig. 3.1. As a whole composed of multiple subsystems, any problem in any subsystem will affect the overall performance and function of the battery energy storage system. Therefore, each subsystem should achieve optimized design and mutual adaptation to avoid system shortcomings. The design of the battery cluster should fully consider the performance and consistency level of the single cell and the battery pack; the configuration of the BMS should consider the composition and monitoring needs of the battery pack; the PCS should take into account the function of AC grid connection and the electrical requirements of the DC battery side; and so on.

X. Tang et al., *Electrical Energy Storage Technologies and Applications*,
https://doi.org/10.1007/978-981-96-5625-7_3

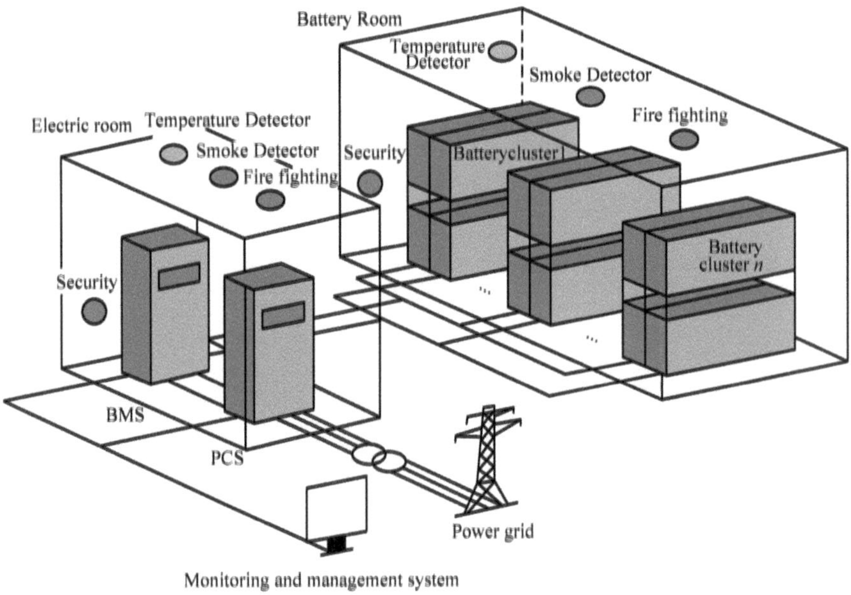

Fig. 3.1 Structure of the battery energy storage system

Furthermore, the power and environment supervision system is also an important part of the battery energy storage system. It realizes centralized monitoring and management of the power system, environmental system, fire system, security system, network system, etc. of the energy storage station. It mainly monitors the operating status and working parameters of various devices, discovers abnormal parameters or faults, takes various alarms and fault handling methods in time, records historical data and alarm events, and has remote monitoring and management functions and WEB browsing functions. The optimization of the power and environment supervision system is an important guarantee for the battery energy storage system. Through the energy storage station monitoring and management system, the state of the power and environment system is closed combined with the operation process of energy storage, which is the premise for the safe, optimized, and efficient operation of the energy storage station.

3.1.1 Lithium Energy Storage System

Since the cell of a lithium-ion battery generally has a small capacity, multiple cells need to be operated in parallel or multiple battery clusters need to be operated in parallel. Regarding the configuration relationship between the battery cluster and PCS, the charging and discharging power and its time (energy) required by the energy storage system are mainly considered. For large-capacity energy storage systems, there are two main technical routes, one is to connect multiple battery clusters in parallel to a high-power PCS; the other is to connect a small number of battery clusters in parallel to a low-power PCS, and then form a high-power system by parallel multiple low-power PCS. For the former, the parallel connection of multiple battery clusters requires high consistency of the batteries, otherwise, there will be large circulating currents. Although there are currently some methods to solve the problem of imbalance between clusters, the operation is difficult. For the latter, it can effectively avoid the problem of circulating current between battery clusters, but the equipment cost will increase, and multiple PCS in parallel also need to solve the problems of parallel balance and output end resonance. With the development of technology, it should be an effective solution to compromise the two routes, that is, a moderate number of battery clusters are connected in parallel to configure an appropriate power PCS, to achieve a reasonable configuration of large-capacity energy storage, of course, this depends on the characteristics and functions of the battery, BMS, and PCS.

BMS is an important part of the lithium energy storage system. BMS measures the electrical and thermal parameters of lithium battery cells, modules, PACK, battery clusters, and battery systems, estimates the state of charge (SOC) of the battery system in real time, realizes the balance of batteries within and between clusters, and carries out various electrical and thermal management and protection. At the same time, the energy storage system also needs to configure supporting facilities such as air conditioning and fire protection.

Taking the 2.2MWh standard energy storage system shown in Fig. 3.2 as an example, the system uses a certain aluminum shell lithium iron phosphate battery, and the single cell is 3.2 V/120 Ah. A battery PACK box is composed of 2 parallel and 15 series (2P15S), and 16 PACKs are connected in series to form a battery cluster, a total of 12 clusters; every 6 clusters are connected in parallel to a 500 kW PCS forms an independent energy storage unit; two units form an energy storage system connected to the medium voltage grid through a double-split transformer, with a capacity of 1 MW/2.2MWh.

3.1.2 Lead-Carbon Energy Storage System

The lead-carbon energy storage system is similar to the lithium energy storage system in composition. Since the single cell capacity of the lead-carbon battery can be larger, there are more choices in capacity configuration, and multiple small-capacity batteries can be avoided by choosing large-capacity single cells. For large-capacity energy storage systems, large-capacity batteries can be selected to be connected in series into a cluster, multiple clusters can be connected in parallel to PCS as an energy storage unit, and multiple units can be connected in parallel to form an energy storage system.

During operation, lead-carbon batteries spend most of their time in a Partial State of Charge (PSOC), requiring no float charging and offering good safety. This is an

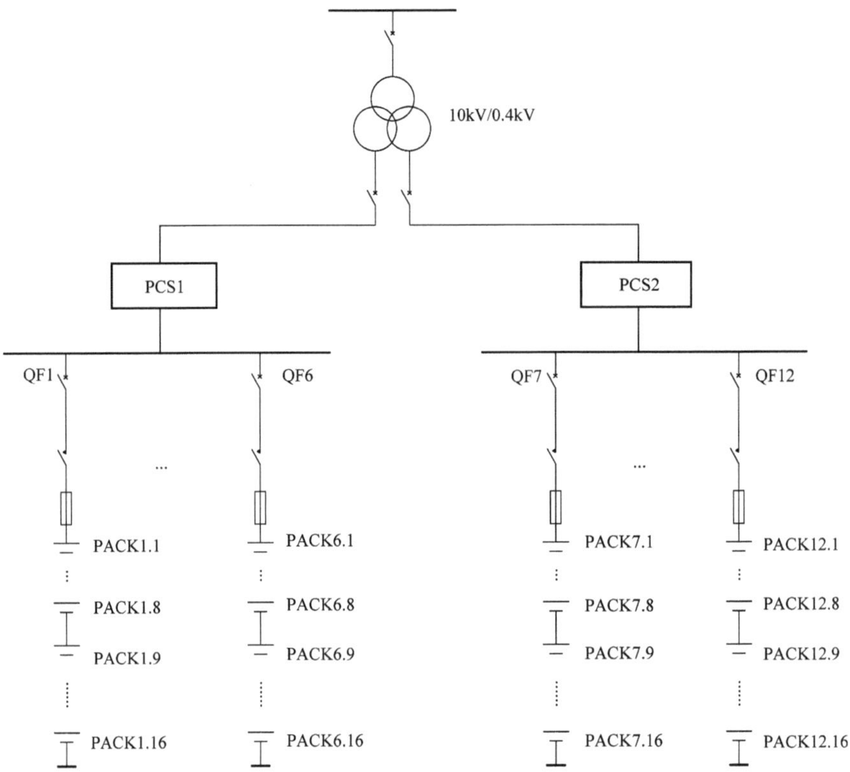

Fig. 3.2 Electrical diagram of lithium battery energy storage system

important feature that makes lead-carbon batteries suitable for energy storage. However, lead-carbon batteries need to be regularly balanced and charged for maintenance (generally every 2–3 months) to fully activate the active material and improve the battery's cycling performance. During the balancing charging process, inconsistencies between batteries can be effectively balanced, and the State of Charge (SOC) and State of Health (SOH) parameters can be calibrated. Therefore, the Battery Management System (BMS) used for lead-carbon energy storage can be appropriately simplified in terms of functional configuration. At the same time, integrating the balancing charging process with the parameter calibration of the BMS, and incorporating it into the energy management of the storage system, is an important measure to improve the overall performance of the system.

Taking a 1.2MWh lead-carbon energy storage system as an example, the system is configured with 600 LLC-1000 lead-carbon batteries, each with a single cell of 2 V/1000 Ah, 300 in series and 2 in parallel (300S2P), connected to a 125 kW or 250 kW PCS; a BMS is configured to monitor battery voltage, temperature, internal resistance, and estimate SOC, SOH, etc. The electrical structure of the lead-carbon battery energy storage system is shown in Fig. 3.3.

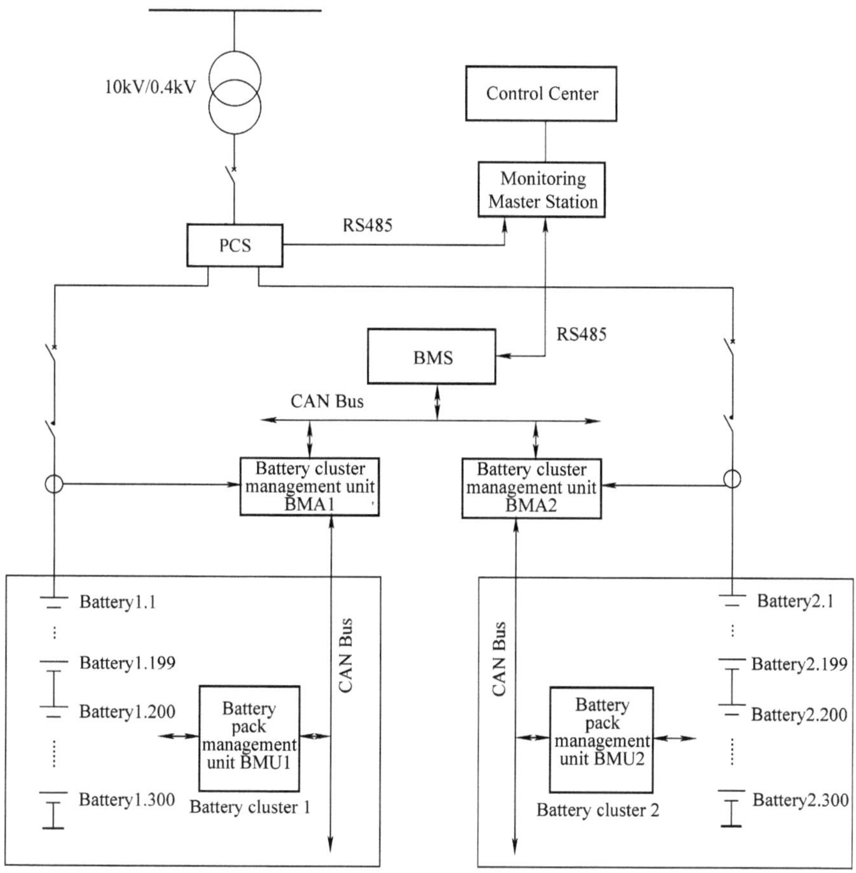

Fig. 3.3 Electrical diagram of the lead-carbon energy storage system

3.1.3 Monitoring Management System

The monitoring management system of energy storage can generally be divided into integrated control layer and spacer layer, as shown in Fig. 3.4. The integrated control layer includes a monitoring host, data server, data communication gateway, etc., to achieve information collection and real-time display, remote control of equipment, data storage, query and statistics, and communication with related systems; the spacer layer includes local monitoring devices, measurement and control devices, etc. The integrated control layer and spacer layer are connected through a communication network. The monitoring of the energy storage system mainly involves the detection and control of various units of the energy storage system such as batteries and modules, Battery Management System (BMS), energy storage converter PCS, electric meter, power and environment supervision system, etc.; monitoring of the grid status at the grid connection point of the energy storage system,

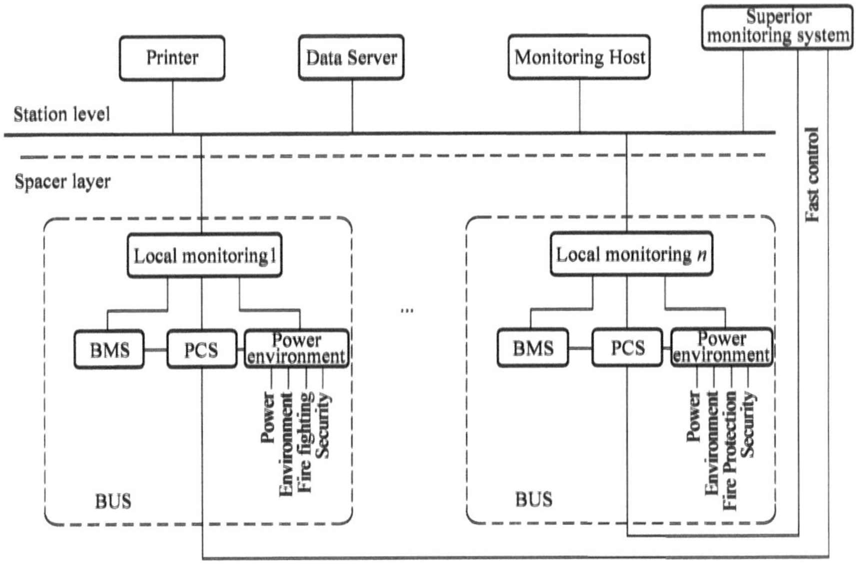

Fig. 3.4 Battery energy storage system monitoring management architecture

including frequency, voltage, exchange power, etc.; including data interaction, command transmission, data storage, early warning and protection, and human-machine interface, etc. At the same time, a communication interface with the host computer is reserved to accept instructions from the superior station or dispatch. The energy management of the energy storage system mainly involves predicting the operating status of various subsystems within the energy storage system, determining the overall SOC and SOH status of the energy storage system based on the SOC and SOH status of each subsystem, and the available capacity under specific charging and discharging power, to provide a decision basis for the operation control of the energy storage system. At the same time, optimize the management of each energy storage subsystem, implement regular and irregular maintenance, optimize the overall status of the energy storage system, and improve operational reliability and availability. For larger-scale energy storage stations, they are often composed of relatively independent energy storage units. Each energy storage unit generally achieves information collection, analysis, control, and upward communication through the configured local monitoring management device. The local monitoring management device uploads specific information and analysis calculation results to the superior monitoring management system, realizing a reasonable division of tasks, resource optimization, and timely control between the integrated control layer and spacer layer.

IEC61970 "Energy Management System Application Interface (EMS-API)" has realized the seamless integration of EMS application software, promoting the development of modern power system automation. For energy storage systems used in power systems, specifying the interface semantics of the energy storage system in the Common Information Model (CIM) in IEC61970 is the basis for the efficient access of the energy storage system to the energy management system.

As a special unit for bidirectional energy flow, the energy storage system cannot inherit from the unidirectional energy output model GenerationUnit. Literature [47] creates a new class EnergyStorageUnit to describe the unit model of bidirectional energy input and output scheduling. The BatteryEnergyStorageUnit of the battery energy storage system can be extended from the EnergyStorageUnit. It is composed of the battery body, battery type, control system, operating status, energy storage station, etc., as shown in Fig. 3.5 and Table 3.1.

The interaction information involved in the above model mainly focuses on the energy storage system status information required by the monitoring system, measurement/metering values, and accepts the setting values and control commands of the monitoring system. The construction of specific status information, settings,

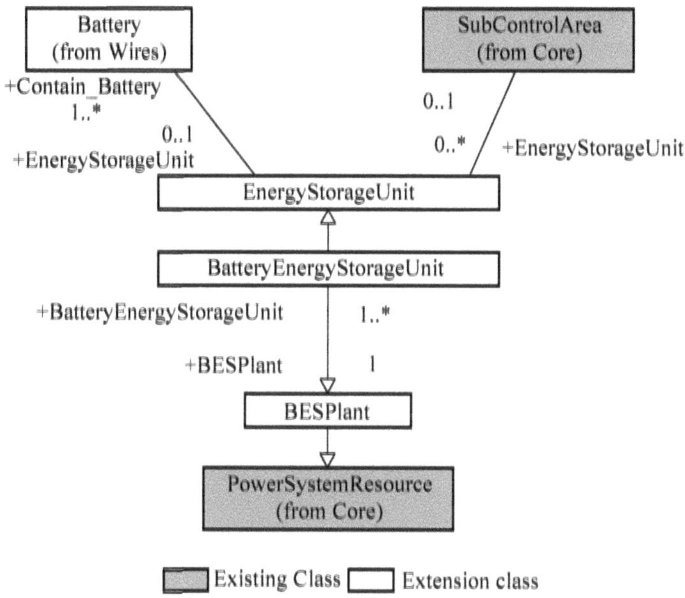

Fig. 3.5 Energy storage system CIM model

Table 3.1 Extended class of battery energy storage system

New Class	Parent Class	Package	Description
Battery	RegulationCondEg	Wires	Battery
Battery type		Domain	Battery type
BatteryOperationModel		Domain	Battery operating status
BESPlant	PowerSystemResource	Production	Battery energy storage plant
EnergyStorageUnit	SubControlArea	Production	Energy storage unit
BatteryEnergyStorageUnit	EnergyStorageUnit	Production	Battery energy storage unit

control, and measurement/metering value information models needs to be set according to actual conditions.

3.2 Energy Storage PCS Main Circuit Topology

The grid-connected equipment for energy storage, namely the power conversion system PCS, can be divided into low-voltage access PCS and medium-high voltage access PCS according to the voltage level of the grid-connected grid. In the selection of topology structure, factors such as on/off-grid, nonlinear/unbalanced load, electrical isolation, etc. need to be considered. Different topology structures have different adaptabilities to the above functions.

PCS needs to adapt to the characteristics of the local power grid, including power supply system, voltage level, grounding mode, relay protection, etc. For general low-voltage power grids, three-phase four-wire system or three-phase five-wire system is often used, which can supply power to three-phase or single-phase loads. Therefore, PCS mostly adopts a three-phase four-wire system topology. The choice of topology structure for the three-phase four-wire converter depends on the formation method of the neutral point of the output AC system [48] .

3.2.1 Based on ▲/Y Transformer Topology

Through the transformer obtains the neutral point of the three-phase PCS main circuit as shown in Fig. 3.6. The secondary side of the transformer uses Y type connection, the neutral point is connected to the one of the grid, providing a path for the zero-sequence current of the unbalanced load.

Due to the existence of PCS output impedance and transformer leakage inductance, unbalanced loads will cause asymmetry in the output voltage. Therefore, the

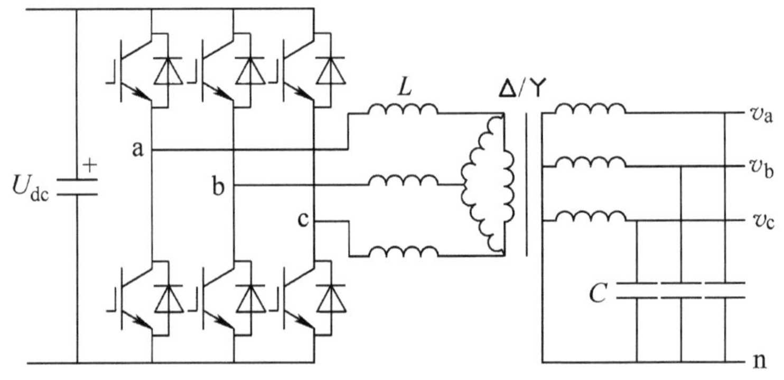

Fig. 3.6 Main circuit topology based on ▲/Y transformer

presence of ▲/Y transformer cannot fundamentally solve the problem of PCS output voltage asymmetry, and can only be used in situations where the three-phase load imbalance is small. Since the energy storage PCS inevitably exists off-grid operation and unbalanced load, to solve the output voltage symmetry problem under this scenario, it is necessary to combine the main circuit topology with the three-phase load imbalance control method.

3.2.2 Three Single-Phase Transformer Combination Topology

The combination topology is composed of three single-phase converters, as shown in Fig. 3.7. By connecting one end of the three single-phase transformers to form the neutral point of the system output voltage, a three-phase four-wire converter is formed. Since the three single-phase converters are independent of each other, conventional single-phase converter control methods can be applied, which have the advantages of strong ability to carry unbalanced loads, simple control, high utilization rate of DC bus voltage, etc., and are more suitable for high-power situations [49,50] .

The main disadvantage of this topology is: it requires three output transformers, which leads to large equipment volume and high cost; moreover, the number of

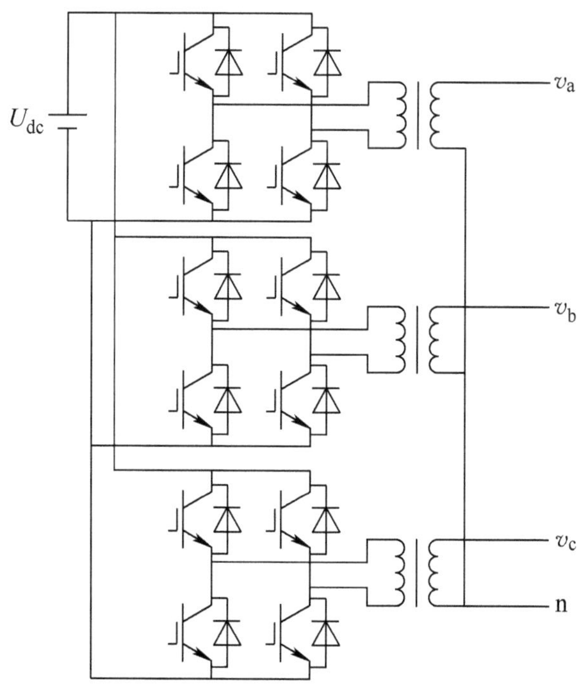

Fig. 3.7 Three Single-Phase Transformer Combination Main Circuit Topology

power switch tubes used in the equipment is as many as 12, which is costly and not suitable for medium and small power applications.

3.2.3 DC Bus Split Capacitor Topology

The main circuit topology based on DC bus split capacitor, by connecting two sets of DC filter capacitors in series, forms a split capacitor on the DC bus side, constitutes the fourth line as the midpoint of the system output voltage, provides a zero-sequence current path, to solve the problem of unbalanced load, as shown in Fig. 3.8.

The main disadvantage of this topology is: the DC capacitor value often needs to be very large in series on the DC bus side to maintain the constancy of the neutral point potential, which increases the volume of the equipment and reduces the reliability; moreover, the DC voltage utilization rate is low, and the peak value of the AC output phase voltage is theoretically only half of the DC bus voltage.

3.2.4 Three-Phase Four-Bridge Topology

The three-phase four-bridge topology is based on the traditional three-phase bridge converter, with an additional bridge added to directly control the neutral point voltage and generate neutral point current to adjust the impact of unbalanced loads, as shown in Fig. 3.9. Because an additional control parameter is added, it can generate independent three-phase voltage outputs, enabling it to maintain the symmetry of the three-phase voltage under unbalanced load conditions.

The three-phase four-bridge topology can achieve a higher utilization rate of DC voltage. To improve the overall filtering effect, a filter inductor can also be connected in series on the neutral line.

Fig. 3.8 Main Circuit Topology Based on DC Bus Split Capacitor

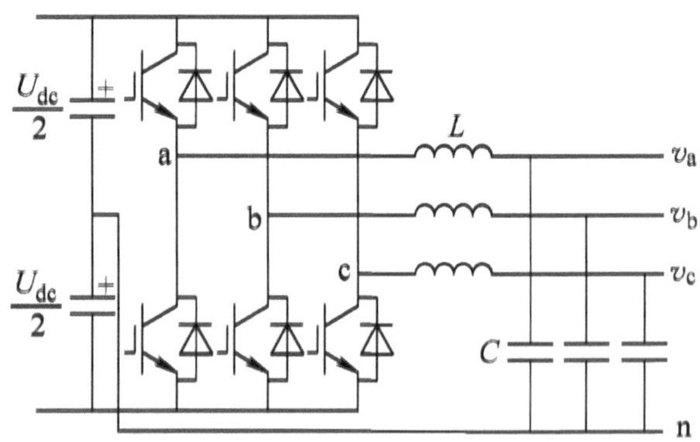

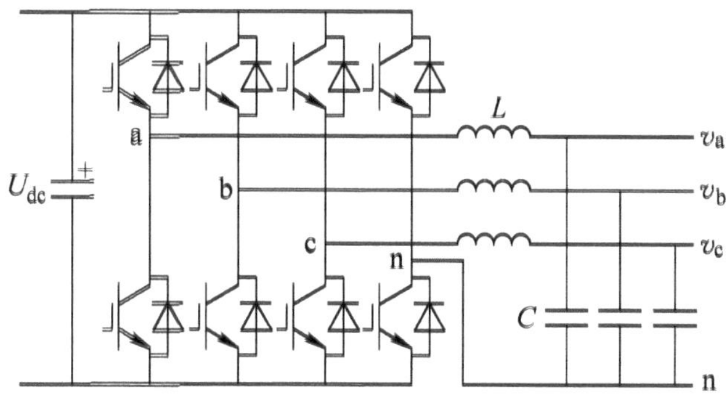

Fig. 3.9 Main circuit topology based on three-phase four-bridge

Of course, the control process of the three-phase four-bridge PCS is relatively complex [51,52], the design and debugging of the controller are relatively difficult, and the cost will increase due to the addition of a pair of power switch bridge.

3.2.5 Medium and High Voltage Topology Based on Cascaded H Bridge

Large-capacity energy storage systems often need to be connected to higher-level power grids of 10 kV and above. Medium and high voltage grid connection can be achieved by using a low-voltage converter in conjunction with a step-up transformer, but the equipment volume and weight of this scheme are large, and the overall efficiency is low.

Medium and high voltage energy storage converters often adopt the following several topology structures. One is to use two-level or three-level topology, and connect power switch devices in series to obtain higher voltage. This scheme requires high uniformity of devices, the control difficulty of simultaneous triggering of devices is high, and the voltage on the DC bus side also needs to be matched very high. Another scheme is to use the cascaded H bridge unit CHB on the AC side to achieve medium and high voltage AC output. The DC side of each H bridge can be directly connected to the battery pack, or it can be connected to the battery pack after an isolated DC/DC converter. In addition, the modular multilevel converter MMC is also a suitable topology structure for medium and high voltage, which can also be used in battery energy storage systems, but it requires more complex control such as circulating current suppression.

Among the above several medium and high voltage access schemes, the CHB medium and high voltage energy storage system, where the energy storage battery pack is directly connected to the H bridge without DC/DC, has the advantages of simple structure and flexible control, and has good practicality, as shown in Fig. 3.10.

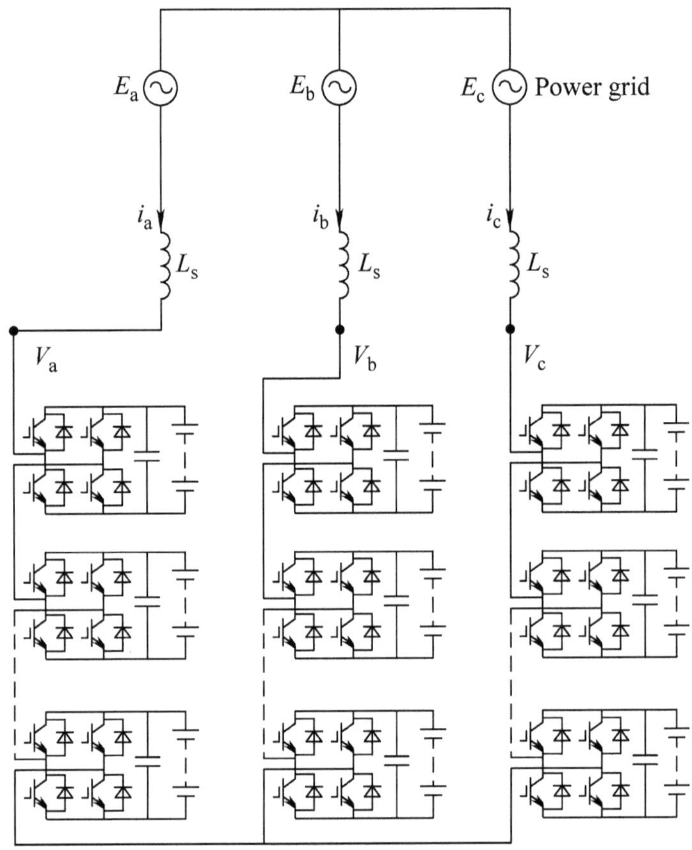

Fig. 3.10 Topology of Medium and High Voltage Energy Storage System Based on CHB

In addition, because the cascaded H bridge unit has many units, when a unit fails and exits operation, it basically does not affect the entire bridge and system operation, which is also an advantage of this topology.

When the medium and high voltage energy storage converter based on CHB performs charging and discharging control, it can use vector control based on grid voltage orientation, or it can use direct power control. Since the energy storage batteries are distributed in each H bridge unit, the state of charge SOC of each battery group needs to be balanced, including inter-phase balance and balance of each battery unit within the phase. The adjustment of inter-phase SOC balance can be achieved by superimposing zero-sequence voltage in the three-phase voltage, and the adjustment of intra-phase SOC balance can be achieved by superimposing an appropriate correction amount on the modulation wave of each unit.

3.3 Energy Storage PCS Control Technology

Island operation and grid-connected operation are the basic operating modes of battery energy storage applied to power systems. In the island operation mode, the function of the energy storage PCS is similar to that of the grid-connected main power source, but it also monitors the state of the grid voltage in real time, and is ready to switch to the grid-connected operation mode when conditions permit. In the island operation mode, the control target of the energy storage PCS is to output three-phase symmetrical sine wave voltage, and its research focus is on the control of the output voltage waveform and the parallel control technology of multiple units.

3.3.1 PCS Mathematical Model

PCS generally uses a three-phase voltage source converter VSC, and the mathematical model of VSC is the basis for analyzing and designing energy storage grid-connected systems. Different forms of mathematical models can be established from different perspectives, and the corresponding control methods are often different. The general mathematical model of three-phase VSC can take the following two forms [53]:

1. General mathematical model described by the switching function;
2. General mathematical model described by the duty cycle.

The general mathematical model described by the switching function is an accurate description of the VSC switching process, which is more suitable for VSC waveform simulation. However, because this switching function model includes high-frequency components during the switching process, it is difficult to use for controller design.

When the VSC switching frequency is much higher than the AC output fundamental frequency, to simplify the general mathematical model of VSC, the high-frequency components in the VSC switching function model can be ignored, that is, only the low-frequency components are considered, thereby obtaining the low-frequency mathematical model described by the duty cycle. This VSC low-frequency mathematical model described by the duty cycle is very suitable for control system analysis and can be directly used for controller design. However, because this model omits the high-frequency components during the switching process, it cannot perform accurate dynamic waveform simulation.

In summary, the general mathematical models of VSC described by the switching function and the duty cycle each play an important role in the design and simulation of VSC control systems. The latter is commonly used to design the VSC control system, and then the former is used to simulate the VSC control system to verify the performance indicators of the control system design.

3.3.1.1 Three-Phase VSC Switching Function Model

For a VSC converter controlled by a PWM pulse signal, the two power switches on the same bridge alternately conduct, defining its switching function S as follows:

$S_{i(i=a,b,c)} = 1$: i The upper switch of the bridge is on, and the lower tube is off;

$S_{i(i=a,b,c)} = -1$: iThe lower switch of the bridge is on, and the upper tube is off.

Taking the energy storage PCS topology based on the split DC bus capacitor as an example, in its main circuit structure, because the reference point of the system is the midpoint of the DC bus filter capacitor, the three-phase bridge output phase voltage is:

$$\begin{bmatrix} u_a \\ u_b \\ u_c \end{bmatrix} = \begin{bmatrix} S_a \\ S_b \\ S_c \end{bmatrix} \cdot \frac{U_{dc}}{2} \tag{3.1}$$

Taking the load current as a disturbance, set the state variables as the three-phase filter capacitor phase voltage v_a, v_b, v_c and the three-phase filter inductor current i_{1a}, i_{1b}, i_{1c}, the input variables are the three-phase bridge output phase voltage u_a, u_b, u_c and the three-phase filter inductor current i_{1a}, i_{1b}, i_{1c}, the output variables are the phase voltage of the three-phase filter capacitor v_a, v_b, v_c and the three-phase load current i_a, i_b, i_c . According to Kirchhoff's law, the current and voltage equations of the filter capacitor and filter inductor are given as:

$$\begin{cases} C\dfrac{dv_a}{dt} = i_{1a} - i_a \\ C\dfrac{dv_b}{dt} = i_{1b} - i_b \\ C\dfrac{dv_c}{dt} = i_{1c} - i_c \\ L\dfrac{di_{1a}}{dt} = u_a - v_a - ri_{1a} \\ L\dfrac{di_{1b}}{dt} = u_b - v_b - ri_{1b} \\ L\dfrac{di_{1c}}{dt} = u_c - v_c - ri_{1c} \end{cases} \tag{3.2}$$

Converted into the form of state space matrix:

$$\begin{bmatrix} \dot{v}_a \\ \dot{v}_b \\ \dot{v}_c \\ \dot{i}_{1a} \\ \dot{i}_{1b} \\ \dot{i}_{1c} \end{bmatrix} = \begin{bmatrix} 0 & 0 & 0 & 1/C & 0 & 0 \\ 0 & 0 & 0 & 0 & 1/C & 0 \\ 0 & 0 & 0 & 0 & 0 & 1/C \\ -1/L & 0 & 0 & -r/L & 0 & 0 \\ 0 & -1/L & 0 & 0 & -r/L & 0 \\ 0 & 0 & -1/L & 0 & 0 & -r/L \end{bmatrix} \begin{bmatrix} v_a \\ v_b \\ v_c \\ i_{1a} \\ i_{1b} \\ i_{1c} \end{bmatrix} + \begin{bmatrix} 0 & 0 & 0 & -1/C & 0 & 0 \\ 0 & 0 & 0 & 0 & -1/C & 0 \\ 0 & 0 & 0 & 0 & 0 & -1/C \\ 1/L & 0 & 0 & 0 & 0 & 0 \\ 0 & 1/L & 0 & 0 & 0 & 0 \\ 0 & 0 & 1/L & 0 & 0 & 0 \end{bmatrix} \begin{bmatrix} u_a \\ u_b \\ u_c \\ i_a \\ i_b \\ i_c \end{bmatrix} \tag{3.3}$$

Substitute eq. (3.1) into eq. (3.3) to get:

$$\begin{bmatrix} \dot{v}_a \\ \dot{v}_b \\ \dot{v}_c \\ \dot{i}_{1a} \\ \dot{i}_{1b} \\ \dot{i}_{1c} \end{bmatrix} = \begin{bmatrix} 0 & 0 & 0 & 1/C & 0 & 0 \\ 0 & 0 & 0 & 0 & 1/C & 0 \\ 0 & 0 & 0 & 0 & 0 & 1/C \\ -1/L & 0 & 0 & -r/L & 0 & 0 \\ 0 & -1/L & 0 & 0 & -r/L & 0 \\ 0 & 0 & -1/L & 0 & 0 & -r/L \end{bmatrix} \begin{bmatrix} v_a \\ v_b \\ v_c \\ i_{1a} \\ i_{1b} \\ i_{1c} \end{bmatrix} + \begin{bmatrix} 0 & 0 & 0 & -1/C & 0 & 0 \\ 0 & 0 & 0 & 0 & -1/C & 0 \\ 0 & 0 & 0 & 0 & 0 & -1/C \\ U_{dc}/2L & 0 & 0 & 0 & 0 & 0 \\ 0 & U_{dc}/2L & 0 & 0 & 0 & 0 \\ 0 & 0 & U_{dc}/2L & 0 & 0 & 0 \end{bmatrix} \begin{bmatrix} S_a \\ S_b \\ S_c \\ i_a \\ i_b \\ i_c \end{bmatrix} \tag{3.4}$$

The structure diagram of the switch function model of the three-phase VSC in the abc stationary coordinate system is shown in Fig. 3.11. It can be seen that the three-phase four-wire VSC can be controlled independently in three phases or unified vector control can be adopted.

3.3.1.2 Three-Phase VSC State Space Average Model

The mathematical model of the three-phase VSC described by the switch function is an accurate description of the switching process of the power device. However, because the switch model contains high-frequency components during the

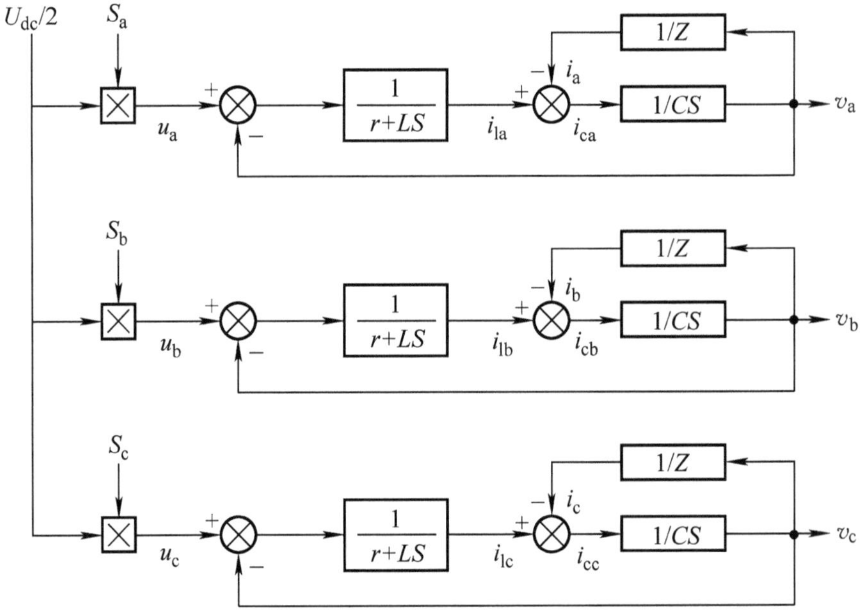

Fig. 3.11 Switch function model of three-phase VSC in abc stationary coordinate system

switching process of the power device, it is difficult to guide the design of the controller. When the output fundamental frequency of the converter and the resonance frequency of the LC filter are sufficiently small compared to the switching frequency of the power device, the VSC can be regarded as a constant gain amplifier, and the state space averaging method can be used to establish the linearized model of the VSC.

The state space averaging method uses the average value of the variables in a switching period to replace their instantaneous values, thereby obtaining a continuous state space average model. Under this model, regular sampling can replace natural sampling. At this time, the PWM switch function waveform in a switching period is as shown in Fig. 3.12.

In the figure, the high-frequency carrier is a bipolar triangle wave U_c, its amplitude is U_{cm}, and its frequency is f_{sw}. The modulated wave is a power frequency sine wave $U_{ir(i=a,b,c)}$, its amplitude is U_{rm}. In Fig. 3.12a, within a switching period, $d_{i(i=a,b,c)}$ is the PWM duty cycle of the corresponding phase, and $d_i \le 1$. The average value of the switch function S_i on a switching period $\overline{S_i}$ is:

$$\overline{S_i} = \frac{1 \times t_{on} + 0 \times (1/f_{sw} - t_{on})}{1/f_{sw}} = \frac{1 \times d_i / f_{sw} + 0 \times (1/f_{sw} - d_i / f_{sw})}{1/f_{sw}} = d_i \tag{3.5}$$

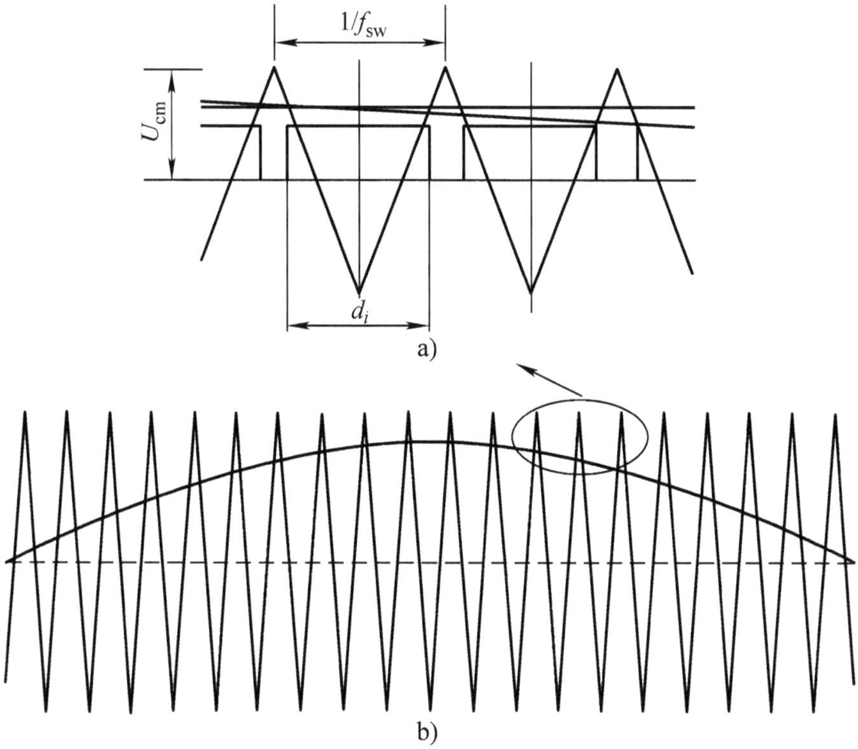

Fig. 3.12 PWM switch waveform diagram

From eq. (3.5), it is known that the PWM duty cycle is actually the average value of the switch function in a switch cycle. In Fig. 3.12a, according to the trigonometric relationship, the PWM duty cycle is:

$$d_i = \frac{1}{2}\left(\frac{U_{ir}}{U_{cm}} + 1\right) \tag{3.6}$$

That is,

$$\begin{cases} d_a = \dfrac{1}{2}\left(\dfrac{U_{rm} \cdot \sin \omega t}{U_{cm}} + 1\right) \\ d_b = \dfrac{1}{2}\left(\dfrac{U_{rm} \cdot \sin\left(\omega t - 2\pi / 3\right)}{U_{cm}} + 1\right) \\ d_c = \dfrac{1}{2}\left(\dfrac{U_{rm} \cdot \sin\left(\omega t + 2\pi / 3\right)}{U_{cm}} + 1\right) \end{cases} \tag{3.7}$$

Define the modulation ratio of PWM as $m = \dfrac{U_{rm}}{U_{cm}}$ and $m \le 1$, then eq. (3.7) is:

$$\begin{cases} d_a = \dfrac{1}{2} m \cdot \sin \omega t + 0.5 \\ d_b = \dfrac{1}{2} m \cdot \sin\left(\omega t - 2\pi / 3\right) + 0.5 \\ d_c = \dfrac{1}{2} m \cdot \sin\left(\omega t + 2\pi / 3\right) + 0.5 \end{cases} \tag{3.8}$$

Take the average of eq. (3.1) in a switch cycle, then

$$\begin{bmatrix} \overline{u}_a \\ \overline{u}_b \\ \overline{u}_c \end{bmatrix} = \begin{bmatrix} \dfrac{d_a}{2} \\ \dfrac{d_b}{2} \\ \dfrac{d_c}{2} \end{bmatrix} \cdot U_{dc} \tag{3.9}$$

where, $\overline{u}_i$ is the average value of u_i in a switch cycle, $i = a, b, c$. Similarly, take the average of eq. (3.4) in a switch cycle, you can get the state space average model of the three-phase VSC:

$$\begin{bmatrix} \dot{v}_a \\ \dot{v}_b \\ \dot{v}_c \\ \dot{i}_{1a} \\ \dot{i}_{1b} \\ \dot{i}_{1c} \end{bmatrix} = \begin{bmatrix} 0 & 0 & 0 & 1/C & 0 & 0 \\ 0 & 0 & 0 & 0 & 1/C & 0 \\ 0 & 0 & 0 & 0 & 0 & 1/C \\ -1/L & 0 & 0 & -r/L & 0 & 0 \\ 0 & -1/L & 0 & 0 & -r/L & 0 \\ 0 & 0 & -1/L & 0 & 0 & -r/L \end{bmatrix} \begin{bmatrix} v_a \\ v_b \\ v_c \\ i_{1a} \\ i_{1b} \\ i_{1c} \end{bmatrix} + \begin{bmatrix} 0 & 0 & 0 & -1/C & 0 & 0 \\ 0 & 0 & 0 & 0 & -1/C & 0 \\ 0 & 0 & 0 & 0 & 0 & -1/C \\ U_{dc}/2L & 0 & 0 & 0 & 0 & 0 \\ 0 & U_{dc}/2L & 0 & 0 & 0 & 0 \\ 0 & 0 & U_{dc}/2L & 0 & 0 & 0 \end{bmatrix} \begin{bmatrix} d_a \\ d_b \\ d_c \\ i_a \\ i_b \\ i_c \end{bmatrix} \tag{3.10}$$

According to eq. (3.10), the structure diagram of the state space average model of the three-phase VSC in the *abc* stationary coordinate system is shown in Fig. 3.13.

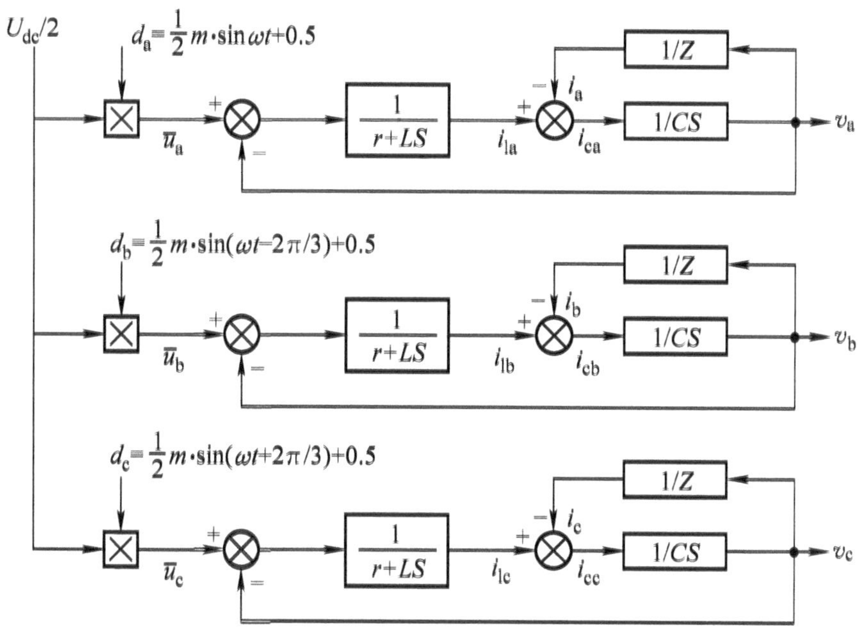

Fig. 3.13 State space average model of three-phase VSC in abc stationary coordinate system

3.3.1.3 Mathematical Model of Three-Phase VSC in dq0 Synchronous Coordinate System

The previous section analyzed the general mathematical model of VSC in the three-phase stationary coordinate system *abc*. The general mathematical model of the three-phase VSC has clear physical meanings and is intuitive. However, in this model, all quantities on the AC side of the VSC are AC variables, which is not conducive to the design of the controller.

In the modeling of three-phase AC motors and three-phase converters, the Park transformation has been widely used. Through the Park transformation, the three-phase AC quantity becomes a two-phase DC quantity, reducing the number of variables, simplifying the system model, and facilitating the design of the controller. The general mathematical model of VSC in the three-phase stationary coordinate system is transformed by Park, and the *dq0* model is obtained [54–56] .

The relationship between the three-phase stationary coordinate system *abc* and the synchronous rotating coordinate system *dq0* is shown in Fig. 3.14, where the *d* axis rotates counterclockwise at an angular speed of ω, $\omega = 2\pi f$, f is the grid frequency of 50 Hz, in the stationary coordinate system, it lags behind the *a* axis at the position of 90° as the phase angle θ at the starting moment, $\theta = \omega t$, *q* axis lags behind the *d* axis 90° .

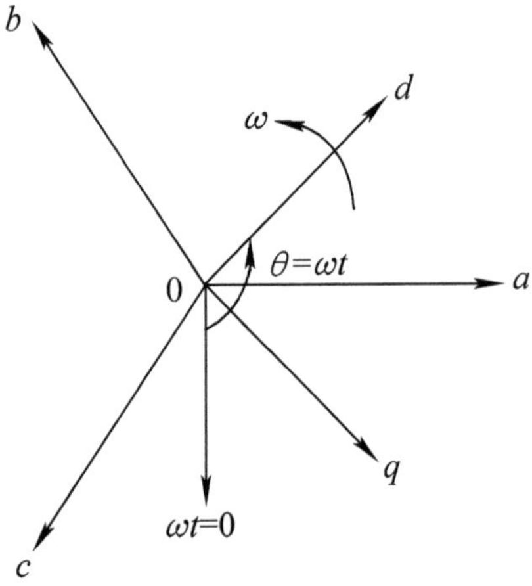

Fig. 3.14 The relationship between *abc* and *dq0* coordinate systems

According to the principle of equal quantity transformation, the above transformation relationship can be described by the transformation matrix T and its inverse matrix T^{-1}:

$$T = \frac{2}{3}\begin{bmatrix} \sin\omega t & \sin\left(\omega t - \frac{2\pi}{3}\right) & \sin\left(\omega t + \frac{2\pi}{3}\right) \\ -\cos\omega t & -\cos\left(\omega t - \frac{2\pi}{3}\right) & -\cos\left(\omega t + \frac{2\pi}{3}\right) \\ \frac{1}{\sqrt{2}} & \frac{1}{\sqrt{2}} & \frac{1}{\sqrt{2}} \end{bmatrix} \quad (3.11)$$

$$T^{-1} = \begin{bmatrix} \sin\omega t & -\cos\omega t & 1/\sqrt{2} \\ \sin\left(\omega t - \frac{2\pi}{3}\right) & -\cos\left(\omega t - \frac{2\pi}{3}\right) & 1/\sqrt{2} \\ \sin\left(\omega t + \frac{2\pi}{3}\right) & -\cos\left(\omega t + \frac{2\pi}{3}\right) & 1/\sqrt{2} \end{bmatrix} \quad (3.12)$$

Therefore, we can obtain the voltage and current quantities under the synchronous rotating coordinate system *dq0*, $\begin{bmatrix} u_d \\ u_q \\ u_o \end{bmatrix} = T \cdot \begin{bmatrix} u_a \\ u_b \\ u_c \end{bmatrix}$, $\begin{bmatrix} v_d \\ v_q \\ v_o \end{bmatrix} = T \cdot \begin{bmatrix} v_a \\ v_b \\ v_c \end{bmatrix}$, $\begin{bmatrix} i_{1d} \\ i_{1q} \\ i_{1o} \end{bmatrix} = T \cdot \begin{bmatrix} i_{1a} \\ i_{1b} \\ i_{1c} \end{bmatrix}$, $\begin{bmatrix} i_d \\ i_q \\ i_o \end{bmatrix} = T \cdot \begin{bmatrix} i_a \\ i_b \\ i_c \end{bmatrix}$.

According to formula (3.11), perform the Park transformation on formula (3.2), and convert the AC quantities to the *dq0* coordinate system, we can get,

$$\begin{cases} C\dfrac{dv_d}{dt} = i_{1d} - i_d - \omega C v_q \\ C\dfrac{dv_q}{dt} = i_{1q} - i_q + \omega C v_d \\ C\dfrac{dv_o}{dt} = i_{1o} - i_o \\ L\dfrac{di_{1d}}{dt} = u_d - v_d - ri_{1d} - \omega L i_{1q} \\ L\dfrac{di_{1q}}{dt} = u_q - v_q - ri_{1q} + \omega L i_{1d} \\ L\dfrac{di_{1o}}{dt} = u_0 - v_0 - ri_{1o} \end{cases} \tag{3.13}$$

From formula (3.13), we can see that in the *S* domain, the output *LC* filter model structure under the synchronous rotating coordinate system is shown in Fig. 3.15, it can be seen that there is a strong coupling relationship between the variables under the *dq0* coordinates.

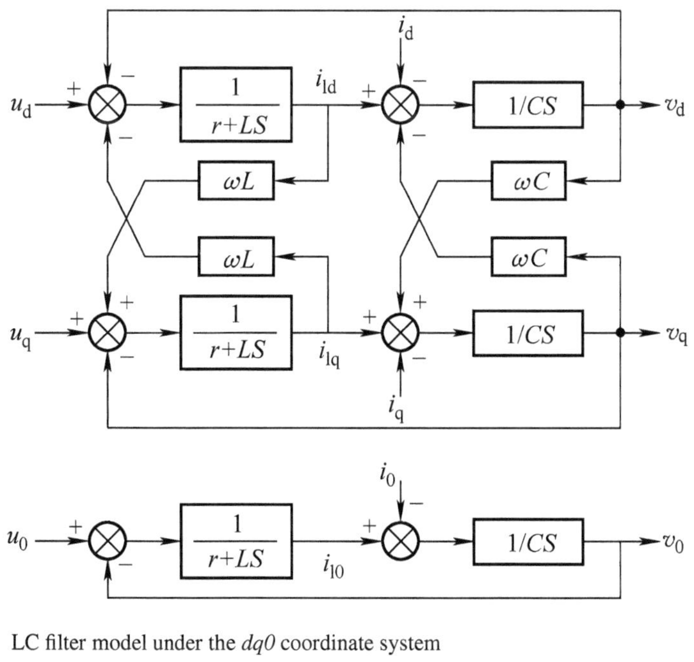

Fig. 3.15 LC filter model under the *dq0* coordinate system

Similarly, using formula (3.11), perform the Park transformation on formula (3.1), so that the switch function is transferred from the three-phase stationary coordinate system *abc* to the synchronous rotating coordinate system *dq0*, that is,

$$\begin{bmatrix} u_d \\ u_q \\ u_o \end{bmatrix} = \begin{bmatrix} S_d \\ S_q \\ S_o \end{bmatrix} \cdot \frac{U_{dc}}{2} \tag{3.14}$$

where, $\begin{bmatrix} S_d \\ S_q \\ S_o \end{bmatrix} = T \cdot \begin{bmatrix} S_a \\ S_b \\ S_c \end{bmatrix}$.

Substituting eq. (3.14) into eq. (3.13), we get,

$$\begin{cases} C\dfrac{dv_d}{dt} = i_{1d} - i_d - wCv_q \\ C\dfrac{dv_q}{dt} = i_{1q} - i_q + wCv_d \\ C\dfrac{dv_o}{dt} = i_{1o} - i_o \\ L\dfrac{di_{1d}}{dt} = \dfrac{U_{dc}}{2} S_d - v_d - ri_{1d} - wLi_{1q} \\ L\dfrac{di_{1q}}{dt} = \dfrac{U_{dc}}{2} S_q - v_q - ri_{1q} + wLi_{1d} \\ L\dfrac{di_{1o}}{dt} = \dfrac{U_{dc}}{2} S_o - v_o - ri_{1o} \end{cases} \tag{3.15}$$

From eq. (3.15), we can see that in the S domain, the structure diagram of the switch function model of the three-phase VSC under the synchronous rotating coordinate system dq0 is as shown in Fig. 3.16.

Equation (3.15) and Fig. 3.16 describe the switch function model of the three-phase VSC under the synchronous rotating coordinate system dq0. Due to the presence of the switch function (S_d, S_q, S_o), this model still exhibits typical nonlinearity, which is not conducive to the design of the controller.

In the three-phase stationary coordinate system abc, using the same method, the Park transformation is applied to another form in the general mathematical model of VSC - the state space average model, and the state space average model of the three-phase VSC under the synchronous rotating coordinate system dq0 is obtained.

Using eq. (3.11), the PWM duty cycle under the three-phase stationary coordinate system abc $[d_a, d_b, d_c]^T$ is transformed into the synchronous rotating coordinate system dq0 $[d_d, d_q, d_o]^T$, that is,

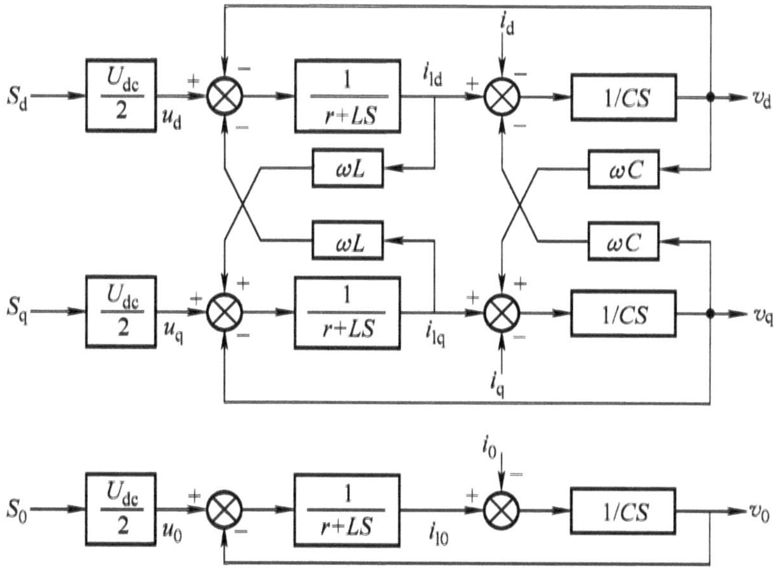

Fig. 3.16 Structure diagram of the switch function model of the three-phase VSC under the dq0 coordinate system

$$\begin{bmatrix} d_d \\ d_q \\ d_o \end{bmatrix} = \begin{bmatrix} \frac{m}{2} \\ 0 \\ \frac{\sqrt{2}}{2} \end{bmatrix} \tag{3.16}$$

Similarly, the Park transformation is applied to eq. (3.9) to obtain the variable of the three-phase bridge output voltage represented by the state space averaging method under the synchronous rotating coordinate system dq0 $\left[\overline{u}_d, \overline{u}_q, \overline{u}_o\right]^T$, that is,

$$\begin{bmatrix} \overline{u}_d \\ \overline{u}_q \\ \overline{u}_o \end{bmatrix} = \begin{bmatrix} d_d \\ d_q \\ d_o \end{bmatrix} \cdot \frac{U_{dc}}{2} \tag{3.17}$$

Substituting eq. (3.17) into eq. (3.13), we get,

$$
\begin{cases}
C\dfrac{dv_d}{dt} = i_{1d} - i_d - \omega C v_q \\
C\dfrac{dv_q}{dt} = i_{1q} - i_q + \omega C v_d \\
C\dfrac{dv_o}{dt} = i_{1o} - i_o \\
L\dfrac{di_{1d}}{dt} = \dfrac{U_{dc}}{2} d_d - v_d - r i_{1d} - \omega L i_{1q} \\
L\dfrac{di_{1q}}{dt} = \dfrac{U_{dc}}{2} d_q - v_q - r i_{1q} + \omega L i_{1d} \\
L\dfrac{di_{1o}}{dt} = \dfrac{U_{dc}}{2} d_o - v_o - r i_{1o}
\end{cases}
\tag{3.18}
$$

According to eq. (3.18), we can see that in the S domain, the structure diagram of the state space average model of the three-phase VSC under the synchronous rotating coordinate system dq0 is as shown in Fig. 3.17.

Equation (3.18) and Fig. 3.17 describe the state space average model of the three-phase VSC under the synchronous rotating coordinate system dq0. Equation (3.18) is obviously a set of linear first-order constant coefficient differential equations, and in Fig. 3.17, in the S domain all quantities are DC quantities, hence, this model is very convenient for the design of VSC controllers.

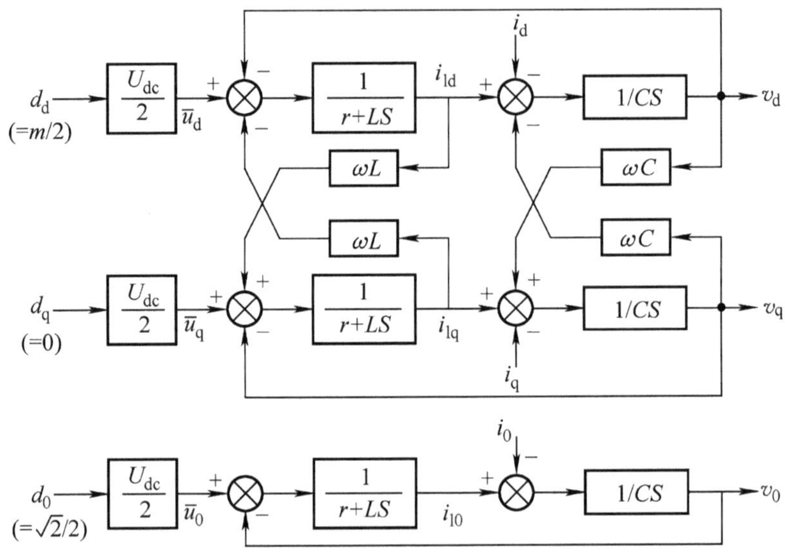

Fig. 3.17 Structure diagram of the state space average model of the three-phase VSC under the dq0 coordinate system

3.3.2 V/f Control

The constant voltage and frequency V/f control of energy storage PCS aims to provide stable voltage and frequency support, and can serve as the balance node for off-grid operation systems. By setting the reference values for voltage and frequency, the voltage and frequency at the PCS output port are detected in real-time as feedback, and the zero-error tracking is achieved under the synchronous coordinate system dq0 through the PI regulator.

Set the d-axis voltage $u_d = U_m$, and the q-axis voltage $u_q = 0$. Under three-phase balanced conditions, $u_0 = 0$. The controller includes a dual closed-loop structure of the output voltage outer loop and the filter inductance current inner loop, as shown in Fig. 3.18. The voltage outer loop is for tracking the output voltage to the set value, and the current inner loop is for improving the control system bandwidth and dynamic characteristics.

In the figure, v_{dref}/v_{qref} is the dq-axis voltage set value, v_d/v_q is the actual value of the dq-axis output voltage, and i_{ld}/i_{lq} is the actual value of the dq-axis inductance current. The PCS under V/f control can be regarded as an ideal voltage source and output impedance series model, as shown in Fig. 3.19.

In the figure, $G(s)$ is the voltage gain, $Z_0(s)$ is the equivalent output impedance, u_{ref} is the voltage set value, and u_0 and i_0 are the actual values of the output voltage and current of the PCS, respectively.

3.3.3 PQ Control

The constant power PQ control of energy storage PCS aims to provide the given active power P_{ref} and reactive power Q_{ref}, and can serve as the PQ node in the system. The implementation idea is to set the power P_{ref}/Q_{ref} Conversion to active current setting and reactive current setting i_{dref}/i_{qref}, as shown in formulas (3.22) and

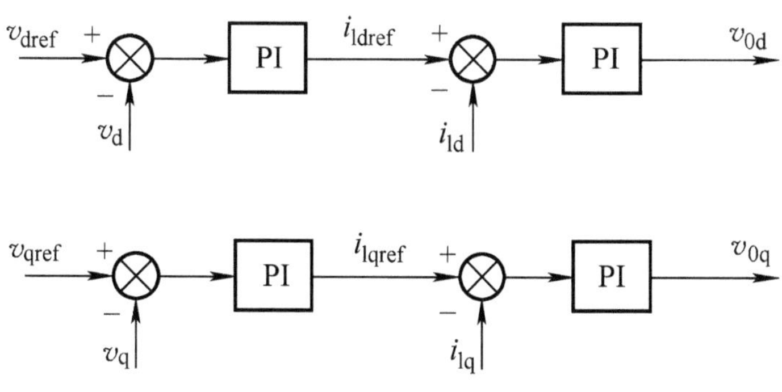

Fig. 3.18 V/f control strategy diagram

Fig. 3.19 Thevenin equivalent model of PCS under V/f control

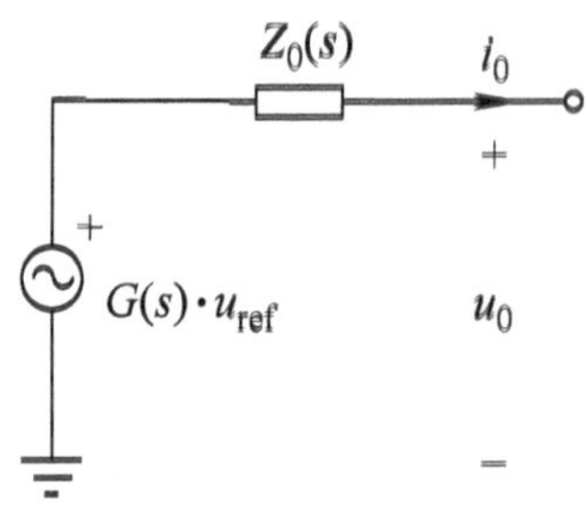

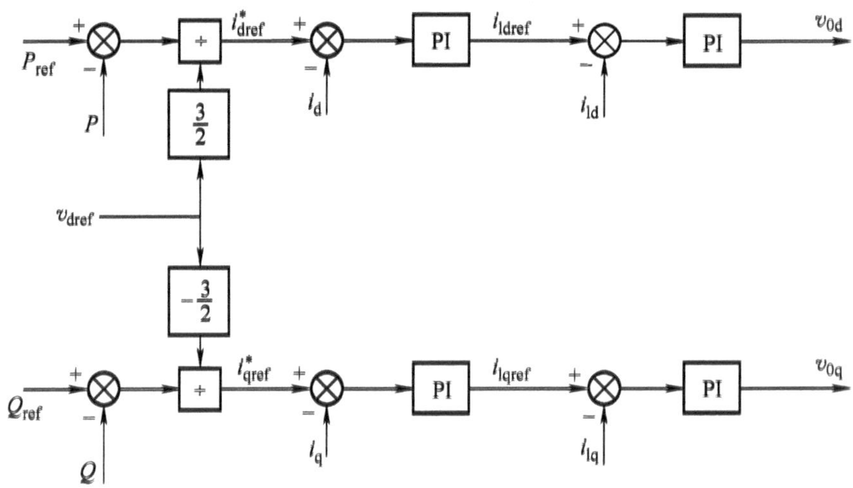

Fig. 3.20 PQ control strategy block diagram

(3.23), then through the *PI* controller for closed-loop control. The control block diagram is shown in Fig. 3.20 [57] .

$$
\begin{aligned}
P_{ref} &= \frac{3}{2}\left(u_d \cdot i_d + u_q \cdot i_q\right) = \frac{3}{2}u_d \cdot i_d \\
Q_{ref} &= \frac{3}{2}\left(u_q \cdot i_d - u_d \cdot i_q\right) = -\frac{3}{2}u_d \cdot i_q
\end{aligned}
\tag{3.22}
$$

Therefore,

$$
\begin{aligned}
i_d &= \frac{2}{3}\frac{P_{ref}}{u_d} \\
i_q &= -\frac{2}{3}\frac{Q_{ref}}{u_q}
\end{aligned}
\tag{3.23}
$$

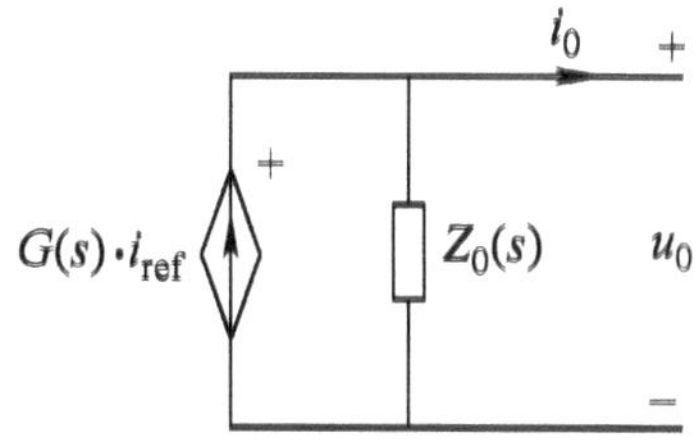

Fig. 3.21 Norton equivalent model of PCS under PQ control

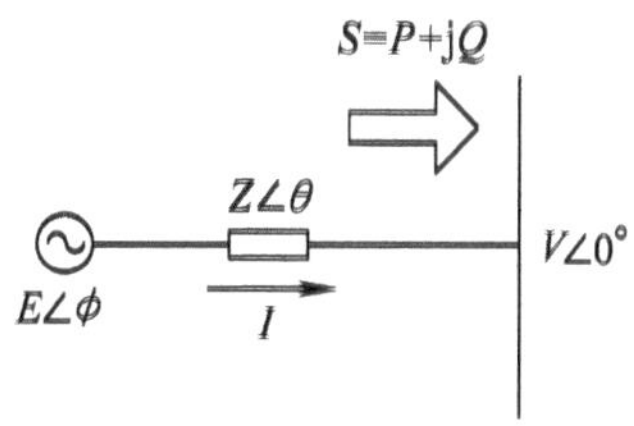

Fig. 3.22 Equivalent model of PCS connected to AC bus

In the figure, P and $Q$are the active power and reactive power values output by the PCS, i_d and i_q are the output current dq axis components, i_{ld} and i_{lq} are the filtered inductance current dq axis components. PQ control allows the PCS to be equivalent to the form of an ideal current source in parallel with the output impedance, as shown in Fig. 3.21.

In the figure, $G(s)$ is the current gain, $Z_0(s)$ is the equivalent output impedance, i_{ref} is the current set value, u_0 and i_0 are the actual voltage and current values at the output end of the PCS.

3.3.4 Droop Control

Droop control simulates the static power-frequency characteristics of traditional synchronous generator sets, making the PCS have similar droop characteristics to synchronous generator sets at the external port. Under this control, the PCS has a primary frequency regulation capability similar to that of conventional synchronous generator sets, which can provide voltage and frequency support for the system alone, or provide voltage and frequency support together through parallel operation of multiple units.

The equivalent model of PCS connected to the AC bus is shown in Fig. 3.22. The traditional droop idea assumes that the connecting line is inductive, i.e., $X >> R$, at this time $\theta = 90°$, the power transmission expression for PCS is:

$$P = \frac{EV}{X}\sin\phi \tag{3.24}$$

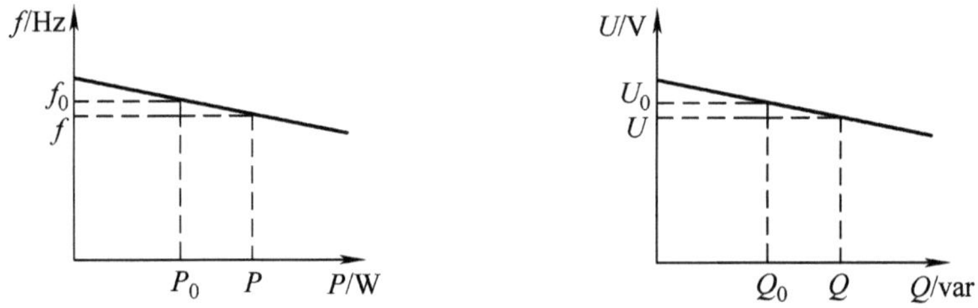

a) Active power/frequency droop characteristics b) Reactive power/voltage droop characteristics

Fig. 3.23 P-ω/Q-v Droop curve

$$Q = \frac{EV\cos\phi - V^2}{X} \tag{3.25}$$

Actually, ϕ is very small, so it is approximately considered that sin$\phi \approx \phi$,cos$\phi \approx 1$, it can be seen that the active power P mainly depends on the voltage phase angle ϕ, while the reactive power Q mainly depends on the voltage amplitude E. Since the phase angle ϕ is the integral of the angular frequency ω ($\omega = d\phi/dt$), it can be dynamically adjusted $\phi$by adjusting ω. Therefore, the droop characteristic is determined as shown in Fig. 3.23, and its relationship is:

$$\omega = \omega^* + m\left(P_0 - P\right) \tag{3.26}$$

$$E = E^* + n\left(Q_0 - Q\right) \tag{3.27}$$

where, $\omega^*$and E^* are the angular frequency and amplitude of the PCS no-load output voltage, m and n are the corresponding droop coefficients. In the case of X> > R, the application of droop method can achieve a good distribution of load current among parallel units [58,59].

However, in the low-voltage grid, the line is resistive [60], if the droop relationship of formulas (3.23) and (3.24) is still adopted, it will lead to the generation of reactive power circulation between parallel PCSs, and in severe cases, some units may even absorb reactive power [61]. In $R >\ > X$ In this case, the power output of PCS is:

$$P = \frac{E\left(E\cos\phi - V\right)}{R} \tag{3.28}$$

$$Q = \frac{EV}{R}\sin\phi \tag{3.29}$$

It can be seen that the active power P mainly depends on the voltage amplitude E, while the reactive power Q mainly depends on the voltage amplitude ϕ. Therefore, the suitable droop characteristic is shown in Fig. 3.24, and its relationship should be:

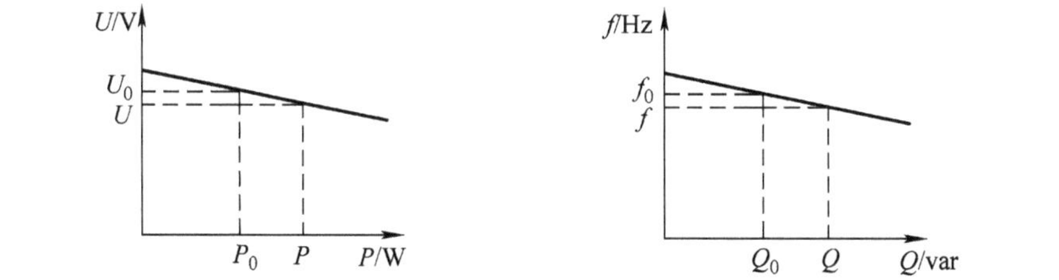

a) Active power/voltage droop characteristics b) Reactive power/frequency droop characteristics

Fig. 3.24 $P-v/Q-\omega$ Droop curve

$$\omega = \omega^* + m(Q_0 - Q) \tag{3.30}$$

$$E = E^* + n(P_0 - P) \tag{3.31}$$

Theoretically, the above $P-v/Q-\omega$ droop can reasonably achieve the balance between parallel PCSs in the low-voltage grid, but there are still many problems in practical applications. First, in low-voltage lines, although the impedance is often greater than the reactance, it is generally difficult to reach the $R>>X$ state, making the equal current effect of using $P-v/Q-\omega$ droop worse; secondly, the active load in the grid is often much higher than the reactive load, so it is more desirable to accurately distribute the active output P between power sources through the global variable ω. And E as a local quantity, it is greatly affected by line differences and is difficult to use to accurately adjust P, which will cause active circulating current, which is far more serious than reactive circulating current; furthermore, because the $P-v/Q-\omega$ droop is opposite to the characteristics of the synchronous generator set, it makes it difficult to control the parallel operation of PCSs and synchronous machines.

In summary, regardless of whether it is $P-\omega/Q-v$ droop or $P-v/Q-\omega$ droop, there are problems in the application in the medium and low voltage power grid. The fundamental reason is that the line impedance characteristics (as shown in Table 3.2) change, causing the power transmission characteristics to be different from those in the high voltage power grid. To solve this problem, virtual impedance technology can be introduced to correct the line impedance characteristics, suppress the circulating current between parallel units, which will be detailed in the following chapters.

Table 3.2 Typical line parameters at different voltage levels

Type	Resistance $r/(\Omega/km)$	Resistance $X/(\Omega/km)$	Impedance Ratio r/X
Low voltage line	0.642	0.101	6.35
Medium voltage line	0.161	0.190	0.85
High voltage line	0.060	0.191	0.31

Chapter 4
Flywheel Energy Storage System

Flywheel energy storage stores energy in the form of mechanical energy in a high-speed rotating rotor. The core technology is the rotor material, support bearing, and electromechanical control system. This chapter mainly introduces the main structure of the flywheel energy storage system, the electromechanical control system, and the charging and discharging control process [62].

4.1 Structure of Flywheel Energy Storage System

The flywheel energy storage system generally consists of a flywheel rotor, support bearing, motor, protective shell, and power electronic conversion equipment. During charging, the power conversion circuit controls the motor to operate in motor mode, driving the coaxially connected flywheel rotor to rotate at high speed, converting electrical energy into mechanical energy. When the flywheel speed reaches the design limit, it is considered that the system is fully charged, stop inputting external electrical energy, and enter standby mode; when energy needs to be released, the power conversion circuit controls the motor to operate in generator mode, converting the mechanical energy stored in the flywheel rotor into electrical energy for external power supply. The specific manifestation of discharge is that the flywheel speed decreases until the designed lower limit, it is considered that the energy is exhausted, the system stops discharging and prepares to re-enter the charging state.

Because flywheel energy storage has high power density and energy density, no environmental pollution, high conversion efficiency, low standby loss, long service life, and wide operating temperature range, it has been applied in certain fields such as uninterruptible power supply, power quality regulation, power system frequency regulation, satellite attitude control, regenerative braking energy recovery in rail transit, and high power load.

X. Tang et al., *Electrical Energy Storage Technologies and Applications*,
https://doi.org/10.1007/978-981-96-5625-7_4

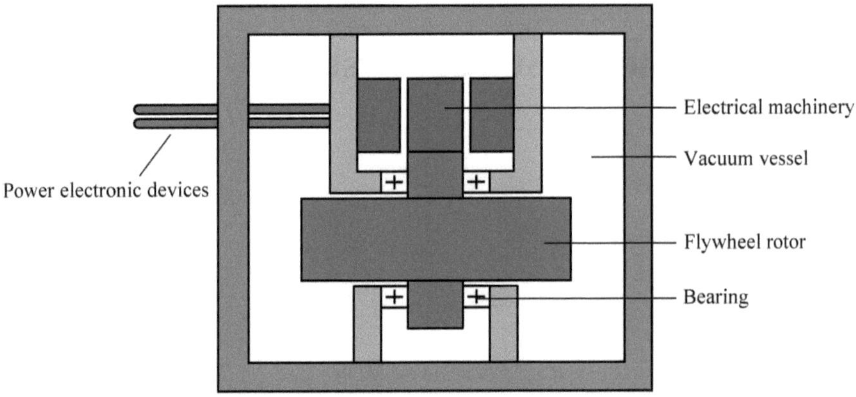

Fig. 4.1 Schematic diagram of the structure of the flywheel energy storage system

A typical main structure of a flywheel energy storage system is shown in Fig. 4.1 [63], its internal motor can operate as both a motor and a generator, the motor is coaxially connected with the rotor, and can also be designed as one. Because of the use of a vacuum container and magnetic levitation bearings, the energy loss of the system in standby mode is very low.

The energy stored in the flywheel can be represented as:

$$\Delta E = \frac{1}{2} J \left(\varpi_{\max}^2 - \omega_{\min}^2 \right) \tag{4.1}$$

where, J is the moment of inertia of the rotor, ω is the angular velocity of the rotor.

4.2 Main Circuit Topology of Flywheel Energy Storage System

The flywheel energy storage system presents certain DC power characteristics through the motor controller, and can therefore be connected to the AC grid through a Voltage Source Converter (VSC).

4.2.1 Main Circuit Topology

Figure 4.2 shows the main circuit topology of the flywheel energy storage system based on the Back-Back dual PWM converter, which consists of a grid-side *LCL* filter, a back-to-back dual PWM converter, a permanent magnet synchronous motor, a flywheel rotor, etc. The *LCL* filter consists of a grid-side inductor *Lg*, a converter-side inductor *Lconv*, and a filter capacitor *C*; The Back-Back dual PWM converter

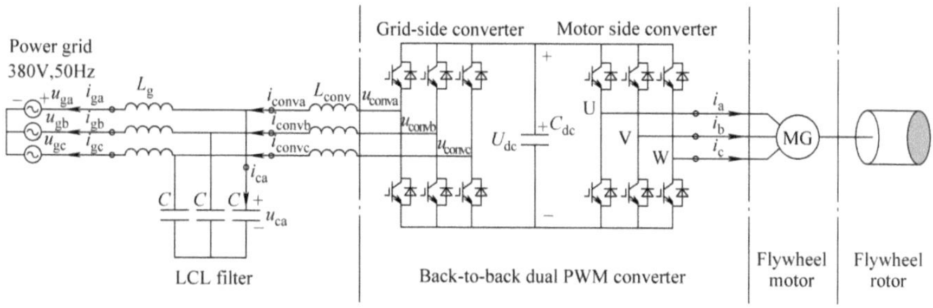

Fig. 4.2 Main Circuit Topology of Flywheel Energy Storage System Based on Back-to-Back Dual PWM

consists of a grid-side converter, a motor-side converter, and a DC bus capacitor C_{dc}, etc.

Based on the above main circuit topology, the grid-connected charging and discharging control of the flywheel energy storage system consists of grid-side converter control and motor-side converter control, and goes through three stages: pre-charging, pre-grid connection, and grid operation. Among them, it is necessary to solve several key links such as flywheel rotor speed control, DC bus voltage control, and grid quality tracking control. In the pre-charging and pre-grid connection stages, the grid-side converter adopts uncontrolled rectification; the motor-side converter adopts speed outer loop control in the pre-charging stage, and voltage outer loop control in the pre-grid connection stage. During the grid operation stage, the grid-side converter control adopts a direct power control strategy based on the grid-side inductor current outer loop and the converter-side inductor current inner loop, controlling the grid-connected active power; the motor-side converter control adopts a double closed-loop control strategy with the DC bus voltage outer loop and the current inner loop, controlling the DC bus voltage.

4.2.2 *Mathematical Model of Permanent Magnet Synchronous Motor*

In the *dq* coordinate system, the voltage-current relationship of the permanent magnet synchronous motor in the flywheel energy storage system is:

$$\begin{cases} u_{md} = R_s i_{md} + L_d \dfrac{di_{md}}{dt} - \omega_e L_q i_{mq} \\ u_{mq} = R_s i_{mq} + L_q \dfrac{di_{mq}}{dt} + \omega_e L_d i_{md} + \omega_e \psi_f \end{cases} \tag{4.2}$$

where u_{md} and u_{mq} are the d-axis and q-axis components of the stator voltage respectively, R_s is the stator resistance, i_{md} and i_{mq} *represent the direct-axis and*

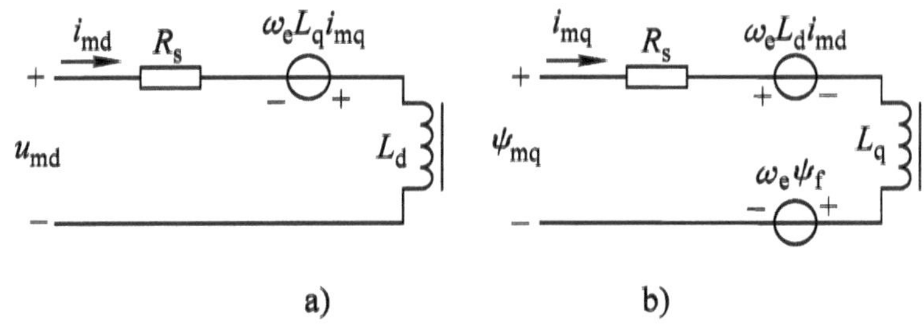

Fig. 4.3 Equivalent circuit diagram of the *dq* axis of the permanent magnet synchronous motor

quadrature-axis armature currents, L_d, L_q represent the d-axis and q-axis inductances, ω_e is the electrical angular speed of the motor, ψ_f is the rotor excitation flux linkage.

The equivalent circuits of the d-axis and q-axis of the permanent magnet synchronous motor are shown in Fig. 4.3a, b respectively:

When the control strategy of the d-axis armature current i_{md} is 0, the q-axis voltage-current relationship is:

$$u_{mq} = R_s i_{mq} + L_q \frac{di_{mq}}{dt} + \omega_e \psi_f \tag{4.3}$$

The motor torque balance equation is:

$$J\frac{d\omega_e}{dt} + B\omega_e = p \times T_e = \frac{3}{2} p^2 \psi_f i_{mq} = K_t i_{mq} \tag{4.4}$$

where, $K_t = 3p^2\psi f/2$, J is the moment of inertia, B is the viscous friction coefficient, p is the number of pole pairs, Te is the electromagnetic torque.

4.2.3 Design and Stability of Motor Side Controller Parameters

In the grid-connected operation control of the flywheel energy storage system, the motor-side controller includes a DC bus voltage outer loop controller and a current inner loop controller [64].

4.2.3.1 Design and Stability of Current Inner Loop Controller

According to eqs. (4.3) and (4.4), the structure of the motor-side quadrature-axis current controller is shown in Fig. 4.4.

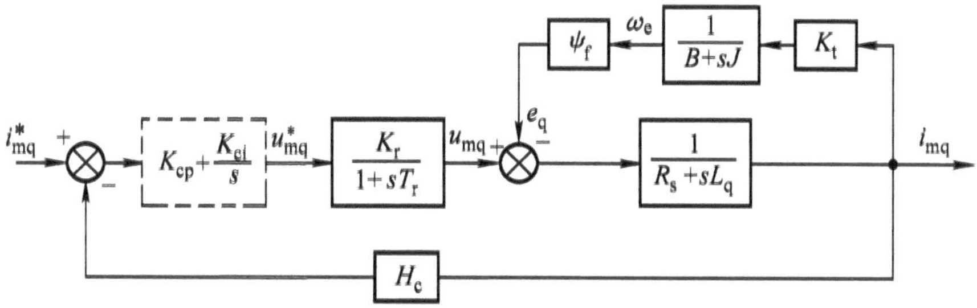

Fig. 4.4 Structure diagram of the motor-side quadrature-axis current controller

The closed-loop transfer function of the quadrature-axis current loop can be derived as:

$$\frac{i_{mq}(s)}{i_{mq}^{*}(s)} = \frac{N_c(s)}{D_c(s)} \tag{4.5}$$

where,

$$N_c(s) = K_a K_r (1+sT_m)(K_{ci}+sK_{cp})$$

$$\begin{aligned} D_c(s) = {} & (1+sT_a)(1+sT_m)(1+sT_r)s + H_c K_a K_{ci} K_r (1+sT_m) \\ & + H_c K_a K_{cp} K_r (1+sT_m)s + K_a K_m K_t \psi_f (1+sT_r)s \end{aligned}$$

And,

$$K_a = 1/R_s,\ T_a = L_q/R_s,\ K_m = 1/B,\ T_m = J/B$$

At the bandwidth frequency, the following conditions approximately hold:

$$(1+sT_r)\approx 1,\ (1+sT_m)\approx sT_m,\ (1+sT_a)(1+sT_r)\approx 1+sT_{ar}$$

where, $T_{ar} = T_a + T_r$.

Based on the above approximate conditions, eq. (4.5) can be simplified as:

$$\frac{i_{mq}(s)}{i_{mq}^{*}(s)} = \frac{N_c'(s)}{D_c'(s)} \tag{4.6}$$

In which,

$$N_c'(s) = K_a K_r T_m K_{ci}\left(1+sK_{cp}/K_{ci}\right)$$

$$\begin{aligned} D_c'(s) &= \left(T_{ar}T_m s^2 + T_m s + K_a K_m K_t \psi_f\right) + H_c K_a K_r T_m K_{ci}\left(1+sK_{cp}/K_{ci}\right) \\ &= K_a K_m K_t \psi_f (1+sT_1)(1+sT_2) + H_c K_a K_r T_m K_{ci}\left(1+sK_{cp}/K_{ci}\right) \end{aligned}$$

$$T_1 = 2T_{ar}T_m / \left(T_m + \sqrt{T_m^{\ 2} - 4T_{ar}T_mK_aK_mK_t\psi_f}\right)$$

$$T_2 = 2T_{ar}T_m / \left(T_m - \sqrt{T_m^{\ 2} - 4T_{ar}T_mK_aK_mK_t\psi_f}\right)$$

Let

$$K_{cp} / K_{ci} = T_i \tag{4.7}$$

Equation (4.6) can be further simplified by cancelling out the zeros and poles to:

$$\frac{i_{mq}(s)}{i_{mq}^*(s)} = \frac{K_i}{1+sT_i} = G_i(s) \tag{4.8}$$

where,

$$K_i = K_aK_rT_mK_{ci} / \left(K_aK_mK_t\psi_f + H_cK_aK_rT_mK_{ci}\right)$$

$$T_i = K_aK_mK_t\psi_fT_2 / \left(K_aK_mK_t\psi_f + H_cK_aK_rT_mK_{ci}\right)$$

It can be obtained,

$$K_{ci} = \frac{K_mK_t\psi_f\left(T_2 - T_i\right)}{H_cK_rT_mT_i} \tag{4.9}$$

After simplification, the d-axis current loop can be reduced from the fourth order system shown in eq. (4.5) to the first order system shown in eq. (4.8). Since the adjustment time of the first order system:

$$t_s = 3T_i \tag{4.10}$$

Therefore, according to the current loop adjustment time t_s required by the system, T_i can be determined, and then according to T_i, K_{ci} can be determined according to eq. (4.9), and then K_{cp} according to eq. (4.7).

From the first-order simplified current closed-loop transfer function (4.8), it can be concluded that the characteristic equation of the current inner loop control system is $1 + sT_i = 0$, its root is a negative number, so it can be known that the current inner loop control system is stable.

4.2.3.2 Voltage Outer Loop Controller Design and Stability

The parameters of the voltage outer loop controller can be designed first during the pre-grid-connected stage, and then fine-tuned in the experiment. In the pre-grid-connected stage, the structure of the voltage control loop is shown in Fig. 4.5, where $K_{Te} = 3\ p\psi_f/2$, from which the voltage open-loop transfer function can be derived as:

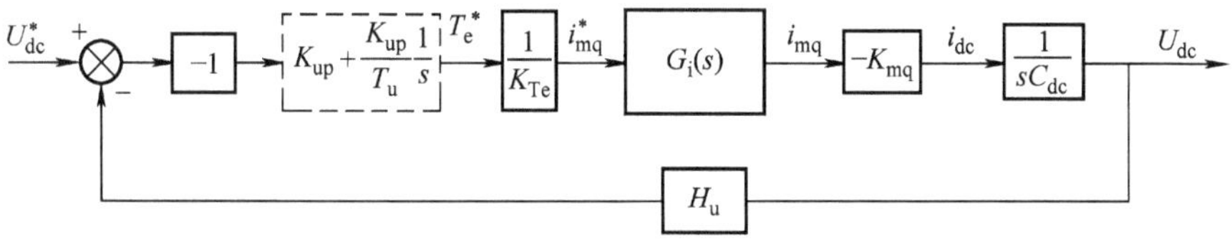

Fig. 4.5 Structure diagram of the motor side voltage control loop

$$G_{u\_ol}=\frac{K_{ug}K_{up}\left(1+sT_u\right)}{T_u s^2\left(1+sT_i\right)} \tag{4.11}$$

The voltage closed-loop transfer function is:

$$\frac{U_{dc}}{U_{dc}^*}=\frac{1}{H_u}\frac{\left(1+sT_u\right)}{\frac{T_iT_u}{K_{ug}K_{up}}s^3+\frac{T_u}{K_{ug}K_{up}}s^2+\left(1+sT_u\right)} \tag{4.12}$$

where, $K_{ug}=\frac{H_uK_iK_{mq}}{K_{Te}C_{dc}}$

The voltage closed-loop transfer function (4.12) is equivalent to the symmetric optimization function shown in formula (4.13) with a damping ratio of 0.707.

$$\frac{U_{dc}(s)}{U_{dc}^*(s)}=\frac{1}{H_u}\frac{\left(1+sT_u\right)}{\left(\frac{1}{16}T_u^3\right)s^3+\left(\frac{3}{8}T_u^2\right)s^2+\left(1+sT_u\right)} \tag{4.13}$$

Comparing formula (4.12) with formula (4.13), we get,

$$K_{up}=\frac{4}{9K_{ug}T_i}$$

$$K_{ui}=\frac{K_{up}}{T_u}=\frac{2}{27K_{ug}T_i^2}$$

$$T_u=6T_i$$

From the voltage closed-loop transfer function (4.13), the characteristic equation of the motor-side converter control system can be obtained as:

$$D_u(s)=\left(\frac{1}{16}T_u^3\right)s^3+\left(\frac{3}{8}T_u^2\right)s^2+T_us+1=0 \tag{4.14}$$

Listing the Routh table, the first column elements are $T_u^3/16$, 3 $T_u^2/8$, 5 $T_u/6$ and 1. Since $T_u=6\ T_i>0$, all the elements in the first column of the Routh table are greater than 0, according to the Routh stability criterion, the motor-side converter control system is stable.

4.3 Operation Control of Flywheel Energy Storage System

The grid-connected control process of the flywheel energy storage system is completed by the cooperation of the grid-side converter and the motor-side converter. The entire control process is implemented through the grid-connected preparation stage (pre-charging, pre-grid-connected) and the grid-connected operation stage.

4.3.1 Grid-Connected Preparation

The grid-connection preparation phase undergoes pre-charging after power-on and the subsequent pre-grid-connection process. During the pre-charging phase, the grid-side converter adopts an uncontrolled rectification method; the motor-side converter uses a speed outer loop and current inner loop control method to accelerate the flywheel motor to the set speed. When the flywheel energy storage system is in the pre-charging phase, the logic switch *S* in Fig. 4.6 is set at *a* position.

When the flywheel motor accelerates to the set speed, the control method of the grid-side converter remains unchanged and still uses uncontrolled rectification; however, the motor-side converter switches from the speed outer loop to the voltage outer loop control method, at which point the flywheel energy storage system enters the pre-grid-connection phase.

Since the DC bus voltage obtained by the uncontrolled rectification of the grid is lower than the DC bus voltage required for the grid-connected operation of the flywheel energy storage, therefore, the DC bus voltage command value U_{dc} * in the voltage outer loop control of the motor-side converter needs to be greater than the

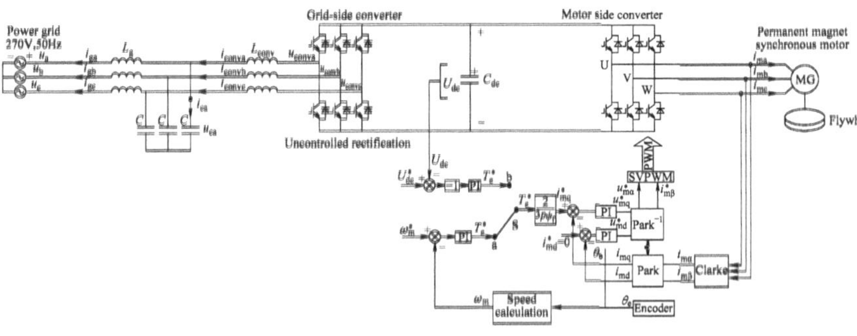

Fig. 4.6 Pre-charging and pre-grid-connection control methods of the flywheel energy storage system

DC bus voltage value obtained by the uncontrolled rectification of the grid. Since the peak value of the grid line voltage is lower than the DC voltage required for the normal operation of the flywheel motor, the grid voltage is reverse blocked by the diodes of the uncontrolled rectification bridge, and the grid-side converter can be equivalent to an open circuit, so the flywheel motor in the pre-grid-connection phase can be regarded as a voltage outer loop no-load discharge. When the flywheel energy storage system is in the pre-grid-connection phase, the logic switch *S* in Fig. 4.6 is set at *b* position.

4.3.2 Grid Operation

After entering the pre-grid-connection phase, enable the PWM control of the grid-side converter, and the control method of the motor-side converter remains unchanged as the voltage outer loop, and the flywheel energy storage system enters the grid operation phase.

The control method of the flywheel energy storage system during grid operation is shown in Fig. 4.7. Among them, the grid-side converter adopts a direct power control strategy based on the grid-side inductor current outer loop and the converter-side inductor current inner loop, controlling the power exchange between the DC bus and the grid. The motor-side converter adopts a dual closed-loop control strategy of the DC bus voltage outer loop and the current inner loop, maintaining a constant DC bus voltage, and indirectly controlling the power exchange between the DC bus and the flywheel motor to meet the power demand of the grid-side converter for the DC bus.

When the grid-connected active power command value P^* is positive, the flywheel energy storage system discharges during grid connection; when P^* is negative and the absolute value is greater than the system loss, the flywheel energy storage system charges during grid connection. When there is no need for the flywheel energy storage system to exchange reactive power with the grid, the grid-connected reactive power command value Q^* can be set to 0.

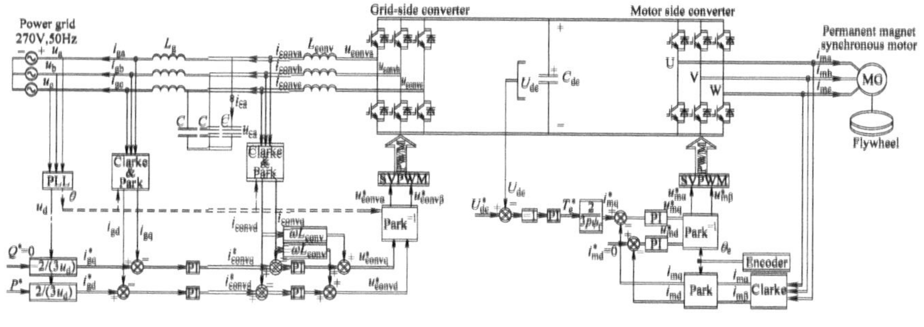

Fig. 4.7 Control method of grid-connected operation of flywheel energy storage system

When P^* is negative, the active power input from the grid to the flywheel energy storage system is $| P^* |$, and the total loss of the system P_{loss} ($| P^* |$, ω_m) is related to $| P^* |$ and the flywheel motor speed ω_m. Thus, when $| P^* | = P_{loss}$ ($| P^* |$, ω_m), the active power input from the grid to the flywheel energy storage system equals the total loss of the flywheel energy storage system, and the flywheel motor speed will stabilize at ω_m. Therefore, when $| P^* | = P_{loss}$ ($| P^* |$, ω_m), the flywheel energy storage system operates in grid-connected standby mode, and the corresponding standby speed is ω_m.

4.3.3 Flywheel Energy Storage Experiment

4.3.3.1 Experimental Platform and Main Parameters

The flywheel energy storage system experimental platform is set up, consisting of an array of three flywheel systems, including the flywheel energy storage controller as shown in Fig. 4.8, and the flywheel energy storage units as shown in Fig. 4.8 (including the permanent magnet synchronous motor and flywheel rotor). The main parameters of the permanent magnet synchronous motor and flywheel rotor are shown in Table 4.1, the main circuit parameters of the system are shown in Table 4.2, and the controller parameters are shown in Table 4.3, Fig. 4.9.

Fig. 4.8 Flywheel Energy Storage Controller

Table 4.1 Permanent Magnet Synchronous Motor and Flywheel Parameters

Number of pole pairs p	2	Rated speed	5000 rpm
Rated current	25 A	Rated torque	14 nm
Stator resistance R_s	0.097 ohm	Excitation flux linkage ψ_f	0.1286 Wb
Direct-axis inductance L_d	1.435 mH	Transverse inductance L_q	2.085 mH
Flywheel rotational inertia J	2.063 kgm ^{2}	Viscous friction coefficient B	0.0005 nm/(rad/s)

Table 4.2 Main circuit parameters of the system

Effective value of grid line voltage	270 V	Grid frequency	50 Hz
L_g	0.3 mH	L_{conv}	0.8 mH
C	60 uF	U^*_{dc}	500 V

Table 4.3 Controller parameters

Control system of grid-side converter		Control system of motor-side converter	
Grid-side current outer loop	Converter-side current inner loop	DC bus voltage outer loop	Current inner loop
$K_{gp} = 0.2$	$K_{ip} = 153$	$K_{up} = 0.8$	$K_{cp} = 48.5$
$K_{gi} = 300$	$K_{ii} = 307230$	$K_{ui} = 210$	$K_{ci} = 2226$

Fig. 4.9 Flywheel Energy Storage Units

Combining the known parameters such as K_r, T_r, K_{mq}, C_{dc}, H_c, H_u, etc., the parameters of the motor-side converter control system can be obtained as shown in Table 4.2.

Set the adjustment time $t_s = 2$ ms required by the current inner loop, according to Sect. 4.2, can obtain $T_i = 0.000667$ s, $T_1 = 0.0218$, $T_2 = 1.9949$, and $K_{ci} = 2226$, $K_{cp} = 48.5$.

Substituting $K_{cp} = 48.5$, $K_{ci} = 2226$ and the data in the table into formulas (4.5) and (4.8), the fourth order current closed-loop transfer function Bode plot and the first order simplified current closed-loop transfer function Bode plot of the motor side can be obtained respectively, as shown in Fig. 4.10. It can be seen that the first order simplified model is relatively close to the fourth order accurate model in amplitude-frequency characteristics and phase-frequency characteristics, indicating that the design parameters based on the first order simplified model are relatively accurate.

After determining the parameters of the current inner loop controller, proceed to design the parameters of the DC bus voltage outer loop controller. According to

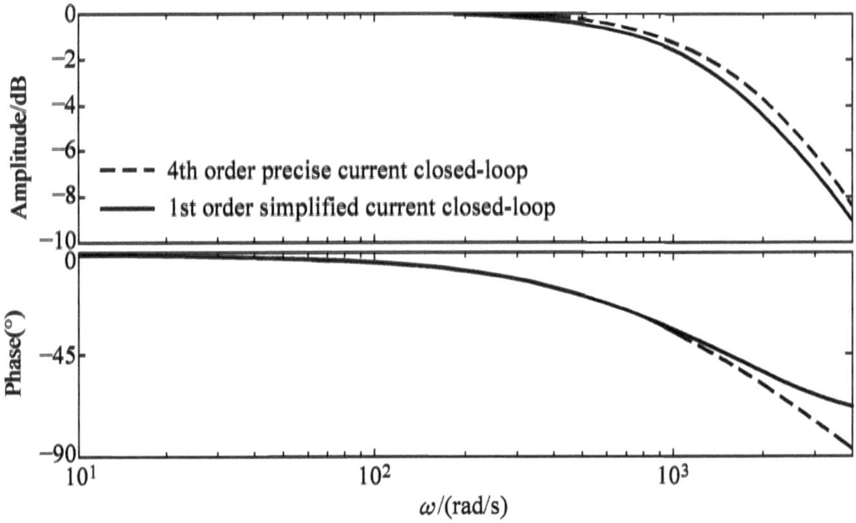

Fig. 4.10 Bode plots of fourth order and first order simplified current loop on the motor side

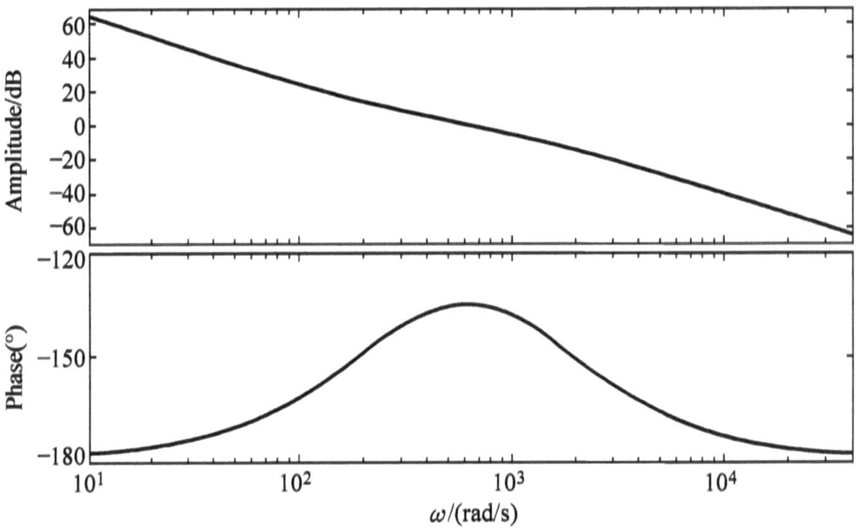

Fig. 4.11 Bode plot of open-loop voltage on the motor side

Sect. 4.2, we can get $K_l = 0.9997$, $K_{ug} = 793.86$, $K_{up} = 0.8$, $K_{ui} = 210$. Figure 4.11 is the Bode plot of the open-loop transfer function of the voltage outer loop.

In the figure, the high-frequency section of the amplitude-frequency curve decays quickly, indicating that the system has a strong ability to resist high-frequency noise. In addition, the phase-frequency curve is always above the −180° line, ensuring the stability of the system.

4.3.3.2 Experimental Results

On the flywheel energy storage system experimental platform, pre-charging, pre-grid connection, and grid-connected operation experiments were conducted to verify the proposed grid-connected control method of the flywheel energy storage system.

Pre-charging and Pre-grid Connection

The waveforms of grid voltage u_a, grid current i_{ga}, DC bus voltage U_{dc}, and flywheel motor speed n during the pre-charging and pre-grid-connection stages are shown in Fig. 4.12.

In the figure, during the pre-charging stage, the flywheel motor accelerates to a preset speed of 4200 rpm with the speed outer loop; at the moment of transition from pre-charging to pre-grid-connection, the DC bus voltage value jumps from about 380 V (obtained from uncontrolled rectification of the grid) during the pre-charging stage to 500 V controlled by the voltage outer loop of the flywheel motor. Moreover, during the pre-grid-connection stage, although the grid-side converter is equivalent to an open circuit, due to the presence of the filtering capacitor C in the LCL filter, the grid current i_{ga} is a certain reactive current.

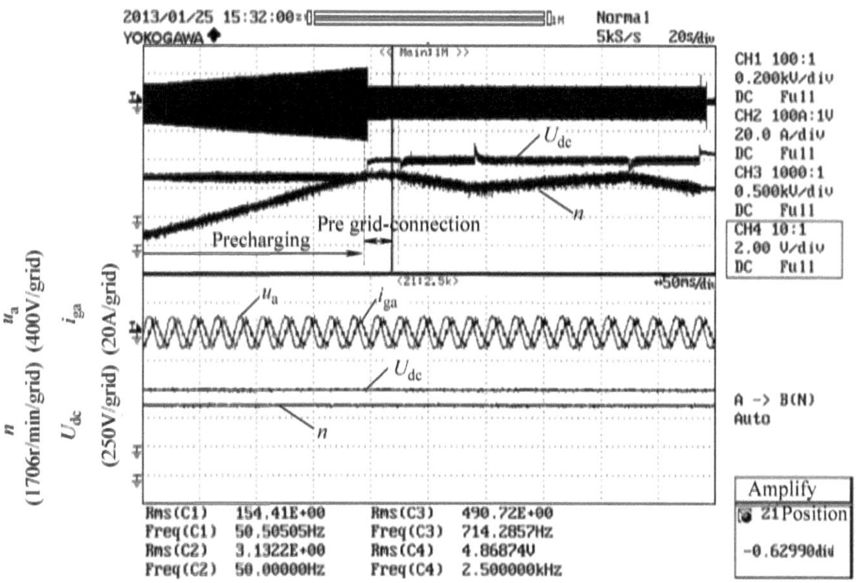

Fig. 4.12 Waveforms of pre-charging and pre-grid-connection experiment

Grid Operation

During the pre-grid-connection stage, the grid power set value P^* is set to $P^* = 1.6$ kW, at this time the PWM pulse of the grid-side converter is enabled, and the system immediately transits from pre-grid-connection to grid-connected discharge stage. At the moment when the system transits from pre-grid-connection to grid-connected discharge, the waveforms of grid voltage u_a, grid current i_{ga}, DC bus voltage U_{dc}, and flywheel motor speed n are shown in Fig. 4.13.

It can be seen that after enabling the PWM pulse of the grid-side converter, the grid current i_{ga} quickly becomes in-phase with the grid voltage u_a, the system transits from pre-grid-connection mode to grid-connected discharge mode, with a fast response speed. Although the DC bus voltage will experience a brief drop during the transition, it can rise back in a short time and maintain at 500 V. The experimental waveform after the system enters the grid-connected discharge is shown in Fig. 4.14.

It can be seen that during grid-connected discharge, the phase angle difference between the grid voltage u_a and the grid current i_{ga} is 0°, and the flywheel energy storage system provides a constant active power to the grid; the DC bus voltage is maintained at 500 V, and the flywheel speed drops from 4200 rpm to 3400 rpm.

When the flywheel speed drops to 3400 rpm, $P^* = 1.6$ kW is changed to P^* At −2 kW, the system switches from grid discharging to grid charging, as shown in the grid charging experiment waveform in Fig. 4.15.

As can be seen, during grid charging, the grid voltage u_a and the grid current i_{ga} are 180° of phase, and the flywheel energy storage system absorbs a constant active power from the grid; the DC bus voltage is maintained at 500 V, and the flywheel

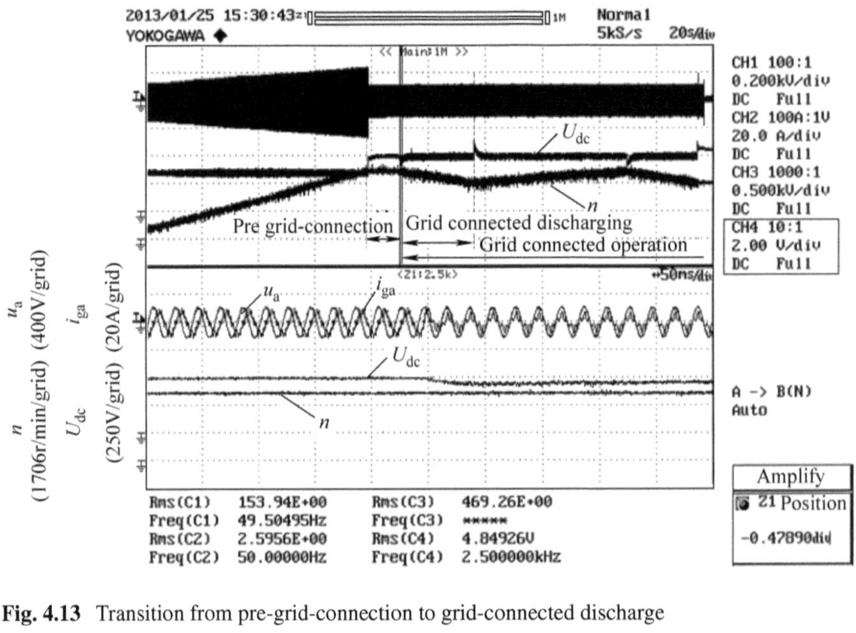

Fig. 4.13 Transition from pre-grid-connection to grid-connected discharge

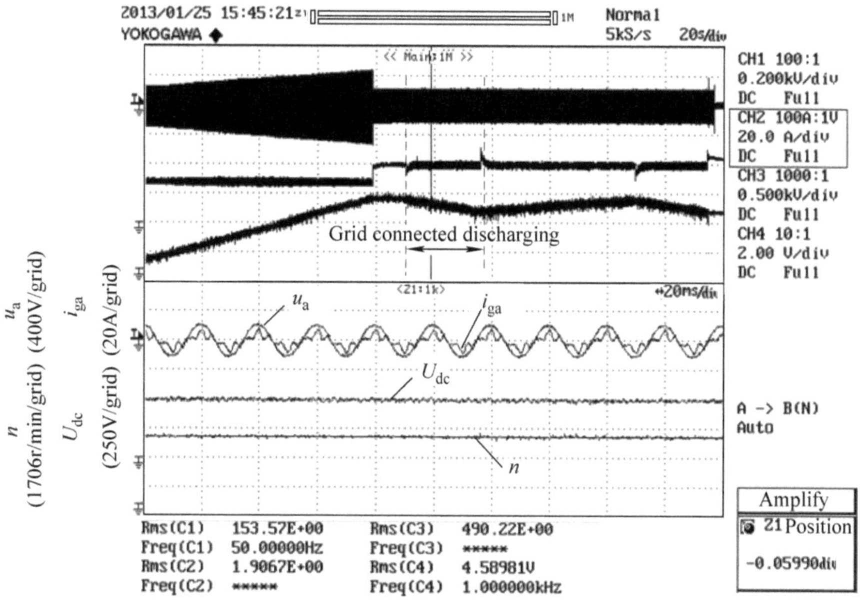

Fig. 4.14 Grid-connected discharge experimental waveform

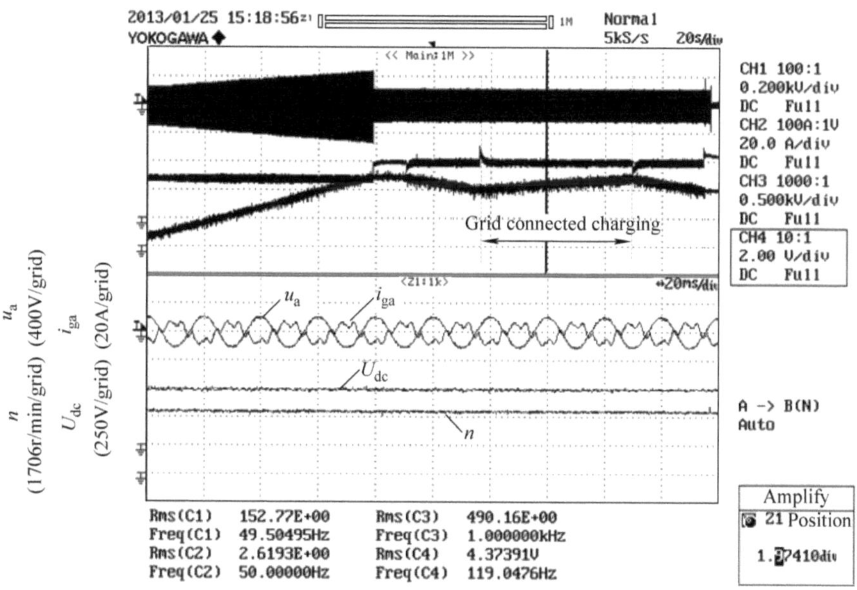

Fig. 4.15 Grid charging experimental waveform

speed rises from 3400 rpm to 4200 rpm, completing a grid charging and discharging cycle. The experimental waveform at the moment of switching from grid discharging to grid charging is shown in Fig. 4.16.

It can be seen that the system can switch from grid discharging to grid charging in a very short time (about 10 ms), with a fast response speed; although the DC bus voltage will briefly rise during the transition, it can fall back and maintain at 500 V in a short time.

When the flywheel speed rises to 4200 rpm, $P^* = -2$ kW is changed back to $P^* = 1.6$ kW, and the system switches from grid charging back to grid discharging. The experimental waveform of the switch from grid charging to grid discharging is shown in Fig. 4.17.

It can be seen that the system can quickly switch from grid charging to grid discharging, with a fast transition process; although the DC bus voltage will briefly drop during the transition, it can rise back and maintain at 500 V in a short time.

As can be seen, the flywheel energy storage, as a typical electromechanical control system, has advantages such as high power density, high operating efficiency, fast charging and discharging switching, high operational stability, and long cycle life. Of course, there are also issues such as complex overall structure, high mechanical requirements, and complex control system. According to the characteristics of the application field of flywheel energy storage, good product design, modular and serial production, and the use of flywheel array technology with multiple units operating in combination can improve the technical and economic performance of flywheel energy storage.

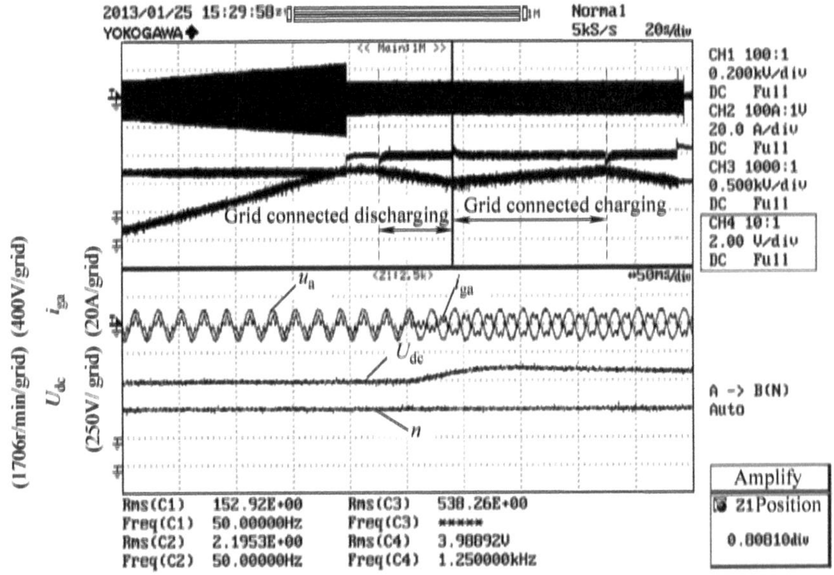

Fig. 4.16 Process from grid discharging to grid charging

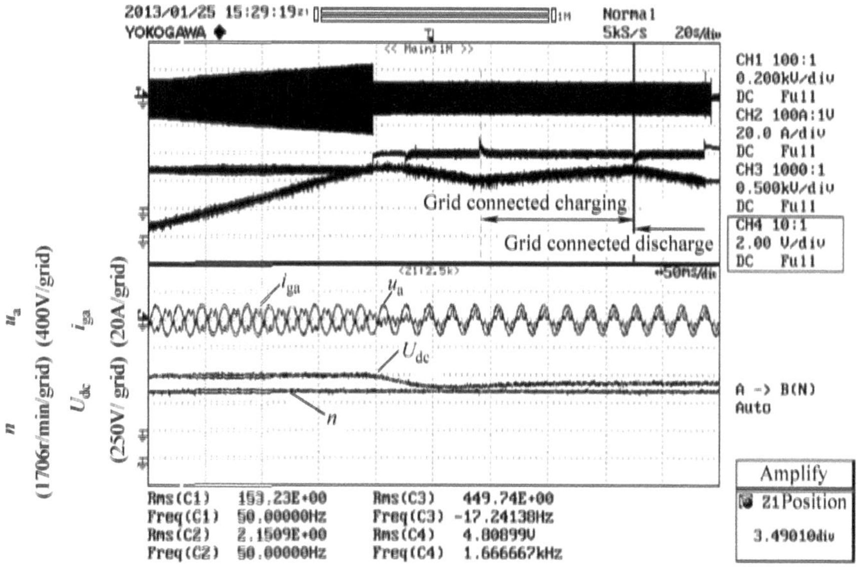

Fig. 4.17 Process from grid charging to grid discharging

Chapter 5
Composite Energy Storage System

Combining two or more complementary energy storage systems according to application requirements is an effective way to solve the current economic insufficiency of single energy storage technology. This chapter analyzes the overall performance improvement of composite energy storage and the corresponding combination control strategies.

5.1 Introduction of Composite Energy Storage

Power supply systems or power systems are facing more and more pulsating load demands, such as electric vehicles, mobile digital devices, directed energy weapons, and various auxiliary services of power systems. Their typical characteristic is that the peak power is very high, but the average power is relatively low. For example, a light passenger car has an average power of about 10 kW during driving, but when the car accelerates, the required peak power is about 60 kW [65], which is 6 times the average power, but the duration is usually only a few seconds.

Various types of electrochemical batteries, as widely used energy storage technologies, are limited in their high power output and input capabilities due to the electrochemical reactions that occur during the energy storage process and the influence of the diffusion speed of the ions involved in the reaction. If a battery is used as the main power source for pulsating loads, a large capacity needs to be configured to meet the peak power demand of the load, resulting in significant capacity waste and cost increase, and leading to a large and cumbersome power system. In addition, when driving high-power loads, the internal loss and heat generation of the battery are large, the terminal voltage experiences a significant drop, and it may even prematurely enter over-discharge protection and stop supplying power.

The advantages of power-type energy storage technologies such as supercapacitors and flywheels are high power density, fast charging and discharging speed, high

X. Tang et al., *Electrical Energy Storage Technologies and Applications*,
https://doi.org/10.1007/978-981-96-5625-7_5

energy storage efficiency, and long cycle life, but the disadvantage is low energy density, which is not suitable for large capacity, long-term power storage applications. It can be seen that power-type energy storage such as supercapacitors and batteries have strong complementarity in technical performance. If the two are used in combination, combining the high energy density of batteries with the high power density, fast charging and discharging speed, high energy storage efficiency, and long cycle life of supercapacitors, it will bring significant performance improvement to the energy storage system.

A large amount of research shows that supercapacitors can be used in combination with batteries in a certain way to give the energy storage device good load adaptability, improve the reliability of power supply, reduce the size of the energy storage device, reduce weight, improve the economic performance of the energy storage device, etc. [66–68]. Using supercapacitors in combination can reduce the peak output current of the battery, reduce internal loss, extend discharge time, and optimize the charging and discharging process of the battery, delaying aging and failure. Literature [69,70] uses supercapacitors as power buffers, used in parallel with batteries, applied to electric vehicles or hybrid vehicles, to filter the output and absorption of instantaneous high power required by the battery during vehicle acceleration and deceleration, reducing the peak power demand of the motor on the battery, reducing the installed capacity of the battery, and extending its life. Literature [71–73] points out that the supercapacitor battery hybrid power source has good pulse power characteristics and has significant performance advantages when used on mobile communication equipment. Literature [74] applies the supercapacitor and Ni-MH battery hybrid energy storage to the independent photovoltaic system of the fixed communication station, and the volume and weight of the battery can be reduced by 50%. In addition, using hybrid energy storage in independent photovoltaic systems can reduce the number of charge and discharge cycles of the battery, optimize the charging process, and improve the energy conversion efficiency of the system [75–78].

It can be seen that the use of hybrid energy storage technology with strong performance complementarity is an important way to solve the shortcomings of single energy storage technology and improve the technical and economic performance of power supply systems or power systems. This chapter takes batteries and supercapacitors as examples to analyze and demonstrate their performance after hybridization.

5.2 Composite Energy Storage System Model and Analysis

5.2.1 Composite Energy Storage System Modeling

To simplify the analysis process, the battery model can be simplified to an ideal voltage source in series with an equivalent internal resistance, and the supercapacitor model can be simplified to an ideal capacitor in series with an equivalent internal resistance. Since the dynamic performance of the system is mainly considered, the parallel internal resistance can be ignored. The direct parallel energy storage model of the supercapacitor and battery is shown in Fig. 5.1a [43].

In the figure, R_s is the equivalent series internal resistance of the supercapacitor, R_b is the equivalent series internal resistance of the battery. i_c is the current of the supercapacitor branch, i_b is the current of the battery branch, i_0 is the load current.

The circuit model in Fig. 5.1a is transformed by Laplace and simplified by Thevenin theorem to obtain the simplified model shown in Fig. 5.1b, where,

$$V(S) = \frac{V_b}{S} + \frac{R_b}{R_b + R_s} \frac{V_{co} - V_b}{S + \frac{1}{(R_b + R_s)C}} \tag{5.1}$$

$$Z(S) = R_b // \left(R_s + \frac{1}{SC} \right) = \frac{R_b R_s}{R_b + R_s} \frac{S + \frac{1}{R_s C}}{S + \frac{1}{(R_b + R_s)C}} \tag{5.2}$$

For pulsating power load, set the pulsation period to T, the duty cycle to D, and the current peak to I_0, then the current $i_0(t)$ can be expressed as:

$$i_0(t) = I_0 \sum_{k=0}^{N-1} \left[\varphi(t - kT) - \varphi(t - (k + D)T) \right] \quad k = 0, 1, 2 \cdots \tag{5.3}$$

Where, $\varphi(t)$ is the standard step function. Transforming Eq. (5.3) by Laplace, we can get the frequency domain expression of the pulsating load current:

$$I_0(S) = I_0 \sum_{k=0}^{N-1} \left[\frac{e^{-kTS}}{S} - \frac{e^{-(k+D)TS}}{S} \right] \tag{5.4}$$

The pulsating current generates a pulsating voltage drop Vz(s) on the equivalent internal resistance of the composite energy storage system:

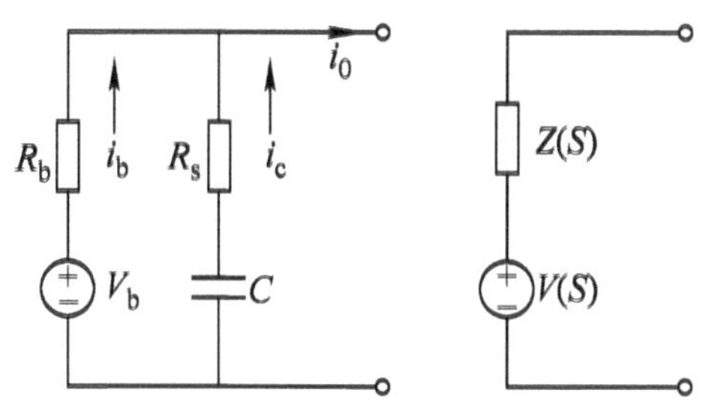

a) Direct parallel structure b) Equivalent circuit

Fig. 5.1 Equivalent circuit model of direct parallel supercapacitor and battery. (**a**) Direct parallel structure (**b**) Equivalent circuit

$$V_Z(S)=\frac{R_bR_sI_0}{R_b+R_s}\sum_{k=0}^{N-1}\left[\frac{S+\frac{1}{R_sC}}{S+\frac{1}{(R_b+R_s)C}}\frac{e^{-kTS}-e^{-(k+D)TS}}{S}\right] \tag{5.5}$$

herefore, the voltage drop $V_0(s)$ generated by the pulsating current on the load is:

$$\begin{aligned}V_0(S)=&\frac{V_b}{S}+\frac{R_b}{R_b+R_s}\frac{V_{co}-V_b}{S+\frac{1}{(R_b+R_s)C}}\\&-\frac{R_bR_sI_0}{R_b+R_s}\sum_{k=0}^{N-1}\left[\frac{S+\frac{1}{R_sC}}{S+\frac{1}{(R_b+R_s)C}}\frac{e^{-kTS}-e^{-(k+D)TS}}{S}\right]\end{aligned} \tag{5.6}$$

Where, V_{c0} is the initial voltage of the ideal capacitor C.

Transforming Eq. (5.6) by inverse Laplace, we can get the time domain expression of the pulsating load voltage drop:

$$\begin{aligned}v_0(t)=&v_b+\frac{R_b}{R_b+R_s}(V_{co}-V_b)e^{-\frac{t}{(R_b+R_s)C}}-\\&R_bI_0\sum_{k=0}^{N-1}\left[\begin{array}{l}\left(1-\frac{R_b}{R_b+R_s}e^{-\frac{t-kT}{(R_b+R_s)C}}\right)\varphi(t-kT)\\-\left(1-\frac{R_b}{R_b+R_s}e^{-\frac{t-(k+D)T}{(R_b+R_s)C}}\right)\varphi(t-(k+D)T)\end{array}\right]\end{aligned} \tag{5.7}$$

From this, the current of the battery branch can be obtained:

$$\begin{aligned}i_b(t)=&-\frac{(V_{co}-V_b)}{R_b+R_s}e^{-\frac{t}{(R_b+R_s)C}}+\\&I_0\sum_{k=0}^{N-1}\left[\begin{array}{l}\left(1-\frac{R_b}{R_b+R_s}e^{-\frac{t-kT}{(R_b+R_s)C}}\right)\varphi(t-kT)\\-\left(1-\frac{R_b}{R_b+R_s}e^{-\frac{t-(k+D)T}{(R_b+R_s)C}}\right)\varphi(t-(k+D)T)\end{array}\right]\end{aligned} \tag{5.8}$$

The current of the supercapacitor branch:

$$i_c(t) = \frac{V_{co} - V_b}{R_b + R_s} e^{-\frac{t}{(R_b+R_s)C}} + I_0 \sum_{k=0}^{N-1} \left[\frac{R_b}{R_b + R_s} e^{-\frac{t-kT}{(R_b+R_s)C}} \varphi(t - kT) - \frac{R_b}{R_b + R_s} e^{-\frac{t-(k+D)T}{(R_b+R_s)C}} \varphi(t-(k+D)T) \right] \tag{5.9}$$

In steady state, the current of the battery branch:

$$i_{bss}(t) = I_0 \sum_{k=0}^{N-1} \left[\left(1 - \frac{R_b}{R_b + R_s} e^{-\frac{t-kT}{(R_b+R_s)C}}\right) \varphi(t - kT) - \left(1 - \frac{R_b}{R_b + R_s} e^{-\frac{t-(k+D)T}{(R_b+R_s)C}}\right) \varphi(t-(k+D)T) \right] \tag{5.10}$$

In steady state, the current of the supercapacitor branch:

$$i_{css}(t) = \frac{R_b I_0}{R_b + R_s} \sum_{k=0}^{N-1} \left[e^{-\frac{t-kT}{(R_b+R_s)C}} \varphi(t - kT) - e^{-\frac{t-(k+D)T}{(R_b+R_s)C}} \varphi(t-(k+D)T) \right] \tag{5.11}$$

Through the above modeling of the composite energy storage of supercapacitors and batteries, we can quantitatively analyze the improvement of the performance of the composite energy storage system, including the increase of peak power capability, the reduction of internal loss, the extension of operation time, etc. [66].

5.2.2 Power Capability Improvement

Based on the above analysis, simulate the current of the battery branch and the supercapacitor branch in steady state. Set $R_b = 200$ mΩ, $R_s = 10$ mΩ, C = 200 F, the pulsation period of the load T = 2 s, duty cycle D = 0.2 s, current amplitude $I_0 = 5$ A.

The operation process of the composite energy storage system is shown in Fig. 5.2. During the load power pulsation, both the battery and the supercapacitor output current, jointly supplying power to the load, but the discharge current of the battery is smaller; during the non-load period, the battery continues to output current, charging the supercapacitor. It can be seen that the use of composite energy storage can greatly reduce the peak output current of the battery during load power pulsation.

For the battery branch, when t = (k + D)T, the output current $i_{bss}(t)$ reaches its maximum value:

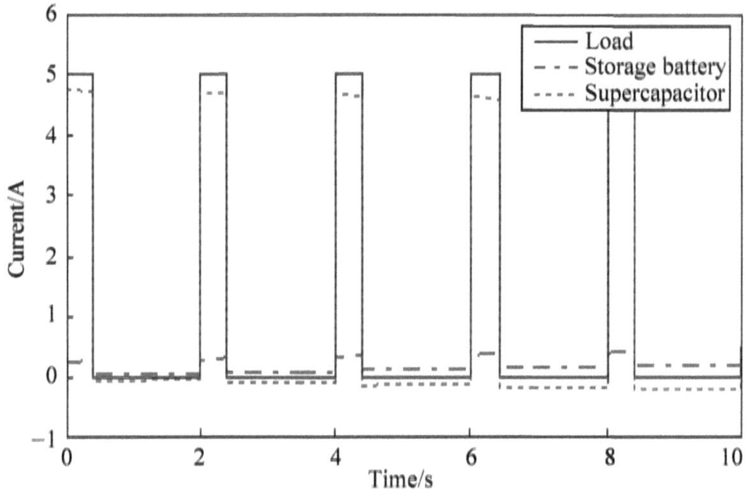

Fig. 5.2 Response of the composite energy storage system during load power pulsation

$$I_{bpeak} = I_0\left(1-\frac{R_b e^{-\frac{DT}{(R_b+R_s)C}}}{R_b+R_s}\frac{1-e^{-\frac{(1-D)T}{(R_b+R_s)C}}}{1-e^{-\frac{T}{(R_b+R_s)C}}}\right)=\frac{I_0}{\gamma} \tag{5.12}$$

Where, γ is always greater than 1. Therefore, when the battery and supercapacitor are paralleled to drive the pulsating load, the maximum output current value of the battery branch is less than the current amplitude of the pulsating load, and most of the load current is shared by the supercapacitor branch. Due to the high power density and strong current output capability of the supercapacitor, the power capability of the composite energy storage system has been improved.

Assuming the working voltage of the battery is V_b, then its maximum output power is:

$$P_{b,peak} = I_{b,peak}\cdot V_b \tag{5.13}$$

If the battery and supercapacitor are directly paralleled according to the structure shown in Fig. 5.1a, the maximum current of the composite energy storage system is determined by the maximum current of the battery, that is,

$$I_{bc,peak} = \gamma\cdot I_{b,peak} \tag{5.14}$$

And the maximum output power of the composite energy storage system:

$$P_{bc,peak} = I_{bc,peak}\cdot V_b = \gamma\cdot I_{b,peak}\cdot V_b = \gamma\cdot P_{b,peak} \tag{5.15}$$

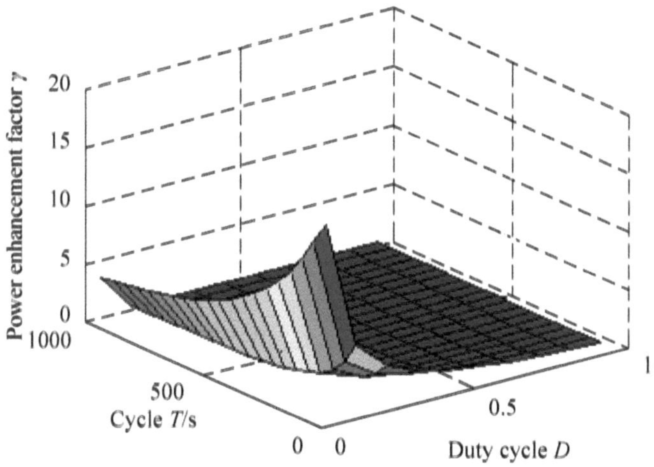

Fig. 5.3 Relationship between power enhancement factor and load parameters

It can be seen that the maximum power of the composite energy storage system is γ times that of the battery working alone, and the power capability has been greatly improved.

Define γ as the power enhancement factor of the composite energy storage system [67]. The larger the γ, the greater the power enhancement effect of the supercapacitor on the composite energy storage system.

$$\gamma = \left(1 - \frac{R_b e^{-\frac{DT}{(R_b+R_s)C}}}{R_b + R_s} \frac{1 - e^{-\frac{(1-D)T}{(R_b+R_s)C}}}{1 - e^{-\frac{T}{(R_b+R_s)C}}} \right)^{-1} \tag{5.16}$$

Among them, γ is related to the parameters of the load, including the pulsation period and duty cycle; γ is also related to the parameters of the supercapacitor bank and battery bank, including the internal resistance of the battery bank, the internal resistance and capacitance of the supercapacitor bank, etc.

Figure 5.3 shows the relationship between the power enhancement factor γ and the parameters of the pulsating load. The smaller the duty cycle of the pulsating load, the larger the γ; the shorter the period, the larger the γ, the greater the power enhancement effect of the supercapacitor on the composite energy storage system. For the limit case of the duty cycle D = 0, it can be known from formula (5.16) that γ has reached the maximum value:

$$\gamma_{\max} = \frac{R_b + R_s}{R_s} = 1 + \frac{R_b}{R_s} \tag{5.17}$$

Equation (5.17) indicates that the maximum power enhancement capability of the composite energy storage system depends on the equivalent series resistance of the battery bank and the supercapacitor bank. The smaller the equivalent resistance of the supercapacitor bank, the greater the power enhancement capability. In this example, the series resistances of the battery bank and the supercapacitor bank are $R_b = 200\,\text{m}\Omega$ and $R_s = 10\,\text{m}\Omega$ respectively, and the maximum value of the power enhancement factor is 21.

For the supercapacitor bank, the equivalent series resistance depends not only on the performance of the individual capacitors but also on the number of series and parallel connections. If the resistance of a single supercapacitor is R_{sc} and the capacitance is C_s, and the combination mode is N_s series and N_p parallel. Define m as the structural coefficient of the supercapacitor bank:

$$m = \frac{N_p}{N_s} \tag{5.18}$$

Then, the combined $R_s = R_{sc}/m$ and $C = mC_s$. Therefore, increasing the structural coefficient m of the supercapacitor bank can improve the power enhancement capability of the composite energy storage system.

Figure 5.4 shows the effect of the equivalent series resistance of the battery bank and the supercapacitor bank on γ. It can be seen that the smaller the equivalent series resistance of the supercapacitor bank, the larger γ, and the greater the power enhancement effect of the supercapacitor on the composite energy storage system.

It can be seen that the supercapacitor battery composite energy storage system is very suitable for pulsating power applications with high peak power but low average power. Of course, in actual design, it is often necessary to use multiple

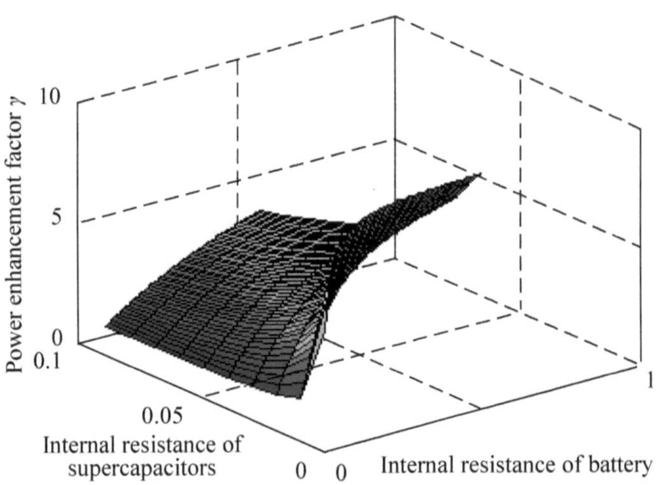

Fig. 5.4 Relationship between power enhancement factor and energy storage resistance ($D = 0.1$, $T = 1s$)

supercapacitor branches in parallel to reduce its equivalent series resistance to a certain extent according to the actual situation of the pulsating load.

5.2.3 Reduction of Internal Losses

As can be seen from Fig. 5.2, during the load power pulsation, most of the load current is provided by the supercapacitor branch with very small internal resistance, thus reducing the internal losses of the composite energy storage system and reducing the heat generation of the battery.

When the battery is powered alone, the total energy that can be output when driving a pulsating load is:

$$W_b = \left(V_b\sqrt{D}I_0 - R_bDI_0^2\right)\tau_{bo} = V_b\sqrt{D}I_0\left(1-\frac{R_b\sqrt{D}I_0}{V_b}\right)\tau_{bo} = V_{bo}I_{o,rms}\tau_{bo} \tag{5.19}$$

In the formula, τ_{bo} is the total discharge time of the battery, V_{bo} is the output voltage of the battery group, $I_{o,\,rms}$ is the effective value of the output current of the battery group.

$$V_{bo} = V_b\left(1-\delta\sqrt{D}\right) \tag{5.20}$$

$$I_{o,rms} = \sqrt{D}I_0 \tag{5.21}$$

Where, δ is the ratio of the load current amplitude to the short-circuit current of the battery bank, defined as the load current rate:

$$\delta = \frac{R_bI_0}{V_b} \tag{5.22}$$

Assuming that the battery operates under ideal A-H characteristics and the discharge depth can reach 100%, the loss of the battery bank during the discharge process is:

$$W_i = R_bI_{o,rms}^2\tau_{bo} = R_bDI_0^2\tau_{bo} \tag{5.23}$$

The battery bank and the supercapacitor bank are directly paralleled as shown in Fig. 5.1. According to the effective value of the current of the battery bank and the effective value of the current of the supercapacitor bank, the total energy loss of the composite energy storage system can be calculated. Formula (5.10) gives the instantaneous value of the current in the battery branch at steady state, where the instantaneous value of the pulsating current in the n cycle is:

$$i_{bss}^{nth}(t)=I_0\sum_{k=0}^{n}\left[\begin{array}{l}\left(1-\frac{R_b}{R_b+R_s}e^{-\frac{t-kT}{(R_b+R_s)C}}\right)\varphi(t-kT)\\-\left(1-\frac{R_b}{R_b+R_s}e^{-\frac{t-(k+D)T}{(R_b+R_s)C}}\right)\varphi(t-(k+D)T)\end{array}\right]\cdot\\\left[\varphi(t-nT)-\varphi(t-(n+1)T)\right] \tag{5.24}$$

Therefore, the effective value of the current of the battery bank is:

$$\begin{aligned}I_{b,rms}&=\sqrt{\frac{1}{T}\int_{nT}^{(n+1)T}\left[i_{bss}^{nth}\right]^2dt}\\&=\sqrt{D}I_0\sqrt{1+\frac{R_b}{R_b+R_s}\frac{2\left(1-e^{-\frac{DT}{(R_b+R_s)C}}\right)}{\frac{DT}{(R_b+R_s)C}}\left(\frac{1-e^{-\frac{DT}{(R_b+R_s)C}}}{1-e^{\frac{T}{(R_b+R_s)C}}}-1\right)+\left(\frac{R_b}{R_b+R_s}\right)^2\frac{\left(e^{\frac{DT}{(R_b+R_s)C}}-1\right)\left(1-e^{\frac{(1-D)T}{(R_b+R_s)}}\right)}{\frac{DT}{(R_b+R_s)C}\left(1-e^{\frac{T}{(R_b+R_s)}}\right)}}\\&=\sqrt{D}I_0\lambda\end{aligned} \tag{5.25}$$

Similarly, for the supercapacitor group, the instantaneous value of the pulsating current in the n cycle is:

$$i_{css}^{nth}(t)=\frac{R_bI_0}{R_b+R_s}\sum_{k=0}^{n}\left[e^{-\frac{t-kT}{(R_b+R_s)C}}\varphi(t-kT)-e^{-\frac{t-(k+D)T}{(R_b+R_s)C}}\varphi(t-(k+D)T)\right]\\\cdot\left[\varphi(t-nT)-\varphi(t-(n+1)T)\right] \tag{5.26}$$

Therefore, the effective value of the current of the supercapacitor bank is:

$$\begin{aligned}I_{c,rms}&=\sqrt{\frac{1}{T}\int_{nT}^{(n+1)T}\left[i_{css}^{nth}\right]^2dt}\\&=\sqrt{D}I_0\sqrt{\left(\frac{R_b}{R_b+R_s}\right)^2\left[\frac{1}{(R_b+R_s)CDT}\frac{e^{\frac{DT}{(R_b+R_s)C}}-e^{\frac{T}{(R_b+R_s)C}}-1+e^{\frac{(1-D)T}{(R_b+R_s)C}}}{1-e^{\frac{T}{(R_b+R_s)C}}}\right]}\\&=\sqrt{D}I_0\mu\end{aligned} \tag{5.27}$$

From this, the energy loss of the composite energy storage system within the discharge time τ_{be} can be obtained:

$$W_{l,bc} = \left(R_b I_{b,rms}^2 + R_s I_{\varepsilon,rms}^2\right)\tau_{bc} = R_b D I_0^2 \left(\lambda^2 + \frac{R_s}{R_b}\mu^2\right)\tau_{bc} = R_b D I_0^2 (1-\varepsilon)\tau_{bc} \quad (5.28)$$

Where,

$$\varepsilon = 1 - \left(\lambda^2 + \frac{R_s}{R_b}\mu^2\right) \quad (5.29)$$

Comparing formula (5.28) with formula (5.29), we get,

$$\varepsilon = 1 - \frac{R_b D I_0^2 (1-\varepsilon)}{R_b D I_0^2} = 1 - \frac{P_{l,bc}}{P_{l,b}} = \frac{P_{l,b} - P_{l,bc}}{P_{l,b}} \quad (5.30)$$

In the formula, $P_{l,bc}$ represents the power loss of the supercapacitor battery composite energy storage system, $P_{l,b}$ represents the power loss of the battery alone energy storage. Therefore, ε represents the power loss saving rate of the composite energy storage system, defined as the power saving factor [66].

Figure 5.5 illustrates the relationship between the power saving factor ε and the pulsating load parameters ($R_b = 200\,\text{m}\Omega$, $R_s = 10\,\text{m}\Omega$, $C = 200\,\text{F}$). It can be seen that the smaller the load duty cycle, the shorter the pulsation period, ε the larger, the smaller the internal loss of the composite energy storage system.

Figure 5.6 shows the effect of the equivalent series resistance of the battery bank ($R_b = 200\,\text{m}\Omega$, $R_s = 10\,\text{m}\Omega$, $C = 200\,\text{F}$) and the supercapacitor bank on the power saving factor ε. The smaller the equivalent series resistance of the supercapacitor bank, ε the larger, the smaller the internal loss of the composite energy storage system.

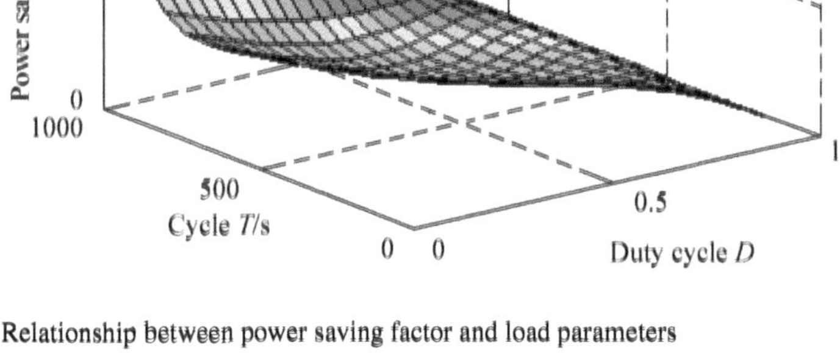

Fig. 5.5 Relationship between power saving factor and load parameters

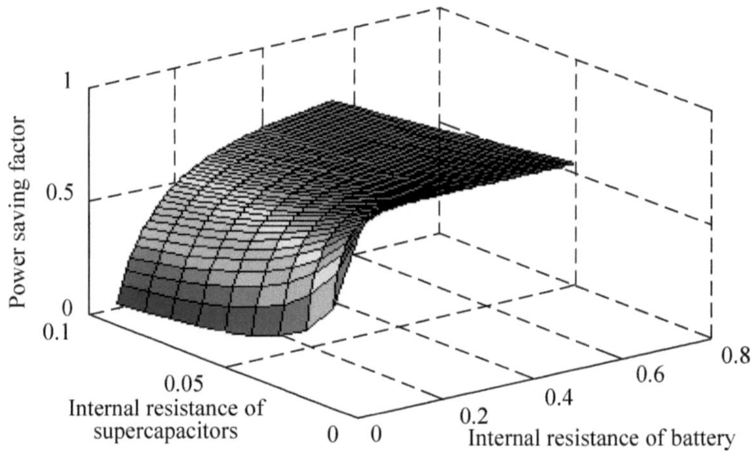

Fig. 5.6 Relationship between power saving factor and storage resistance

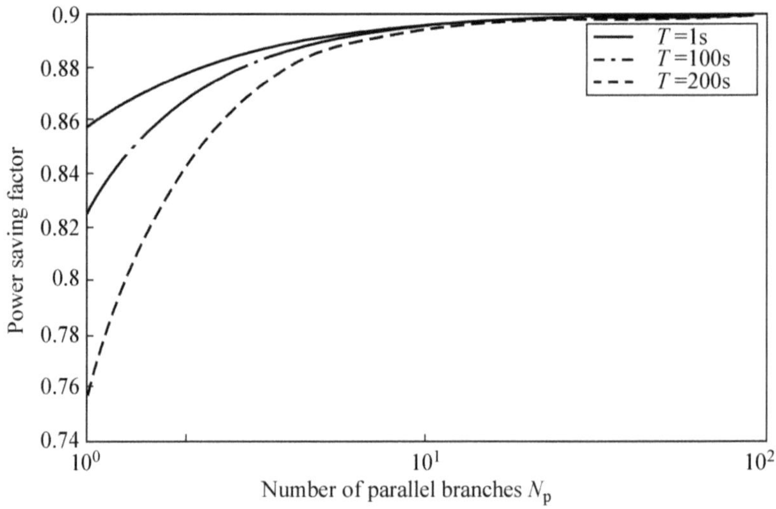

Fig. 5.7 Relationship between power saving factor and the number of supercapacitor parallel branches

Figure 5.7 shows the relationship between the power saving factor ε and the number of parallel branches of the supercapacitor group ($D=0.1$, $T=1s$, $C=200F$), single supercapacitor series branch parameters: $R_s = 10\,\text{m}\Omega$, $C = 200F$. When the number of parallel branches increases, ε increases and tends to flatten when the number of parallel branches increases to a certain extent. Figure 5.7 can be used as a basis for configuring the supercapacitor of the hybrid energy storage system.

5.2.4 Extension of Operating Time

The supercapacitor and battery composite energy storage system can extend the operation time of the battery because the maximum working current of the battery is reduced during operation, which reduces internal losses.

Compared to the battery operating alone, the total energy saved by the hybrid energy storage system is:

$$\Delta W = W_i - W_{i,bc} = R_b D I_0^2 \tau_{bo} - R_b D I_0^2 (1-\varepsilon)\tau_{bc} = R_b D I_0^2 \left[\tau_{bo} - (1-\varepsilon)\tau_{bc}\right] \quad (5.31)$$

Therefore, the extended operation time is:

$$\Delta\tau = \frac{\Delta W}{P_o} \quad (5.32)$$

Where, P_o is the average power output of the composite energy storage system (W_T is the total energy stored by the battery):

$$P_o = \frac{W_T - W_{i,bc}}{\tau_{bc}} = \frac{V_b\sqrt{D} I_0 \tau_{bo} - R_b D I_0^2 (1-\varepsilon)\tau_{bc}}{\tau_{bc}} \quad (5.33)$$

From this, the ratio of the extended operation time of the composite energy storage system to the operation time of the battery (defined as the time extension rate) can be obtained:

$$\zeta = \frac{\Delta\tau}{\tau_{bo}} = \frac{\tau_{bc} - \tau_{bo}}{\tau_{bo}} = \frac{\varepsilon\delta\sqrt{D}}{1 - \varepsilon\delta\sqrt{D}} \quad (5.34)$$

From Eq. (5.34), it can be seen that compared to the battery operating alone, the operation time of the composite energy storage system has been extended. The time extension rate is related to three factors, including the power saving factor ε, the duty cycle of the pulsating load D, and the load current rate δ.

Figure 5.8 shows the relationship between the time extension rate of the composite energy storage system and the load parameters (R_b = 200 mΩ, R_s = 10 mΩ, C = 200 F, V_b = 24 V, I_0 = 5 A). It can be seen that the shorter the pulsation period of the load, the larger the time extension rate ζ. In addition, ζ is also related to the duty cycle of the pulsating load. When the duty cycle is too large or too small, ζ becomes smaller and tends to zero. The maximum value of ζ should be a point between the duty cycle range (0, 1), which depends on the pulsation period of the load and the load current rate δ.

From the above analysis, it can be seen that by paralleling supercapacitors, the peak output current of the battery during load power pulsation is reduced, the drop of terminal voltage is suppressed, and the effect is equivalent to the reduction of the equivalent source impedance of the battery. The reduction of equivalent source

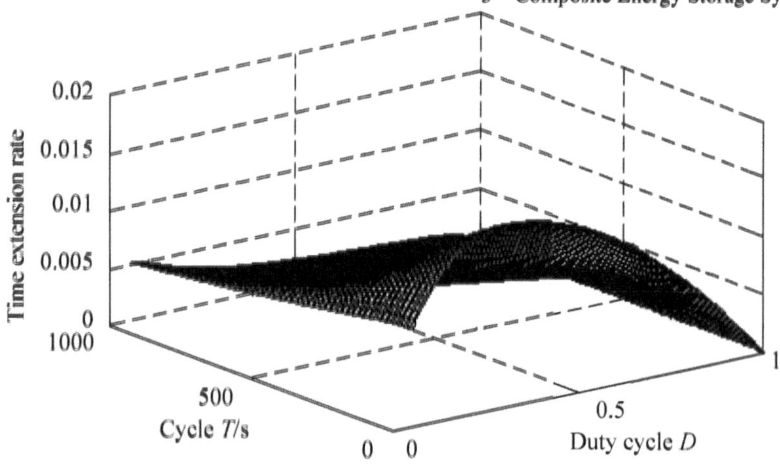

Fig. 5.8 Relationship between time extension rate and load parameters

impedance improves the dynamic response capability of the battery, reduces its internal loss when driving pulsating loads, improves discharge efficiency, extends discharge time, and effectively prevents abnormal shutdown of the power source.

5.3 Control Strategy of Composite Energy Storage

5.3.1 Directly Parallel Composite Energy Storage System

The direct parallel of supercapacitors and batteries is the simplest hybrid energy storage structure, as shown in Fig. 5.1a. In this structure, because the terminal voltage of the battery bank is forced to be equal to the terminal voltage of the supercapacitor bank, the combination method of the supercapacitor bank is strictly required in the design, and the structural parameters of the supercapacitor bank should be reasonably configured according to the working voltage range of the battery bank and the load [79, 80].

This section sets up an experimental platform for direct parallel energy storage of supercapacitors and batteries. The parameters of the supercapacitor group are: 200 F, 10 mΩ; the parameters of the battery bank are: 12 Ah, 200 mΩ; the parameters of the pulsating load are: period 5 s, duty cycle 25%, pulsating power 70 W. According to formula (5.16), it can be calculated that $\gamma = 3.37$, that is, under this application condition, the peak power output capability of the composite energy storage system is 2.37 times higher than that of the battery operating alone.

Figure 5.9 shows the working process of the composite energy storage system driving the pulsating load. In each cycle, when the pulsating power occurs, the peak current output of the battery bank accounts for about 35% of the peak load current, and the remaining 65% is provided by the supercapacitor bank. After the pulsating power ends, the battery bank continues to output current to charge the

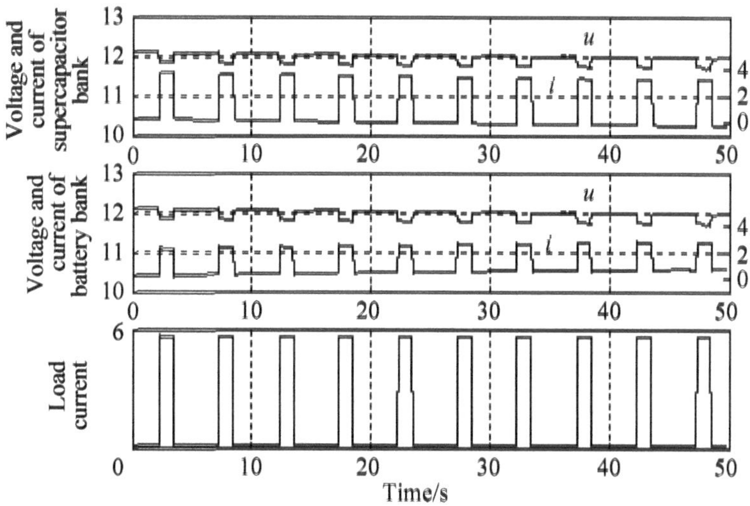

Fig. 5.9 Working process of the directly parallel composite energy storage system

supercapacitor bank. The voltage ripple of the composite energy storage system is about 0.3 V, which is about 2.5% of the output voltage. As the operation process continues, the charge of the composite energy storage system continues to decrease, the output current of the battery bank gradually increases, and the terminal voltage also continues to drop.

The direct parallel structure can greatly reduce the maximum output current of the battery during load power pulsation, and improve the power output capability of the composite energy storage system, but there are also obvious shortcomings in energy storage using the direct parallel method.

First, the terminal voltage of the supercapacitor bank must be consistent with that of the battery bank. The change range of the terminal voltage of the battery during charging and discharging is small, while the change range of the terminal voltage of the supercapacitor during charging and discharging can be large. Therefore, the direct parallel structure will lead to a decrease in the utilization rate of supercapacitor capacity.

Secondly, the power enhancement capability of the composite energy storage only depends on the parameters of the supercapacitor bank and the battery bank itself (mainly the equivalent series internal resistance). If the working condition of the pulsating load changes, it often needs to redesign the capacity configuration and combination structure of the composite energy storage system, which lacks flexibility in practical applications.

Thirdly, compared with batteries, supercapacitors have a higher self-discharge rate, and the direct parallel structure leads to an increase in the overall self-discharge rate of the composite energy storage system.

5.3.2 Through Inductor Parallel Composite Energy Storage System

Based on the direct parallel structure, an inductor is arranged between the battery bank and the supercapacitor bank, and the supercapacitor bank is directly connected to the load, as shown in Fig. 5.10. The role of the inductor is to filter the output current of the battery bank, reduce the output current ripple, and further reduce its internal energy loss and heat generation.

Figure 5.11 shows the working process of the composite energy storage system driven by the pulsating load through the parallel inductor. The experimental conditions are the same as the direct parallel structure in the previous section, and the inductance value is 6.5 mH. It can be seen that when the load power pulsates, the

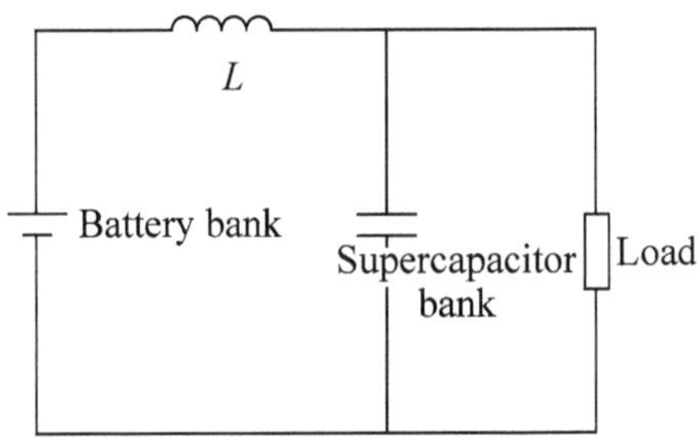

Fig. 5.10 Through Inductor Parallel Structure Diagram

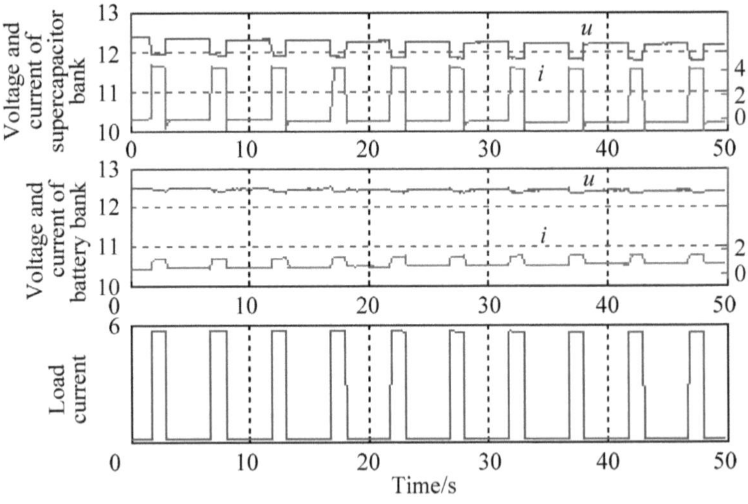

Fig. 5.11 Working Process of Composite Energy Storage System through Inductor Parallel

output current of the battery bank in this structure is smoothed, and the pulsation is significantly weakened. The output current ripple of the battery bank is 0.75 A, which is only 13% of the amplitude of the load pulsating current, and is 22% less than the direct parallel structure. The terminal voltage of the battery bank generates about 0.1 V ripple during the operation of the pulsating load, which is about 0.8% of its working voltage, and only 32% of the direct parallel structure.

This structure has a good effect on improving the power output capability of the energy storage system and optimizing the charging and discharging process of the battery. However, similar to the direct parallel structure, it also has disadvantages such as inflexible system configuration and large load terminal voltage ripple. Moreover, the inductor itself also brings losses and increases in equipment volume and weight.

5.3.3 Active Composite Energy Storage System

The battery is combined with the supercapacitor through the power electronic power converter, and the converter control function of the power converter can greatly improve the performance of the energy storage device and further optimize the working process of the battery, extending its service life.

5.3.3.1 Main Circuit Model

In this structure, the battery bank is paralleled with the supercapacitor bank through the DC/DC converter, and the supercapacitor bank is directly connected to the load, as shown in Fig. 5.12. Because the energy flow process of the battery is controlled by active power devices, it is also called an active composite energy storage system. Correspondingly, direct parallel composite energy storage and composite energy storage through inductor parallel can be called passive composite energy storage.

The DC/DC converter is located between the battery bank and the supercapacitor bank. With its flexible converter control capability, it manages the process of outputting energy from the battery bank to better play the advantages of the supercapacitor, optimize the working process of the battery, and improve the overall technical and economic performance of the composite energy storage system.

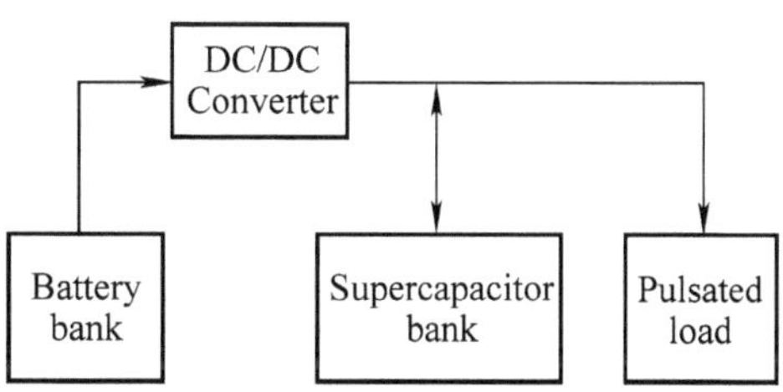

Fig. 5.12 Active Composite Energy Storage System Structure Diagram

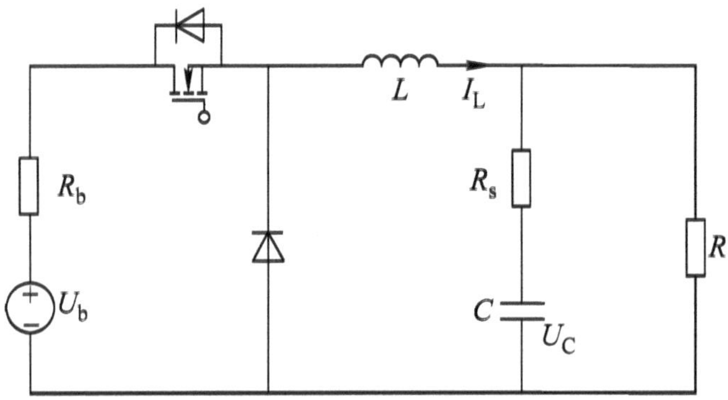

Fig. 5.13 Equivalent circuit model of active composite energy storage system

Taking the step-down DC/DC converter (BUCK) as an example, the equivalent circuit model of the supercapacitor battery composite energy storage system is shown in Fig. 5.13. Where, U_c is the voltage of the ideal capacitor, R As the equivalent impedance of the load.

Assuming the BUCK converter operates in a continuous inductor current state, with the inductor current I_L and ideal capacitor voltage U_c as state variables, applying the state average method, the state average equation can be obtained:

$$\begin{cases} L\dfrac{dI_L}{dt} = -\left(DR_b + \dfrac{RR_s}{R+R_s}\right)I_L - \dfrac{R}{R+R_s}U_c + DU_b \\ C\dfrac{dU_c}{dt} = \dfrac{R}{R+R_s}I_L - \dfrac{U_c}{R+R_s} \end{cases} \tag{5.35}$$

When the system operates in a steady state, apply perturbations to the state variables, let the instantaneous values $d = D + \hat{d}, i_L = I_L + \hat{i}_L, u_c = U_c + \hat{u}_c, u_b = U_b + \hat{u}_b$, obtain the transient equation, and transform to the s domain, ignoring high-order components, the control-output transfer function of the system can be obtained:

$$\left.\frac{\hat{I}_L(s)}{\hat{d}(s)}\right|_{\hat{U}_b(s)=0} = \frac{(U_b - R_bI_L)}{L}\frac{s+\dfrac{1}{C\alpha}}{s^2 + s\dfrac{L+C\beta}{LC\alpha} + \dfrac{DR_b+R}{LC\alpha}} \tag{5.36}$$

$$\left.\frac{\hat{U}_c(s)}{\hat{d}(s)}\right|_{\hat{U}_b(s)=0} = \frac{(U_b - R_bI_L)}{LC\alpha}\frac{R}{s^2 + s\dfrac{L+C\beta}{LC\alpha} + \dfrac{DR_b+R}{LC\alpha}} \tag{5.37}$$

Where, $\alpha = R + R_s$, $\beta = DRR_b + DR_bR_s + RR_s$.

The system is second-order, its poles:

$$s = -\frac{L + C\beta}{2LC\alpha} \pm \sqrt{\left(\frac{L - C\beta}{2LC\alpha}\right)^2 - \frac{R^2}{LC\alpha^2}} \tag{5.38}$$

In Eq. (5.38), since $\frac{L + C\beta}{2LC\alpha} > \frac{L - C\beta}{2LC\alpha}$ always holds, therefore, over the entire frequency range:

$$R_e(s) < 0 \tag{5.39}$$

That is, the two poles of the system are always on the left side of the *s* coordinate plane, so the open-loop system is stable.

5.3.3.2 Control Link Design

The control objective of the active composite energy storage system is to provide instantaneous power from the supercapacitor, while the battery charges and discharges in an approximately constant current manner through the power converter. The control objective of the DC/DC converter is to output a constant current, the current value is equal to the average working current of the pulsating load. The parameters are set as shown in Table 5.1.

The working current of the pulsating load is sampled over a certain period of time, its average value is calculated as the set value of the constant current output of the parallel controller, and compared with the actual output current to generate an error. Through the *PI* regulator, the PWM signal controlling the power switch tube is finally generated. The closed-loop control model of the system is shown in Fig. 5.14, where, $G_p(s)$ is the control-output transfer function of the system, $G_c(s)$ is the transfer function of the PI regulator.

Therefore, the closed-loop transfer function of the system is:

$$G(s) = \frac{\hat{I}_L(s)}{\hat{I}_{ref}(s)} = \frac{G_c(s)G_p(s)}{1 + G_c(s)G_p(s)} \tag{5.40}$$

Table 5.1 Circuit parameter settings

Parameter	Value	Unit
Power converter inductance L	0.15	mH
Supercapacitor Bank capacitance C	200	F
Supercapacitor Bank internal resistance R_s	0.01	Ω
Supercapacitor Bank voltage U_c	12.5	V
Battery Bank internal resistance R_b	0.2	Ω
Battery Bank voltage U_b	25	V

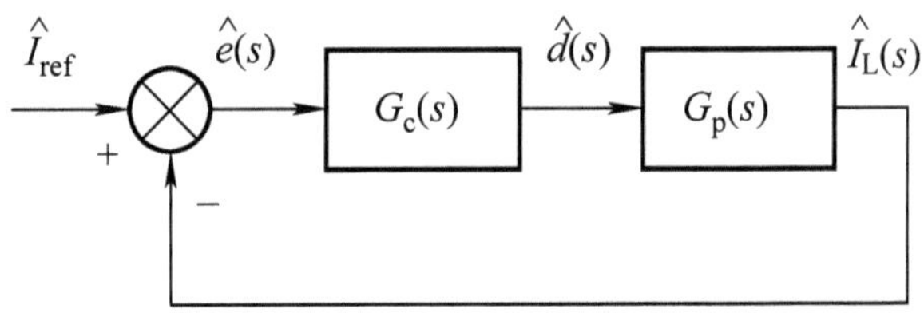

Fig. 5.14 System Control Diagram

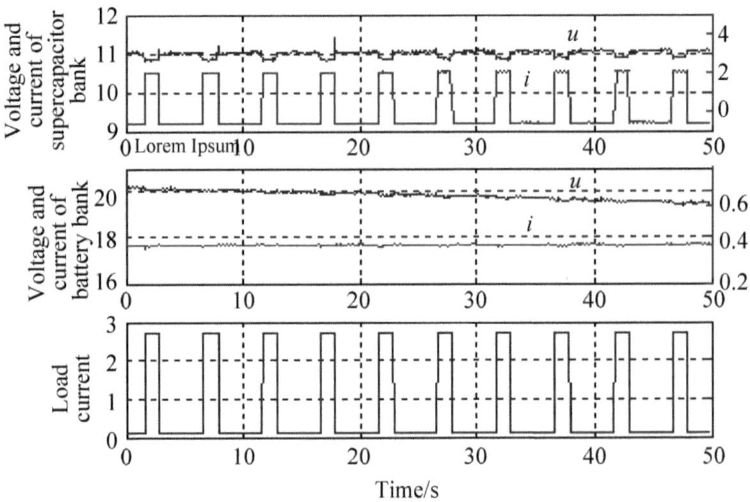

Fig. 5.15 Active Composite Energy Storage System Working Process

5.3.3.3 Composite Energy Storage System Experiment

A supercapacitor-battery active composite energy storage system experimental platform was built, with a pulsating power load applied, with a period of 5 s and a duty cycle of 25%. Figure 5.15 shows the working process. It can be seen that the battery bank discharges in a constant current mode, and its discharge current is about 15% of the peak load current. The supercapacitor bank provides most of the remaining current during the load power pulsation and is charged from the battery group through the parallel controller at other times.

As the pulsating load continues to work, the state of charge of the battery bank continues to decrease, and the terminal voltage continues to decrease. Because it needs to provide the energy required for the DC/DC converter's constant current output, the output current of the battery bank gradually increases. Under the control of the power converter, the terminal voltage of the supercapacitor bank and the load can still maintain good stability.

Compared with the direct parallel structure and the inductive structure and other passive composite energy storage, the active composite energy storage has significant advantages:

Firstly, the terminal voltages of the battery bank and the supercapacitor bank can be different, thus they have great flexibility in capacity configuration and structural design.

Secondly, by utilizing the control function of the active controller, the output current of the battery bank can be limited within a reasonable range, thereby greatly improving the power output capability of the energy storage system.

Thirdly, the battery bank always discharges in an approximately constant current mode, which optimizes the discharge process, reduces internal losses, improves power supply reliability, and extends the service life.

Of course, due to the use of the DC/DC converter, system control becomes complex, and it also consumes a certain amount of energy. However, the development of this type of converter is very mature at present, the energy conversion efficiency is also very high, and the bidirectional flow converter can also serve as a charger for the battery or supercapacitor. Therefore, active composite energy storage has received more applications.

5.4 Composite Energy Storage Application Cases

For the application of energy storage in power systems, a supercapacitor battery composite energy storage system as shown in Fig. 5.16 was designed based on the active parallel structure. The battery bank is connected to the DC bus of the AC/DC converter through a bidirectional DC/DC converter, the supercapacitor bank is directly connected to the DC bus, and the AC/DC converter is connected to the AC

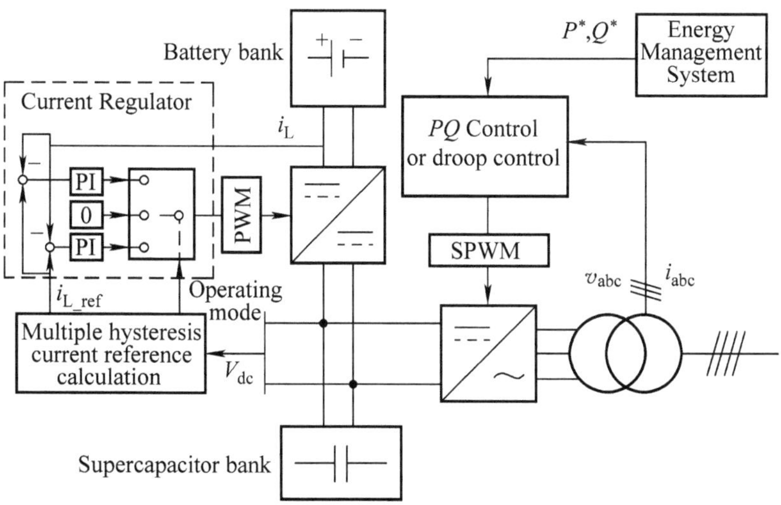

Fig. 5.16 Composite energy storage system applied to power systems

bus of the distribution network or microgrid. It can achieve functions such as smoothing renewable energy power generation fluctuations through active and reactive power regulation, or as a means of instantaneous power balance adjustment for microgrids [81–83].

The bidirectional DC/DC converter is the key to composite energy storage energy management. A BUCK/BOOST bidirectional power converter (BUCK/BOOST Bi-directional Converter, BBBC) is used. This topology has the advantages of small size, fewer power devices, and high working efficiency [84]. For the modeling of BBBC, the BUCK and BOOST working states can be modeled independently [85], and the equivalent circuit of the composite energy storage is shown in Fig. 5.17.

The control scheme of BBBC adopts a multi-hysteresis regulation control strategy, which can flexibly set the charging and discharging current of the battery and its conversion process at multiple levels according to the actual situation of the system, and has good practical operability.

The multi-hysteresis regulation control strategy consists of a multi-hysteresis current set calculation unit and a current regulator. The control logic of the multi-hysteresis current set calculation unit consists of two 3-state hysteresis loops, its input is the DC bus voltage V_{dc}, and the output is the current set value i_{L_ref} and the working mode instruction of BBBC, as shown in Fig. 5.18.

The current set calculation unit first determines the positive and negative values of i_{L_ref} based on the value of V_{dc}, that is, which half of the hysteresis loop it is in, then determines the working mode of the next beat $k + 1$ in combination with the current working mode of BBBC, and finally obtains the current output instruction i_{L_ref} according to the hysteresis curve, and send it to the current regulator. The current regulator determines the operating mode of BBBC according to the work instructions. When BBBC operates in BUCK mode, the drive signal of switch tube

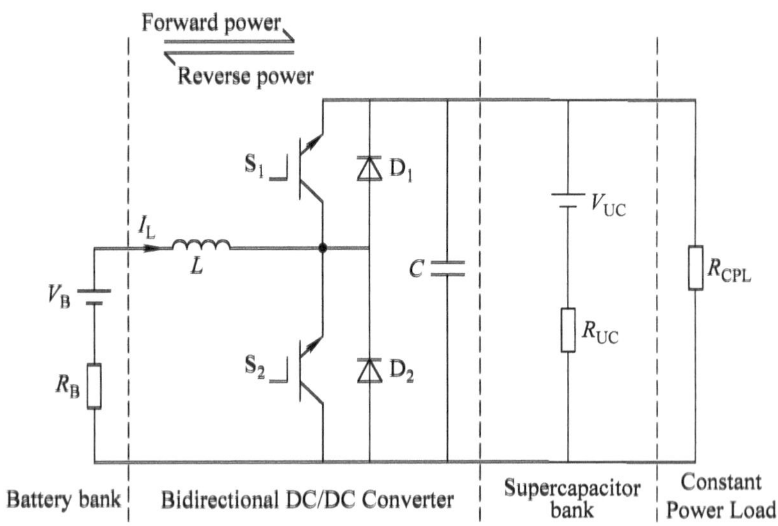

Fig. 5.17 Equivalent circuit diagram of composite energy storage system

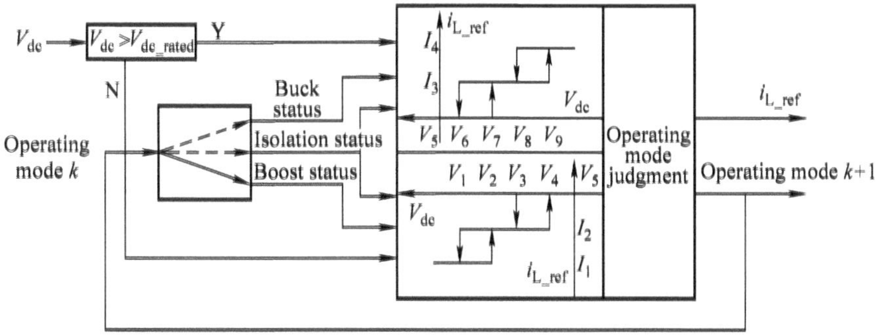

Fig. 5.18 Current set calculation unit

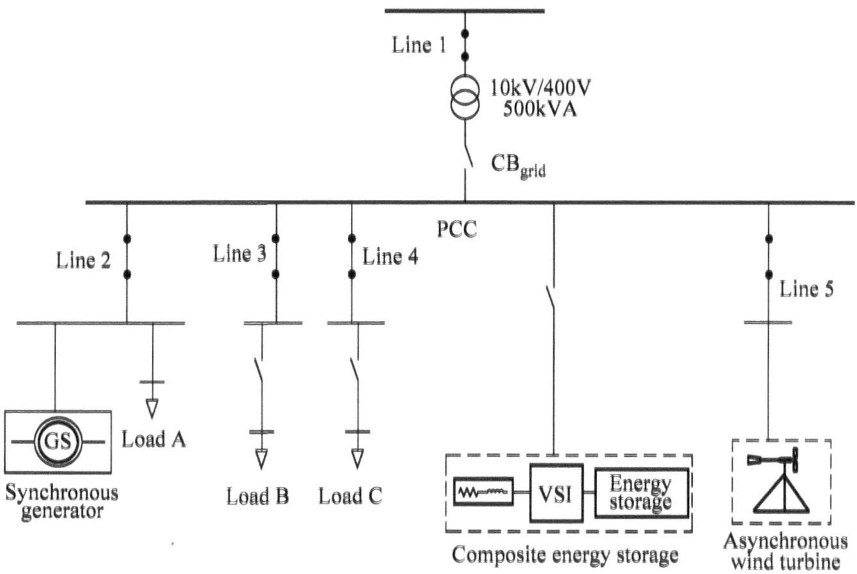

Fig. 5.19 Microgrid simulation model with composite energy storage [Change "Energy Storage" in the figure to "Composite Energy Storage"]

S_2 is blocked, and S_l is enabled; on the contrary, if it operates in BOOST mode, the drive signal of switch tube S_l is blocked, and S_2 is enabled. The *PI* regulator makes the inductor current i_L track the given value i_{L_ref} without static difference.

In order to verify the effectiveness of the composite energy storage and multi-hysteresis regulation control strategy, a microgrid simulation platform based on *Matlab/Simulink* was established, as shown in Fig. 5.19.

The microgrid simulation model includes a fixed pitch stall type asynchronous wind turbine generator AWT, a synchronous generator equipped with excitation regulation and speed control, energy storage and load, etc. Among them, the synchronous generator has a capacity of 30 kVA, AWT has a capacity of 15 kVA, and the energy storage system has a capacity of 15 kVA (where the battery bank has a capacity of 100 Ah and a rated voltage of 240 V; the supercapacitor bank has a capacity of 15 F and a rated voltage of 360 V); the active power of the load P = 15 kW, reactive power Q = 20 kvar; system frequency f = 50 Hz, phase voltage 220 V.

The simulation results are shown in Fig. 5.20. It can be seen that when the battery energy storage operates alone, due to the limited output power, it cannot meet the power demand of the stable operation of the microgrid, resulting in larger fluctuations in the voltage and frequency of the microgrid. When using composite energy storage, the high power characteristics of the supercapacitor greatly improve the

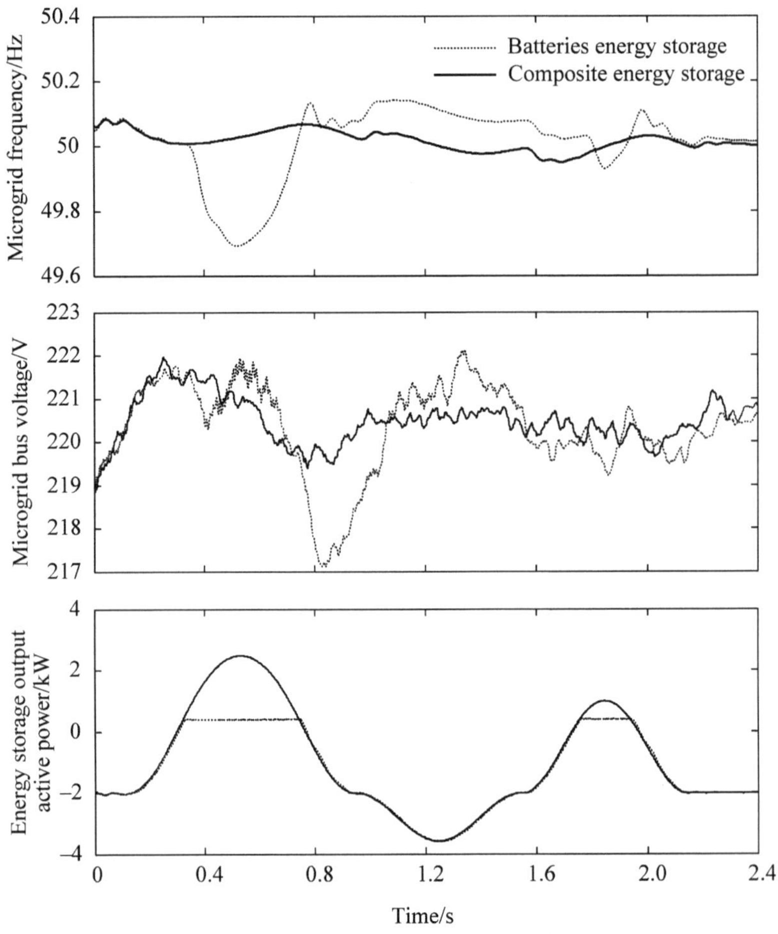

Fig. 5.20 Simulation results of composite energy storage applied to microgrid

Fig. 5.21 Experimental results of composite energy storage applied to microgrid

power capacity of the energy storage system, effectively achieving the instantaneous power balance of the microgrid and improving the operational stability.

In addition, a 10 kW composite energy storage experimental platform was built, with the experimental parameters being: supercapacitor bank capacity 20 F, rated voltage 270 V, internal resistance 0.1 Ω; battery bank rated voltage 120 V, internal resistance 0.5 Ω; the inductance value of the bidirectional DC/DC converter is 0.5 mH, switching frequency 10 kHz.

Simulating the state changes during the operation of the microgrid, controlling the working process of the composite energy storage, the discharge current of the battery group is set to 10 A and 20 A. The experimental results are shown in Fig. 5.21.

During 0–38 s and 463 s–500 s, the microgrid is in a grid-connected operation state, and the energy storage does not exchange energy with the microgrid and is in standby mode. Between 38 s–463 s, the microgrid operates in island mode. As can be seen from Fig. 5.21, the power absorbed/emitted by the energy storage unit during island operation has the characteristics of large fluctuations and high peak power. The supercapacitor can quickly respond to the power changes of the microgrid, meeting most of the power demand. The battery charges and discharges in a segmented constant current mode according to the state of the DC bus and system commands, and its operation process is greatly optimized compared to when the battery operates alone.

Chapter 6
Big Data Analysis of Energy Storage Systems

Battery energy storage systems have broad application prospects, but energy storage systems composed of a large number of individual batteries in series and parallel are easily affected by the premature aging of individual battery health, resulting in unstable overall performance and low efficiency. This chapter introduces the data characteristics of battery energy storage systems, uses big data analysis methods to analyze the aging rules of battery banks, and provides a basis for the intelligent operation and maintenance and optimized design of energy storage systems.

6.1 Big Data Technology and Its Application in Energy Storage

In recent years, with the rapid development of the Internet, the Internet of Things, and information technology, the concept of big data has gradually expanded from a few fields such as finance and IT to all sectors of the national economy. China has clearly proposed the strategic direction of "promoting the deep integration of the Internet, big data, artificial intelligence and the real economy". Big data technology mainly includes big data processing systems and big data analysis algorithms [86].

The big data processing system refers to a new system that is significantly different from traditional data processing systems in architecture and logic, including the Hadoop file system suitable for large-scale clusters and its supporting batch processing programming architecture MapReduce [87, 88]. Storm for real-time processing of streaming data [89], and new interactive information processing systems such as Hive [90], MongoDB [91], etc. The development of these systems is characterized by specialization, diversity, and real-time. In specific application scenarios, they can provide flexible, high-performance services, greatly improving the processing capacity and response speed of information systems, and are the underlying foundation of big data technology.

X. Tang et al., *Electrical Energy Storage Technologies and Applications*,
https://doi.org/10.1007/978-981-96-5625-7_6

Big data analysis algorithms are mainly based on deep learning, and also include knowledge computing, visualization, etc. [86]. The proposal of deep learning can be traced back to the BP neural network algorithm in the 1980s. After entering the twenty-first century, with the improvement of a series of algorithms and the increase of computing power, neural networks have made breakthroughs in fields such as speech, language, and image recognition.

With the continuous advancement of the intelligentization of the energy system, especially the development of energy Internet technology, there are also many instances of the application of big data architecture and analysis technology in the power system [92]. Big data analysis technology was applied early in power equipment fault monitoring, using genetic algorithms, classification algorithms, time series analysis, etc. [93–95]. Literature [96] calculates the transition probability sequence of a single state quantity of a device based on the time axis, and then describes the correlation between multiple state quantities of the device through unsupervised clustering, establishing a rapid detection model for abnormal states, to monitor the abnormal state of power equipment in real time. With the development of plant-level monitoring system and distributed control system, the operational data that thermal power units can collect has achieved exponential growth. Literature [97] proposes a method to determine the benchmark values of thermal power units according to working data using an improved k-means clustering algorithm; the difficulty lies in that the parameters of the thermal power unit can reach more than ten, including steam pressure/temperature, flue gas temperature, feed water temperature, flue gas oxygen content, and other multi-dimensional data. By using an algorithm combining generalized neural networks and average impact values for dimension reduction, a data model with strong correlation is constructed, and then the improved k-means clustering analysis is used for benchmark classification. In the intelligent operation and maintenance of photovoltaic power stations, deep learning is used to extract key operating parameters of photovoltaic power stations and optimize parameter combinations to improve the overall performance of the power station. For commercial and industrial parks, through smart energy management systems or electricity information collection systems, "full coverage, full collection" of information management and interactive services based on user pattern understanding are realized. Through data-driven management models, lean production and energy-saving and consumption reduction of enterprises are achieved [98]. The Energy Data Platform (EDP) developed by AutoGrid in the United States can collect and process data covering energy generation, transmission and distribution, and end-user electricity, constructing a comprehensive and dynamic picture of the energy system.

As a complex system containing a large number of components and diverse operating processes, the energy storage system is inevitably a massive data source. With the large-scale deployment of energy storage systems worldwide, an era of massive growth in energy storage data is coming. How to better collect, process, and utilize these data to provide effective solutions for intelligent operation and maintenance and optimized control is a worthwhile research direction, and related R&D and demonstration work is also rapidly underway.

To share data from energy storage projects and promote the development of energy storage technology and the deployment of energy storage systems, the U.S. Department of Energy has established a Global Energy Storage Database (GESDB) based on the PostgreSQL relational database [99, 100]. As of August 2016, the database included 1591 operational energy storage projects in 69 countries, mainly divided into five categories: electrochemical energy storage, mechanical energy storage, hydrogen energy storage, pumped storage, and thermal storage. The nature of this project is closer to that of a traditional data center, disclosing and displaying basic data of energy storage stations, data on the smoothing effect of renewable energy generation fluctuations, and visualizing these data. A unified energy storage data platform is the basis for conducting big data research on energy storage. Of course, what kind of promotion these data will have on the analysis of energy storage systems in the future remains to be tested.

Due to problems such as the difficulty in accurately measuring battery parameters or the inaccuracy of battery models, the research on battery health status has encountered bottlenecks. Literature [101] uses battery data to train a neural network, setting up a system for online analysis of battery status. Deep neural networks have certain advantages in judging ambiguous states, but they require a large amount of complete training data. Due to the diverse working modes of energy storage systems, the variety of batteries, and the large differences between individual batteries, most research currently does not have the conditions to collect complete data.

It is clear that the application of big data technology in the field of energy storage is still in its infancy, and its research methods and ideas need to be further expanded. However, with the explosive growth of data volume in energy storage systems, and further expansion in the depth and breadth of data, new solutions have been provided for the health diagnosis and intelligent operation and maintenance of energy storage systems. This chapter starts from this point, draws on the application experience of big data technology in the corresponding fields, takes the lead-carbon battery energy storage system with a higher degree of commercialization as an example, analyzes the business and data characteristics of the energy storage system, further improves the clustering analysis method, and explores the regularity of the health status of the energy storage system.

6.2 Data Analysis of Energy Storage System

The battery energy storage system integrates battery bank, Battery Management System (BMS), Power Conversion System (PCS), Energy Management System (EMS) and local monitoring management system [102, 103]. Its data comes from various modules of the system, among which the data of single battery voltage,

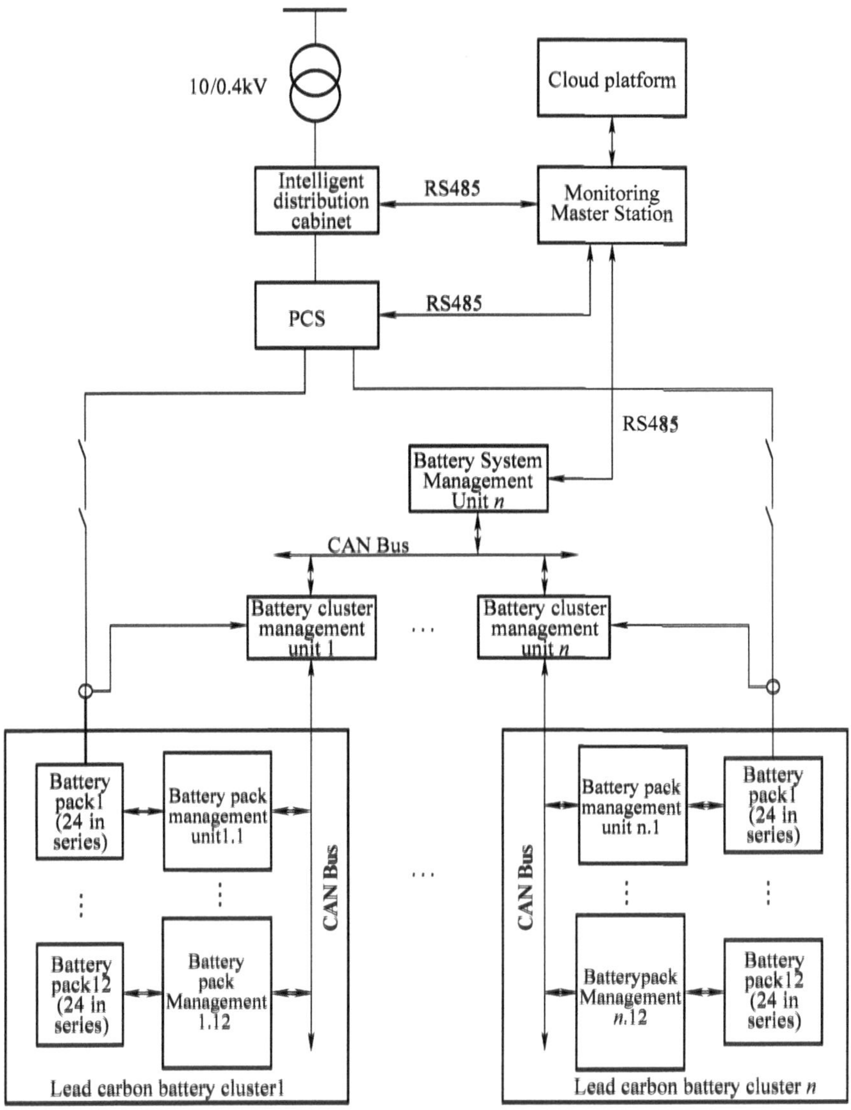

Fig. 6.1 Main components of the lead-carbon battery energy storage system

temperature, etc. comes from the BMS system, PCS provides the overall working status information of the entire energy storage system, and the monitoring management system mainly realizes the collection of the above system-level information, operation status adjustment, and alarm information processing, etc. Figure 6.1 takes the lead-carbon battery energy storage system as an example to illustrate the main components of the energy storage system.

6.2.1 Lead-Carbon Battery

The lead-carbon battery is a new type of lead-acid battery, which improves the utilization rate of active lead material and can inhibit the growth of lead sulfate crystals by incorporating high specific surface carbon materials (such as activated carbon, activated carbon fiber, carbon aerogel or carbon nanotubes, etc.) into the negative lead electrode, taking advantage of the high conductivity and dispersibility of high specific surface carbon materials to lead-based active materials. The addition of carbon materials to the negative electrode plate can also play its advantage of instantaneous high current, buffer during high-rate charge/discharge, effectively protect the negative electrode plate, and inhibit the "sulfation" phenomenon [104]. Therefore, the cycle life of lead-carbon batteries is significantly higher than that of ordinary lead-acid batteries, suitable for applications of energy storage.

Taking a certain type of LLC-1000 lead-carbon battery as an example for design analysis, the rated voltage of a single battery is 2 V, and the rated capacity is 1000 Ah. A 1.2 MWh energy storage system contains 600 LLC-1000 lead-carbon batteries, 300 in series and 2 in parallel, which can be configured with 125 kW or 250 kW PCS. The main configuration of the system is shown in Table 6.1, and the technical parameters of the lead-carbon battery are shown in Table 6.2.

Table 6.1 Main parameters of the energy storage system

Serial number	Project description		Parameter	Remark
1	Energy storage system capacity		1.2 MWh	
2	Rated charge/discharge power		125/250 kW	
3	Rated output voltage		AC 380 V	
4	Rated output frequency		50 Hz	
5	Output wiring method		Three-phase four-wire	
6	External communication manner		Ethernet	Modbus (TCP/IP)
7	Response time		<200 ms	Time from receiving dispatch command from local monitoring to 90% full load power output of PCS
8	Container	Power port	Three-phase four-wire copper bus interface	1 channel
		Distribution port	Single-phase 220 V, 50 Hz	1 channel
		Communication port	Ethernet Modbus (TCP/IP)	1 channel
		Grounding port		1 channel

Table 6.2 LLC-1000 Lead-carbon Battery Parameters

Rated voltage	2 V
Self-discharge (25 °C)	7%/90 days
Operating temperature range	Discharge: −40 °C–50 °C Charging: −20 °C–40 °C Storage: −20 °C–35 °C
Recommended operating temperature	15–25 °C
Maximum charging current	0.3C
Recommended charging current	0.15C
Charging voltage (25 °C)	2.30–2.35 V/cell
Effect of temperature on capacity	105% @ 40 °C 90% @ 0 °C 40% @ −20 °C

Table 6.3 LLC-1000 60%DOD discharge cut-off voltage

Cycle number/N	Discharge cut-off voltage/V
$N \leq 1500$	1.95 V
$1500 < N \leq 2000$	1.92 V
$2000 < N \leq 2200$	1.89 V
$2200 < N$	1.84 V

6.2.1.1 Lead-Carbon Battery Charging and Discharging Strategy

1. Charging: Constant voltage 2.35 V, current limit $0.15C_{10}$ A charging, until the current is $\leq 0.01C_{10}$ A.
2. Generally, the depth of discharge (DOD) should not exceed 60–70%, and the discharge current should be controlled below $0.25C_{10}$. As the number of cycles increases, the cut-off voltage for discharge should also be adjusted accordingly. As shown in Table 6.3.
3. The amount of charge after discharge should generally be controlled at 101% to 103% of the total discharge amount of the last time.
4. Perform an equalizing charge every 2–3 months, constant voltage 2.35 V, current limit $0.15C_{10}$ A charging, until the current is $\leq 0.01C_{10}$ A, and continue for 8–10 h.
5. The above parameters are for standard ambient temperature. According to the actual environmental temperature of the battery system, the charging voltage should be temperature compensated, with a coefficient of -3 mV/(cell·°C).

6.2.1.2 Factors Affecting Battery System Performance

Lead-carbon batteries are essentially lead-acid batteries. There are many conditions affecting their performance and health status, and the process is very complex. However, two obvious characteristics occur during the aging process of the battery, namely the reduction of effective capacity and the increase of internal resistance.

During the discharge process, when the charge is insufficient, lead sulfate gradually precipitates out, and the precipitated lead sulfate tends to crystallize and gradually forms larger crystal salts. This part of the substance can no longer participate in the chemical reaction, reducing the active material of the battery and lowering the effective capacity. During the charging process, in the absence of metal ions that can participate in the reaction in the internal electrolyte, long-term overcharging will cause the electrolysis of water side reaction, generating hydrogen and oxygen, leading to the loss of electrolyte water. These two irreversible processes are the main reasons for the loss of internal reaction materials in the battery [105].

For energy storage systems, which are often composed of hundreds of batteries connected in series and parallel, the capacity difference of individual batteries can greatly affect the overall performance of the energy storage system. Therefore, it is necessary to avoid the following factors affecting the operation control process of the energy storage system [104]:

1. Depth of discharge. Lead-carbon batteries generally have an optimal discharge depth (such as 60%–70%). If the discharge is too deep, the concentration of lead sulfate in the electrolyte will saturate and gradually precipitate, causing the active material of the positive and negative electrodes to sulfate, affecting its cycle life and total discharge amount during the whole life period.
2. Ambient temperature. Lead-acid batteries have an optimal operating temperature range. Exceeding this temperature range will affect the battery's operating life. The physical space of the battery energy storage system is often large, so how to control the ambient temperature within a reasonable range, especially to reduce the temperature difference, is very critical to eliminate the impact of temperature.
3. Thermal runaway. Excessive charging and discharging currents can cause the working temperature of the battery to rise rapidly, so it is necessary to control the maximum charging and discharging current of the battery to prevent the battery temperature from being too high and causing thermal runaway.
4. Overcharging. Overcharging will accelerate battery water loss, affecting battery life; long-term float charging will also cause the battery to accelerate water loss, affecting battery life.
5. Long-term idleness. Long-term idleness will cause the battery to be in a long-term self-reaction process. Due to the parameter differences between individual batteries, this self-reaction will eventually make the voltage and capacity consistency of the entire battery group worse, thereby affecting the overall performance of the energy storage system.

6.2.2 BMS Data

The Battery Management System (BMS) is an important functional unit of the energy storage system, which realizes the collection of battery current, voltage, temperature, internal resistance and other data, evaluates the State of Charge (SOC) and State of Health (SOH), and serves as the basis for system operation control. It also provides early warnings of relevant parameters to prevent faults.

For a battery energy storage system composed of multiple battery clusters in parallel, the BMS can be divided into a Battery Management Unit (BMU) that manages multiple batteries in groups, a Battery Management Array (BMA) that manages battery clusters, and a BMS that manages the entire battery system. The main structure of the BMS is shown in Fig. 6.2.

Data is exchanged between the various levels of the BMS. The BMU aggregates data to the BMA via the CAN bus. Multiple BMAs transmit the data of the battery cluster to the BMS via the CAN bus, and then use the RS485 protocol to package the data and send it to the local monitoring management system. The BMS data frame is shown in Table 6.4.

The BMS data frame is composed of data from various units. When transmitting data to the remote database server, each data content occupies a register address. All data are integers, and the parameters are accurately described through conversion relationships and set units. From the data frame, it can be seen that the information of the lower BMU mainly includes the voltage monitoring data of each individual battery. The middle-level BMA and the BMS will count the voltage, current, temperature, internal resistance, and SOC of the battery group to manage the battery status, warnings, and fault handling.

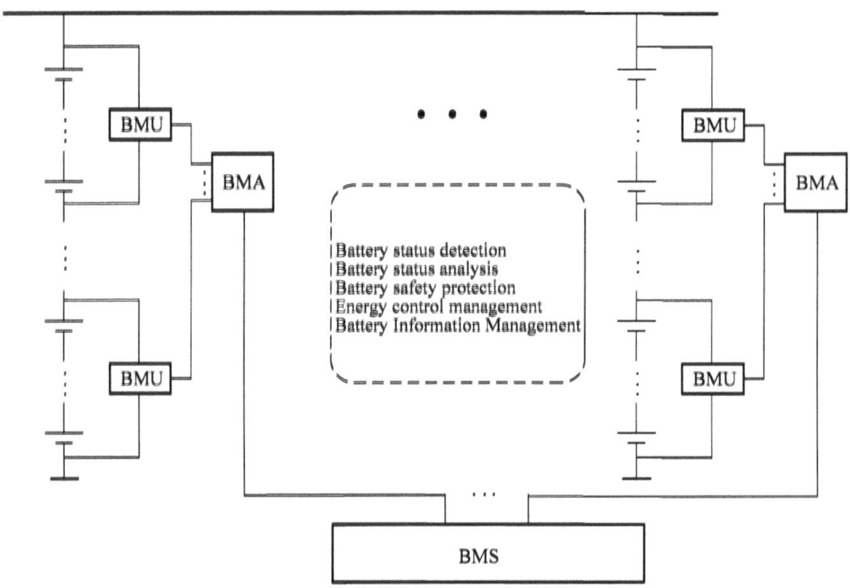

Fig. 6.2 Schematic diagram of BMS structure

Table 6.4 BMS data frame

Monitoring item
Name
Conversion relationship
Format description
BMS
Cumulative discharge quantity of battery system (high position)
*1
Cumulative discharge quantity (high position) H
Cumulative discharge quantity of battery system (low position)
*0.01
Single cell battery voltage low limit alarm value
*0.01
Low limit alarm value of battery cell voltage
High limit protection value of battery cell voltage
*0.01
High limit protection value of battery cell voltage
Low limit protection value of battery cell voltage
*0.01
Low limit protection value of battery cell voltage
High limit alarm value of battery internal resistance
*0.1
High limit alarm value of battery internal resistance
High limit protection value of battery internal resistance
*0.1
High limit protection value of battery internal resistance
High limit protection value of battery internal resistance
*0.1
High limit protection value of battery internal resistance
High limit protection value of battery internal resistance
*0.1
High limit protection value of battery internal resistance
BMA1
Number of batteries in cluster 1
*1
Number of batteries in cluster 1
Voltage of battery bank in cluster 1
*0.1
Voltage of battery bank in cluster 1
Current of battery bank in cluster 1
*0.1
Current of battery bank in cluster 1
Average voltage of single cell in cluster 1
*0.01

(continued)

Table 6.4 (continued)

Average voltage of single cell in cluster 1
Serial number of battery with maximum cell voltage in cluster 1
*1
Serial number of battery with maximum cell voltage in cluster 1
Maximum cell voltage in cluster 1
*0.01
Maximum cell voltage in cluster 1
Serial number of battery with minimum cell voltage in cluster 1
*1
Serial number of battery with minimum cell voltage in cluster 1
Minimum cell voltage in cluster 1
*0.01
Minimum cell voltage in cluster 1
Battery cluster 1 individual SOC minimum battery number battery cluster 1 minimum SOC battery sequence number
Battery cluster 1 minimum SOC value
*0.1
Battery cluster 1 minimum SOC value
Battery cluster 1 rated capacity
*0.1
Battery cluster 1 rated capacity
Battery cluster 1 voltage
*0.1
Battery cluster 1 voltage
Battery cluster 1 chargeable capacity
*0.1
Battery cluster 1 chargeable capacity
Battery cluster 1 dischargeable capacity
*0.1
Battery cluster 1 dischargeable capacity
Battery cluster 1 maximum chargeable power
*0.1
Battery cluster 1 maximum chargeable power
Battery cluster 1 maximum dischargeable power
*0.1
Battery cluster 1 maximum dischargeable power
SOC1
*10
SOH1
*10
Single cell battery information
Voltage of the first cell of the battery
*0.01

(continued)

Table 6.4 (continued)

Battery voltage
Voltage of the second cell of the battery
*0.01
Battery voltage
Voltage of the third cell of the battery
*0.01
Battery voltage
Voltage of the fourth cell of the battery
*0.01
Battery voltage
Voltage of the fifth cell of the battery
*0.01
Battery voltage
Voltage of the sixth cell of the battery
*0.01
Battery voltage
Voltage of the seventh cell of the battery
*0.01
Battery voltage
Battery cell 8 voltage
*0.01
Battery voltage
Battery cell 9 voltage
*0.01
Battery voltage
Battery cell 10 voltage
*0.01
Battery voltage
Battery cell 11 voltage
*0.01
Battery voltage
Battery cell 12 voltage
*0.01
Battery voltage

6.2.3 PCS Data

The Power Conversion System (PCS) realizes the energy transfer between the DC battery group and the grid, and is the carrier of the energy storage system function and performance. The PCS generally uses a Voltage Source Converter (VSC) based on PWM modulation technology. Figure 6.3 shows the main circuit schematic of the PCS.

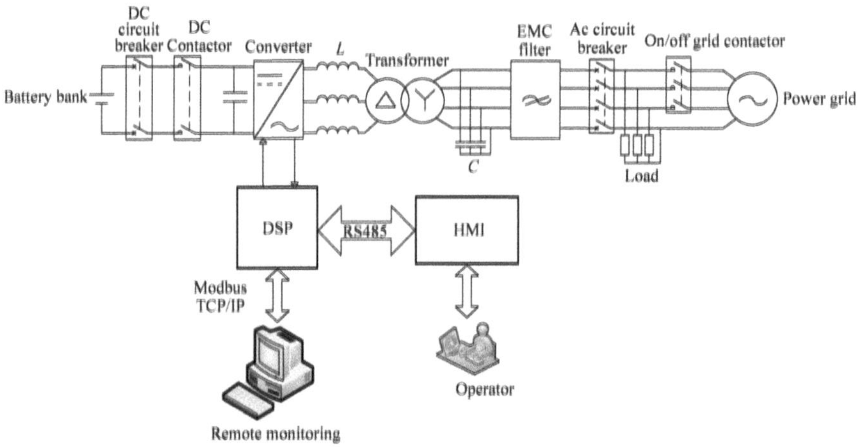

Fig. 6.3 PCS Main Circuit Structure Diagram

The PCS can be started and stopped, parameter setting and status monitoring through the human-machine interface or remote monitoring system, such as setting off-grid/on-grid mode conversion, power setting, alarm threshold setting, etc. The main operating data are battery group current, power, bus voltage, radiator temperature, cumulative charge and discharge amount, local/grid status, etc. The data frame structure provided by the PCS is shown in Table 6.5.

Each data content in the PCS data frame occupies a register address, and all data are integers. The accurate description of parameters is achieved through setting conversion relationships and units. From the perspective of data analysis, the PCS is the main data source for obtaining bus voltage, working current, and charging and discharging status. In addition, the PCS can also implement the control and protection functions of the entire energy storage system, including controlling the grid switch, the system's working parameters and operating logic, equipment and AC/DC side abnormal handling, and fault protection, etc.

6.2.4 Energy Storage System Data Collection and Data Characteristics

The overall data collection of the energy storage system is generally realized through the monitoring management system, mainly monitoring the battery bank, BMS, PCS, metering electric meter and other main equipment, realizing data interaction, command transmission, data storage, and human-machine interface, etc. At the same time, a communication interface with the host computer or external server is reserved.

The monitoring management system mainly communicates with various terminal devices through the local monitoring host (such as PLC), and monitors various

Table 6.5 PCS Data Frame

Name	Conversion relationship	Format description
Battery Voltage	V*0.1	Voltage Range
Local Phase A Voltage	V*0.1	Voltage Range
Local Phase B Voltage	V*0.1	Voltage Range
Local Phase C Voltage	V*0.1	Voltage Range
Grid Side Phase A Voltage		P*1000
Grid reactive power exchange (charging/discharging)	Q*1000	Power range
Grid apparent power exchange (charging/discharging)	S*1000	Power range
Operating status	0 = Off-grid, 1 = On-grid	Operating status
Contactor status	0 = Off, 1 = On	Operating status
Synchronization status	0 = Not synchronized, 1 = Synchronized	Operating status
Grid switch status	0 = Off-grid, 1 = On-grid	Operating status
Grid connection permission flag	0 = Not allowed, 1 = Allowed	Operating status
Grid fault shutdown waiting for automatic recovery and grid connection	0 = Not allowed, 1 = Allowed	Operating status
Power setpoint over limit alarm	0 = Normal, 1 = Alarm	Operating status
Device fault flag	0 = Normal, 1 = Faulty	Device fault status
IGBT_A_OC (Overcurrent)	0 = Normal, 1 = Faulty	Device fault status
IGBT_B_OC	0 = Normal, 1 = Faulty	Device fault status
IGBT_C_OC	0 = Normal, 1 = Faulty	Device fault status
Grid_A_OC (Grid Overcurrent)	0 = Normal, 1 = Faulty	Device fault status
Grid_B_OC	0 = Normal, 1 = Faulty	Device fault status
Grid_C_OC	0 = Normal, 1 = Faulty	Device fault status
IGBT_A_OT (Overtemperature)	0 = Normal, 1 = Faulty	Device fault status
IGBT_B_OT	0 = Normal, 1 = Faulty	Device fault status
IGBT_C_OT	0 = Normal, 1 = Faulty	Device fault status
Bat_OV (Battery Overvoltage)	0 = Normal, 1 = Faulty	Device fault status
Bus_OV (DC Bus Overvoltage)	0 = Normal, 1 = Faulty	Device fault status
Emergency_State (PCS Emergency Stop Switch Status)	0 = Normal, 1 = Faulty	Device fault status
Bat_Charge_OC (Overcurrent during Charging)	0 = Normal, 1 = Faulty	Device fault status
Bat_Discha_OC (Overcurrent during Discharging)	0 = Normal, 1 = Faulty	Device fault status
L_OT (Reactor Overtemperature)	0 = Normal, 1 = Faulty	Device Fault Status
Tran_OT (Transformer Overheat)	0 = Normal, 1 = Fault	Device Fault Status
Hardware_Fault (Hardware Fault)	0 = Normal, 1 = Fault	Device Fault Status
SPD_Fault (Surge)	0 = Normal, 1 = Fault	Device Fault Status
VacOut_OV (AC Side Overvoltage)	0 = Normal, 1 = Fault	Device Fault Status
VacOut_LV (AC Side Undervoltage)	0 = Normal, 1 = Fault	Device Fault Status

(continued)

Table 6.5 (continued)

Name	Conversion relationship	Format description
Phase A IGBT Temperature	*0.1	Phase A IGBT Temperature
Phase B IGBT Temperature	*0.1	Phase B IGBT Temperature
Phase C IGBT Temperature	*0.1	Phase C IGBT Temperature
PCS Charging/Discharging Status	0 = Discharging, 1 = Charging	Operating Status
Spare	Spare 1	Spare 1
Spare	Spare 2	Spare 2
Spare	Spare 3	Spare 3
Active Power Setpoint	*1	Active Power Setpoint
Reactive Power Setpoint	*1	Reactive Power Setpoint
Off-grid Voltage Setpoint	*1	Off-grid Voltage Setpoint
Off-grid Frequency Setpoint	*1	Off-grid Frequency Setpoint
Device Switch	0 = Off, 1 = On	Device Switch
Grid Switch	0 = Off, 1 = On	Grid Switch
Fault Reset	0 = Off, 1 = Reset	Fault Reset
Power Statistics Reset	0 = Off, 1 = Reset	Power Statistics Reset
ModbusID	*1	ModbusID
Allow Automatic Grid Connection	0 = Not Allowed, 1 = Allowed	Allow Automatic Grid Connection
Float Charge Voltage Setting	*0.1	Float Charge Voltage Setting
Equalization Charge Voltage Setting	*0.1	Equalization Charge Voltage Setting
Discharge Under-voltage Setting (Discharge Warning Voltage)	*0.1	Discharge Under-voltage Setting
Over-voltage Shutdown Setting	*0.1	Over-voltage Shutdown Setting
Under-voltage Shutdown Setting (Discharge Stop)	*0.1	Under-voltage Shutdown Setting
Maximum Charging Current	*0.1	Maximum Charging Current
Maximum Discharge Current	*0.1	Maximum Discharge Current
Charging Quantity (High Bit)		
Charging Quantity (Low Bit)		
Discharge Quantity (High Bit)		
Discharge Quantity (Low Bit)		

devices and dynamic environment systems (including power, environment, fire protection, security, network, etc.). It also reserves a communication interface with the host computer or external server, such as Ethernet, and supports various standard power communication protocols, such as IEC61850, IEC60870D, etc.

As shown in Fig. 6.4, the PLC controller is used as the local monitoring management system platform. The PLC communicates with the PCS using Modbus RTU/TCP, collecting data and issuing commands to the PCS; the PLC communicates with the BMS using Modbus RTU protocol RS485, collecting data and monitoring the BMS; the PLC communicates with the electric meter using Modbus RTU protocol RS485, collecting data and monitoring the electric meter.

Data storage is divided into local data storage and remote data storage. Local data storage mainly relies on the memory card configured by the PLC itself, and the PLC can store data in Excel format on the memory card. Remote data storage can be transmitted to the server through an external communication interface; or data can be transmitted to the cloud server via Ethernet, and accessed via Web or mobile APP.

For the above energy storage system, structured data mainly comes from BMS and PCS. Among them, the data of BMS mainly includes the voltage, temperature, etc. of each individual battery, as well as the total voltage and working status of the battery bank; the data of PCS mainly includes the bus voltage, current, power, etc. at each time point. The form of the above data in the database is a table of a certain length, each row represents the instantaneous state of each electrical unit at a specific time point.

To fully grasp the status of an energy storage station, in addition to structured data, many unstructured data outside the database are also needed. This includes battery selection, control logic, operating mode, dynamic environment system, as

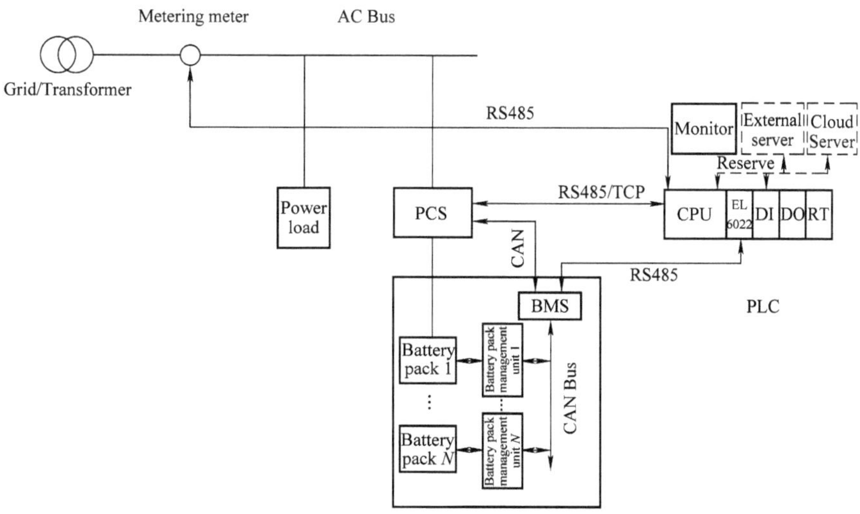

Fig. 6.4 Energy Storage System Information Collection System

well as working logs, operation process, maintenance status, environmental factors, etc. Only by considering enough influencing factors during analysis can we more accurately approach the real state of the system. These unstructured data are generally difficult to obtain from the database, and need to be obtained from technical documents, operation and maintenance records, or even field investigations.

The structured and unstructured data of the energy storage system meet the characteristics of big data, that is, large data volume, fast data acquisition speed, and diverse data types. Only through complete data collection, deep understanding of the operation process of the energy storage system, and effective data processing and analysis methods, can the role of data be fully utilized to solve related problems in the energy storage system.

6.3 Cluster Analysis Method of Energy Storage System

6.3.1 Energy Storage Big Data Analysis Approach

An important factor affecting the operating efficiency and overall lifecycle cost of the energy storage system is the individual aging that occurs during the operation of the battery and the inconsistency of capacity it brings. Aging refers to the phenomenon where the main performance indicators of the battery deviate from the nominal value due to side reactions and material loss during operation, mainly manifested as a decrease in effective capacity, an increase in internal resistance, a rapid rise in terminal voltage during charging, changes in temperature characteristics, etc. Since the energy storage system is often connected by a large number of individual batteries in series and parallel, a few batteries with reduced performance often greatly affect the working state of the entire battery bank.

Although aging of the battery can be observed from the aforementioned manifestations, the aging process is very complex. Factors or phenomena related to aging are either difficult to measure under normal working conditions (such as internal resistance, reactant concentration, etc.), or their relevance is not obvious enough to be quantitatively expressed by mathematical models, because the health status is obtained by comprehensive statistical evaluation of multiple time-varying parameters of the battery. In addition, due to the complexity of the internal electrochemical reactions of the battery, and the increasing complexity of the positive and negative electrode materials and processes in recent years, it is becoming increasingly difficult to analyze the changes in health status by establishing mathematical models.

However, when aging occurs, the degree of aging will be reflected to a certain extent in the data. Based on a large amount of battery operation data, by observing the changes in the data, the pattern of changes in health status can theoretically be discovered. Therefore, by continuously expanding the breadth and depth of energy storage system data, supplemented by effective data processing and analysis methods, a method for judging the aging state of the battery can be formed.

Therefore, according to the idea of clustering algorithms in fuzzy sample classification, methods such as k-means or DBSCAN can be used to perform cluster analysis on battery data to classify the health status of the battery. This chapter models by extracting appropriate parameter information from battery data, selecting parameter weights and the number of clusters, and classifying the health status of each battery in the battery bank through clustering, and positioning it in combination with actual application conditions, thereby exploring the pattern of changes in the health status of the energy storage system.

6.3.2 Introduction to Clustering Analysis Algorithms

Cluster analysis is a type of unsupervised learning in machine learning and a statistical analysis method for studying classification problems [106]. In fact, the obtained sample objects may not have target attributes, and these samples may not be assigned to any known categories, which often occurs in databases with massive data. Clustering algorithms have unique advantages in dealing with these problems. They measure sample similarity through specific algorithms, divide data into various categories based on similarity, and divide different categories according to the data features that exist internally, so that the data within the category is more similar.

In fact, clustering analysis algorithms divide the input data set into several clusters, so that under a specific similarity measurement method, the similarity of samples in the same category is high, and the correlation between elements in different categories is weak [107]. Clustering analysis algorithms are widely used, including pattern learning in psychology, biology, statistics, information retrieval, machine learning, and data mining, etc., and these cross-applications in turn promote the development of clustering algorithms [97]. According to different requirements for performance, efficiency, generalization ability, data form, data dimension, etc., clustering analysis algorithms have derived various efficient and special-purpose algorithms from the earliest k-means algorithm. These algorithms have their own advantages and disadvantages. In practical applications, it is necessary to choose the appropriate algorithm according to the research object and application conditions.

As shown in Fig. 6.5, clustering analysis algorithms can be divided into partition clustering, hierarchical clustering, density clustering, grid clustering, and model-based clustering methods according to their implementation methods.

1. Partition-Based Clustering
   Partition-based clustering methods are the most typical clustering methods, placing data points in a sample space, judging the distance between points through a specific similarity determination method, and then dividing the close points together. The result of the algorithm must ensure that the objects in the same cluster are close, and the objects in different clusters are different. Typical ones include k-means, k-mediods algorithms, etc. These algorithms are simple and

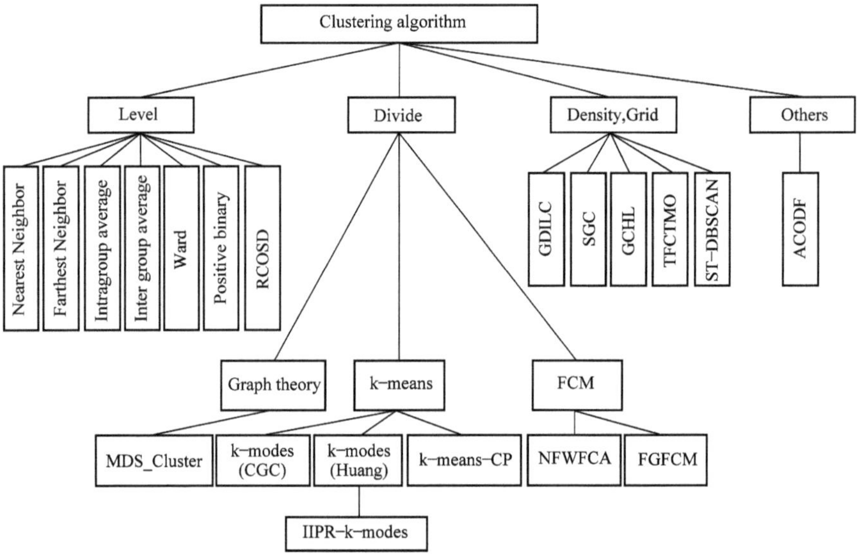

Fig. 6.5 Classification of clustering analysis algorithms

efficient, but as the scale of the dataset expands, the possibility of falling into local minima will also increase.

2. Hierarchical Clustering
   Hierarchical Clustering is different from partition-based clustering, as it will only form one cluster in the end. Its process is: each sample point is regarded as a cluster, and then some similarity measure is used to calculate the similarity between each point, and then the two clusters with the closest similarity are aggregated into a new cluster. This process is repeated until there is only one cluster in the end. In the output results, a tree diagram will be formed based on the distance between clusters. Classic hierarchical clustering methods include CHEMALOEN, BIRCH, ROCK, DIANA, CUBE, AGNES, etc.
3. Density Clustering
   The logic of density clustering is: given the neighborhood radius and density value, when traversing the sample points, if the number of other points in the neighboring area is greater than the density value, the algorithm will continue to traverse the next neighboring point. Typical density clustering algorithms include DBSCAN, OPTICS, DENCLUE, DBCLSD, GDBSCAN, etc. Most classic partitioning methods work better with convex sample set data of spherical distribution, while density algorithms are more suitable for discovering clusters of any shape in the dataset.
4. Grid-based Clustering
   Grid-based clustering methods use a space-driven approach to divide the embedded space into units independent of the distribution of input objects. Grid-based clustering methods use a multi-resolution network data structure, which quantizes the object space into finite units, these units form a grid structure, and the

entire clustering is performed on this grid structure. The advantage of this method is that it is fast and the processing time is independent of the number of data objects, it only depends on the number of units in each dimension of the quantized space. A typical grid-based clustering method is STING.

5. Model-Based Clustering
Model-based clustering methods assume that the data is generated by a model and try to recover the original model from the data. Then, the clusters are defined using the model recovered from the data, and the samples are assigned to the clusters. A common criterion for estimating model parameters is the maximum likelihood method, typical algorithms include COBWEB, CLASSIT, OM, ART, etc.
Considering the actual conditions, the k-means algorithm and DBSCAN algorithm are chosen as the main methods for battery clustering. K-means is a classic clustering algorithm, simple and intuitive, easy to debug, superior performance, suitable for large-scale data mining, but it is sensitive to parameter settings and noise, so it needs to adjust the input parameters based on experience. DBSCAN has good noise point recognition function and can be used as a supplement and verification of the k-means algorithm.

6.3.2.1 K-Means Clustering Algorithm

K-means was independently proposed by Steinhaus in 1955, Lloyd in 1957, Ball & Hall in 1965, and Mc Queen in 1967 in their respective different academic fields. In the more than 50 years since the k-means algorithm was proposed, it has been widely used in different fields and is one of the most widely used clustering algorithms at present [108].

As a real-time unsupervised hard clustering algorithm, the goal of the k-means algorithm is to find undiscovered classifications in the data, with the number of preset classifications represented by the variable k. This algorithm, constrained by the minimization of the error function, iteratively assigns each data point to one of the k groups based on the characteristics provided by the sample data. Since the center points are randomly selected during the initialization of the k-means algorithm, the clustering results are sometimes unstable and the computation time is relatively long.

The input data for the k-means algorithm is a data set X containing n d-dimensional data points = $\{x_1, x_2, x_3 \ldots\ldots x_n\}$ and the number of clusters to be divided, k. The output result is k partition subsets y = $\{y_k\}$ of the data set X, where each partition y_k has a corresponding cluster centroid z_k. If Euclidean distance is chosen as the basis for similarity judgment, then the sum of the squares of the Euclidean distances between the sample points in each class and the center point is the objective value of the algorithm, and the goal of clustering is to minimize the total square distance within each cluster. The process of the k-means algorithm is as follows (Fig. 6.6):

1. Initialization: Randomly initialize k clustering center points in the sample space, and assign each point in the sample to the nearest center point according to the Euclidean distance;
2. Iteration: Calculate the centroids of each cluster after the previous step's allocation, and use these centroids as new centers to reassign each sample point according to Euclidean distance;

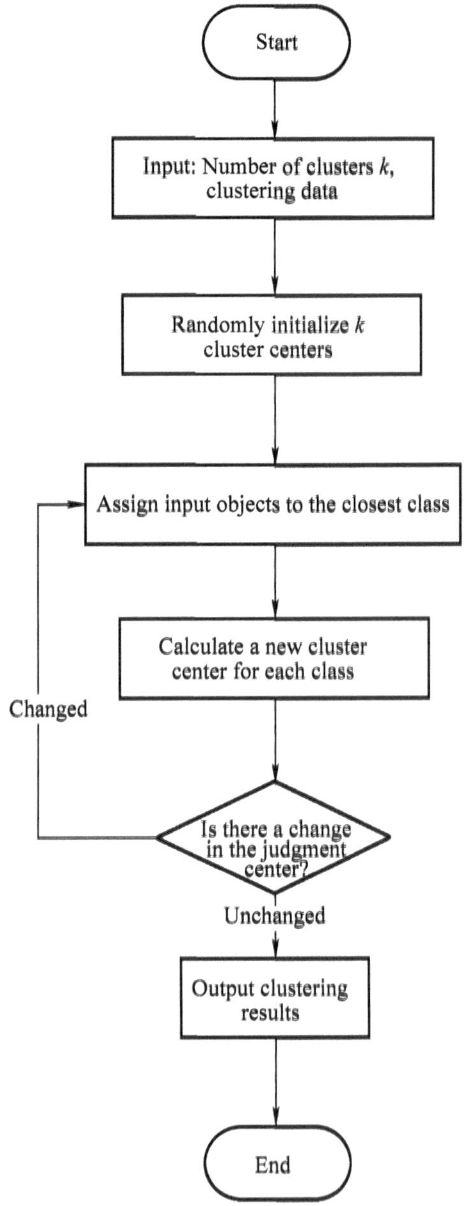

Fig. 6.6 k-means clustering algorithm flowchart

3. Termination: The number of iterations reaches the set limit, or the iteration converges (the result of sample point allocation does not change compared to the previous iteration).

6.3.2.2 DBSCAN Clustering Algorithm

The Density-Based Spatial Clustering of Applications with Noise (DBSCAN), a density clustering method with noise recognition function, was proposed by Martin Ester et al. in 1996 [109]. The basis for density clustering algorithm to classify is the density of sample points in spatial distribution, classifying closely connected samples into different classes to get the final clustering result. The definitions involved in the DBSCAN algorithm are as follows:

1. e-neighborhood: The area within a given object's radius e is called the e-neighborhood of that object;
2. Minimum number of samples MinPts: The threshold for the number of samples in the e-neighborhood of a sample;
3. Core object: If the number of sample points in the given object's E neighborhood is greater than or equal to MinPts, then the object is called a core object;
4. Directly density-reachable: For a sample set D, if sample point q is in the E neighborhood of p, and p is a core object, then object q is directly density-reachable from object p.
5. Density-reachable: For a sample set D, given a string of sample points p_1, p_2 ..... p_n, p= p_1, q= p_n, if object p_i is directly density-reachable from p_{i-1}, then object q is density-reachable from object p.
6. Density-connected: If there is a point o in the sample set D, and object o is density-reachable from both objects p and q, then p and q are density-connected.
7. Non-core objects: Include edge points and noise points.
8. Noise: In the given dataset D, if object p does not belong to any cluster, then object p is a noise point.

When executing the DBSCAN algorithm, all data objects are first marked as unprocessed. Then a random unprocessed point A is selected, and the density of the points around it is used to determine whether this point is a core object or a non-core object. The specific method is to judge whether there are more than MinPts objects within its E domain. If it is a core object, all objects density-connected to point A are traversed according to the above judgment method, and these points are marked as processed and classified into a new cluster; if point A is not a core object, it is marked as processed and a noise point. The termination condition of the algorithm is that all points in the sample set are marked as processed.

The input of the DBSCAN algorithm includes a database of n objects, radius e, and minimum number MinPts; the output is all generated clusters, and the process of meeting the density requirements. The logical flow of the DBSCAN algorithm is as follows (Fig. 6.7):

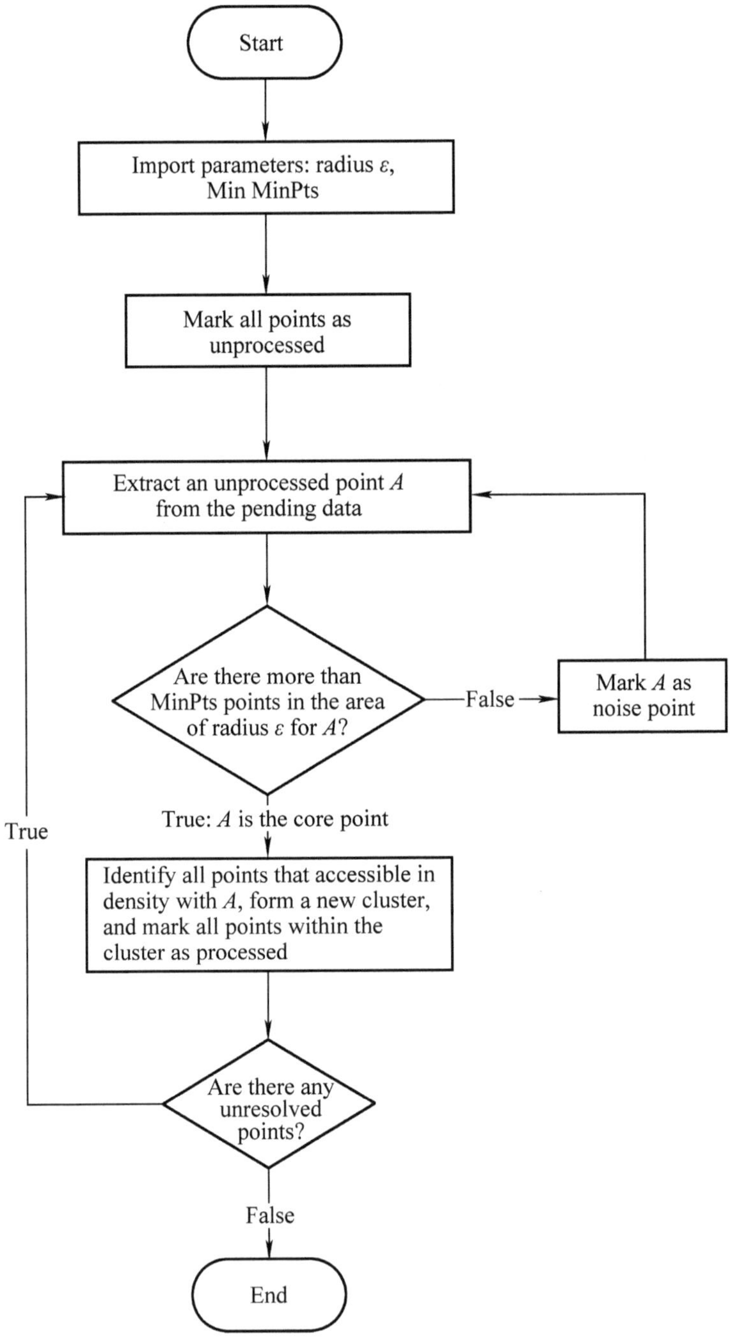

Fig. 6.7 DBSCAN algorithm flowchart

1. Extract an unprocessed point from the data;
2. If the extracted point is a core object, find all objects that are density-reachable from this point, form a new cluster, and mark all points in the cluster as processed;
3. If the extracted point is an edge point (non-core object), mark this point as a noise point, exit this loop, and find the next point;
4. Repeat the loop until all points have been processed.

The key to the DBSCAN algorithm lies in the selection of the two parameters, the region radius E and the density parameter MinPts. Their values will affect the efficiency and results of clustering. In energy storage applications, these two parameters mainly rely on the understanding and experience of specific data, because the main function of DBSCAN is to accurately locate the outliers in battery classification through its noise point screening function, that is, batteries with a high degree of aging, and there are no higher requirements for other effects of clustering. Therefore, the selection of these two parameters will not have a significant impact on the results. In addition, the use of normalized input datasets also reduces the difficulty of determining parameters.

6.4 Energy Storage Big Data Application Case

To address the issues of the k-means clustering analysis algorithm, such as the need to pre-determine the data dimension and the number of clusters k, and its sensitivity to noise, the algorithm is improved in conjunction with the actual characteristics of the battery energy storage system. After extracting the battery features, considering the characteristics of multiple clustering of the battery bank, the battery data of different charging cycles are allocated with a unified normalized weight, supplemented by the adaptive weight method and Hopkins statistics for verification. The elbow method is used to preliminarily determine the number of k for the prepared data, and the value of k is considered comprehensively in combination with the characteristics of the battery itself and the characteristics of multiple clustering. For the sensitivity of the k-means algorithm to noise, the DBSCAN density clustering algorithm with a special screening mechanism for noise points is used to assist in handling noise points.

6.4.1 Data Preprocessing

A case study is conducted in conjunction with a certain lead-carbon battery energy storage system, which is mainly used for demand-side peak-valley shaving. The system contains 336 2 V/1000 Ah lead-carbon batteries, 24 batteries are connected in series to form a battery bank (connected to a battery management unit), and 14 battery bank are connected in series to form a battery cluster, forming the entire

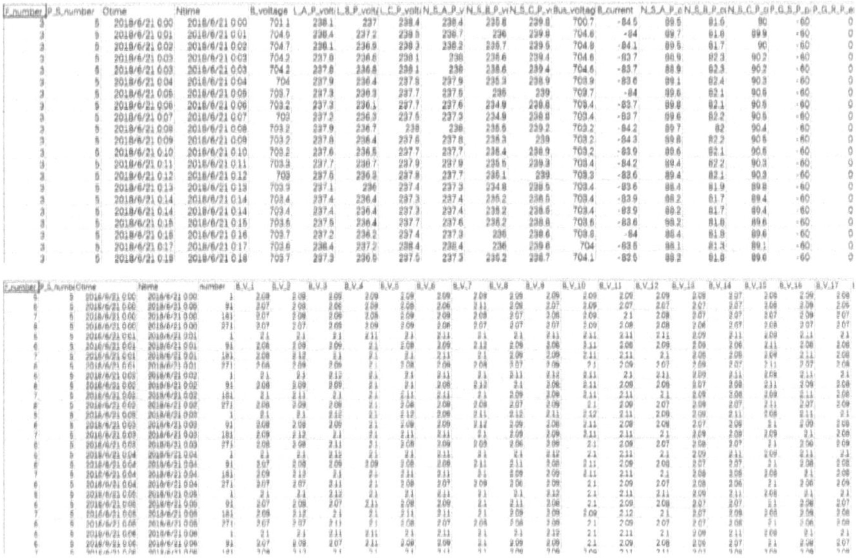

Fig. 6.8 Battery voltage and temperature collection data

battery system. The data of the energy storage system mainly comes from the BMS and the PCS. Among them, the BMS mainly collects the temperature and voltage of each battery, and the PCS mainly collects data such as bus current, total voltage, and working status.

Considering the periodicity and data integrity of the operation status of the energy storage system, the continuous operation data of a certain month is selected for analysis, as shown in Fig. 6.8.

The form of energy storage data is the state quantities at many time points over a period of time. As the input data of the clustering algorithm, it needs to clearly and accurately represent the multi-faceted status of each battery. And the study of data tends to divide the data into charging and discharging cycles on the time scale. Therefore, the purpose of data preprocessing is to divide the time block according to the charging and discharging and the resting period from the voltage, current, temperature, etc. of each single battery continuously sampled over a period of time, and to count the changes in each cycle. The amount, rate of change, overall dispersion, etc. of battery power and voltage are prepared for subsequent feature extraction, parameter allocation, and data analysis.

6.4.2 Data Cleaning

Based on the Dataframe data structure in the Python's Pandas library, a two-dimensional data table structure is imported as an object as a whole, and operations such as filtering, calculation, statistics, and merging are carried out from each row and column. The voltage, current, temperature, etc. data of the battery exported from various data tables at the same time point are imported into various DataFrame data structures as data objects, with timestamps as the index. Then, the integration operation is carried out with their own index as the indicator, and finally a unified time data table is formed, as shown in Fig. 6.9.

The structure of the final integrated table (taking the charging and discharging data of a certain two days as an example) is shown in Fig. 6.10.

In order to analyze the data over a period of time according to each charging and discharging cycle, and extract the corresponding indicators, to get the DataFrame format data table, the indicators shown in Table 6.6 are proposed as the target of data statistics.

According to the statistical table and the operation and maintenance records of the power station, the power station operates in a peak-valley shaving mode, with 1 day as a cycle. A constant power-constant voltage charge begins at 0:00 every day and ends around 7:00 in the morning, with the current during the constant power period being about 0.1C; discharges are carried out at 9:00 and 17:00 each day, each lasting for 3 h. After removing the source data with sampling errors, data loss, and parsing errors from the database, a preliminary analysis of the daily operation of the

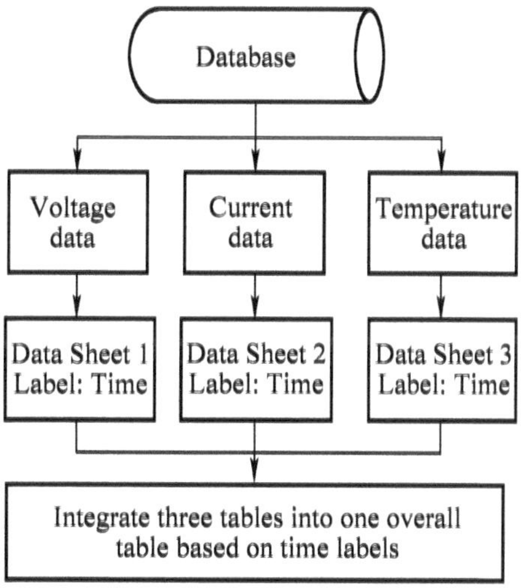

Fig. 6.9 The integration process of battery data

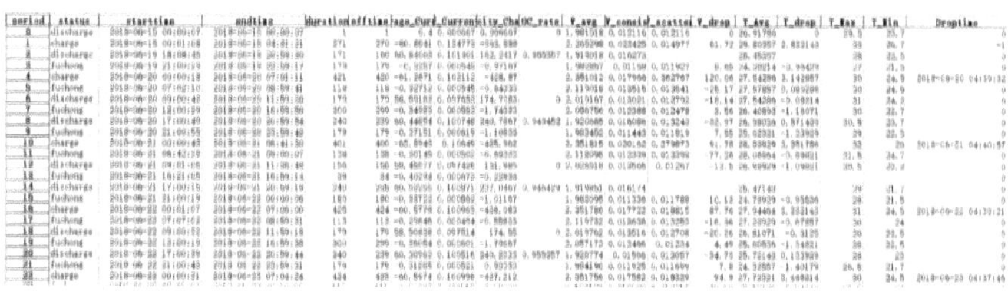

Fig. 6.10 Integrated feature data of each cycle

Table 6.6 Data statistics indicators

Indicator name	Remarks
Status	Charging/discharging/resting
Duration	
Average current	The magnitude of the current within the charge-discharge cycle
Current ratio	Average current/battery rated capacity Determine whether the battery exceeds the maximum current based on the rated ratio
Charge-discharge energy	Calculated by the ampere-hour integration method
Maximum temperature	The highest temperature of all individual cells at the end of charging/discharging
Minimum temperature	The lowest temperature of all individual cells at the end of charging/discharging
Average temperature difference	The change in average temperature within a single charge-discharge cycle
Voltage drop	The change in the total voltage of the battery bank
Cell voltage dispersion	The standard deviation of the individual cell voltages, reflecting the consistency of the battery
Over-discharge rate	At the end of discharge, the proportion of batteries with a single cell voltage less than 1.95 V The proportion of batteries with a DOD exceeding 60%, reflecting the overall discharge depth

power station was conducted. After removing some cycles with unexpected termination of charging and discharging, different working hours, and different working modes, 10 charging cycles with a working duration of 7 h and basically the same working mode and battery status were selected as the original data for one clustering analysis. The numbers of these 10 cycles are 16, 22, 30, 36, 57, 75, 81, 102, 112, 124.

6.4.3 Preliminary Statistics

The original energy storage monitoring data is often real-time information collected by the BMS and PCS during operation, mainly consisting of data such as voltage, temperature, and total current. When integrating data statistics, the time periods need to be divided according to the charging and discharging states, and then the working status of each battery in each period, such as average current, current ratio, charge-discharge energy, highest/lowest temperature, voltage drop, etc., which help to intuitively understand the state of the battery bank, can be obtained through data statistics. Although these data are detailed and abundant, they involve too many variables, some of which reflect the same properties of the battery, and some interference factors may affect the results of clustering analysis. Moreover, too many variables can make the adjustment and evaluation of clustering results difficult. Therefore, before modeling, it is necessary to analyze the original data variables,

Table 6.7 Three Battery Feature Dimensions

Dimension	Meaning
Voltage drop during state switching	The sharp drop in single cell battery voltage when switching from constant power charging to constant voltage charging
Sum of voltage standard deviations	The sum of the absolute differences between the voltage of each battery cell and the average voltage at each moment
Temperature difference	The temperature difference between the end and the beginning of charging and discharging

construct and extract data dimensions that are relatively independent and have a high correlation with battery properties.

Based on a thorough analysis of battery performance characteristics and actual operating rules, and considering the visualization of clustering algorithm results, three dimensions are selected for battery clustering: the voltage drop during battery state switching, the cumulative dispersion of single cell voltage, and the temperature difference in each cycle. Their definitions are shown in Table 6.7.

6.4.3.1 Voltage Drop During State Switching

The electromotive force of a lead-carbon battery is related to the concentration of the acid in its internal electrolyte. The terminal voltage of the battery is directly related to the concentration of hydrogen ions in the electrolyte. Within a certain range, the higher the concentration of hydrogen ions, the higher the voltage, and vice versa, the voltage is the lower.

When a lead-carbon battery is charged at constant power, the current is generally stable at around 0.1C. At this time, the reduction reaction is continuously occurring inside the battery, and $PbSO_4$ is continuously reduced to hydrogen ions. Due to the lag of the concentration diffusion of the sulfuric acid solution relative to the current charging, in a battery, the part closer to the electrode always reacts more intensely and has a higher concentration of hydrogen ions, while the part far from the electrode has a lower concentration of hydrogen ions. In this case, if the charging is switched from constant power to constant voltage, the current flowing through the battery will instantly become smaller, and the impact on the concentration of hydrogen ions near the electrode will also become smaller. At this time, the concentration diffusion inside the battery will dominate the change in the concentration of hydrogen ions, that is, the hydrogen ions inside the battery will rebalance, from the previous state of high concentration near the electrode and low concentration in the middle to a more average state, thereby causing the concentration of hydrogen ions near the electrode to drop rapidly. From the outside, the battery voltage has a sharp drop, as shown in Fig. 6.11. The degree of the sharp drop in voltage is determined by factors such as the concentration of the internal electrolyte, the amount of active substances participating in the electrochemical reaction, and the condensed state structure of the positive and negative electrode materials. It is closely related to the

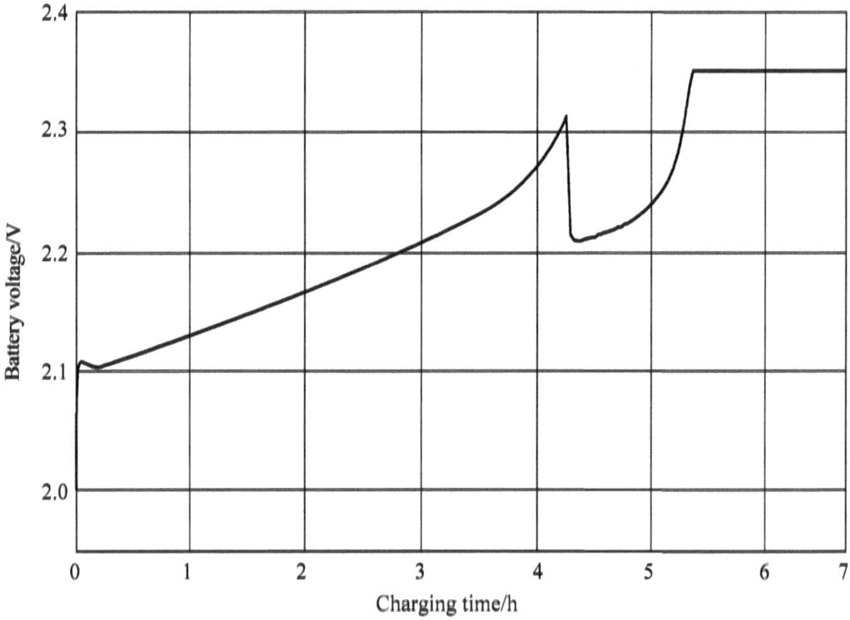

Fig. 6.11 Voltage Drop During State Switching

inherent properties and health status of the battery and can be used as one of the dimensions for clustering analysis.

6.4.3.2 Sum of Battery Voltage Standard Deviations

Figure 6.12 shows two voltage change curves during a 7-h charging process. The red line represents the average voltage of all batteries in the battery bank, and the blue line is the voltage curve of a single cell battery (in this case, battery No. 38). The sum of the standard deviations of the battery voltage is equivalent to the area of the gray shadow between the two curves in the figure, and the expression is:

$$SVSD_i = \int_O^{t_0} \left| v_{average}(t) - v_i(t) \right| dv \tag{6.1}$$

Where, $v_{average}(t)$ represents the overall average voltage change curve over time, $v_i(t)$ represents the change curve of the voltage of the i-th battery over time. The sum of the standard deviations of the battery voltage is the integral of the absolute value of the difference between the battery voltage and the average voltage over the entire charging period, representing the degree to which the individual battery deviates from the whole. It is worth noting that in actual data collection, voltage measurement is not continuous, often with intervals of a few seconds to a few minutes, so this parameter value is obtained by adding the absolute values of the voltage differences at all time points.

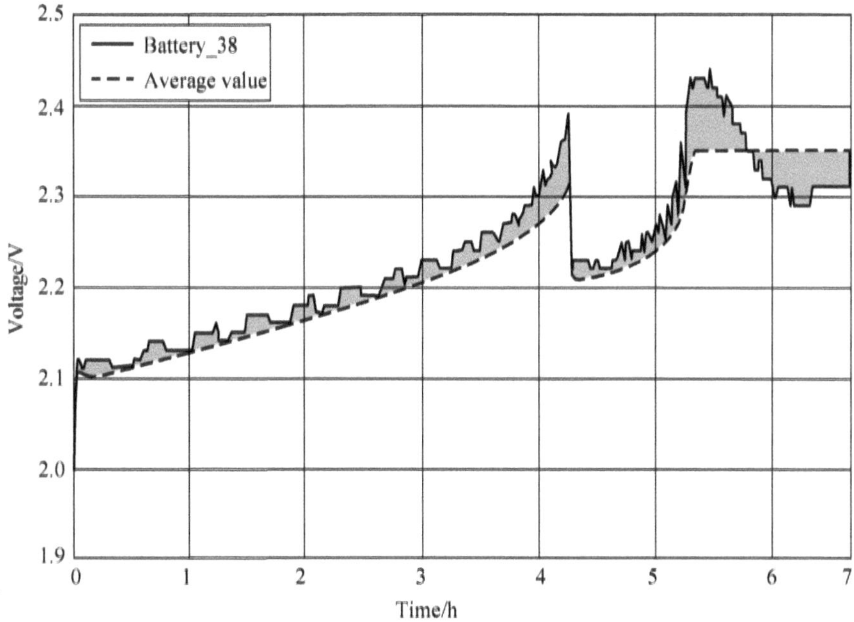

Fig. 6.12 Sum of standard deviations of battery voltage

6.4.3.3 Temperature Difference

The thermal changes that occur during the charging and discharging process of the battery are closely related to the health status of the battery itself. The heat generated by the battery during operation mainly comes from four sources: heat generated by the battery's ohmic internal resistance, heat generated by the internal chemical reactions of the battery, heat generated by the polarization reactions of the battery, and heat generated by the side reactions of the battery [110]. The health of the battery is strongly related to the amount of electrochemical reaction substances inside it. The different concentrations of these substances lead to different amounts of heat generated by the ohmic internal resistance, polarization reactions, and electrochemical reactions when the battery is working. The older and less efficient the battery, the greater the temperature rise during the charging and discharging cycle. Therefore, the change in temperature is also somewhat related to the health status of the battery, so the temperature difference of the battery is listed as the third data dimension.

After calculating and integrating the original data of the ten selected cycles according to these three data dimensions, ten new data tables are formed and saved in the database as input data for the clustering algorithm.

6.4.4 Preprocessing of Clustering: Weight Allocation

Since the operation of the battery energy storage system is multi-cycle, data characteristics are distributed in each working cycle. Therefore, one feature of the clustering analysis of the energy storage system is multiple clustering. This requires extracting and integrating data from multiple working cycles into a unified data feature model, so the input data for each clustering needs to be unified. In the example, data from 10 cycles were selected for unified weight allocation, and the data of the three dimensions were normalized, limiting all data between (0, 1).

In the clustering algorithm, the scale for measuring similarity is the Euclidean distance. If the same weight is assigned to all attributes on the dimensions, it may lead to distortion of distances in the sample space. That is, if two points in the space are close in important attributes, but due to the influence of other irrelevant attributes (noise points) on the distance, these two points may be measured as far apart in Euclidean space. Therefore, by assigning different weights to each dimension, the algorithm performance can be improved and the classification results can be made more reasonable.

The three-dimensional data model constructed here is based on three parameters that are strongly related to the health status of the battery, extracted from existing battery data. Since the impact of attributes on each dimension on the battery does not have very rigorous, quantified conclusions, if the weight parameters are chosen improperly, the clustering process and results will lack interpretability.

Unlike other clustering, the clustering analysis of the energy storage system is a separate clustering for the data of each charging cycle, and the results are then summarized and analyzed. Since the battery data of each cycle is different, if an adaptive algorithm is used for calculation, the weight parameters of each cycle obtained will also be different. Therefore, in order to make the results of each clustering comparable, a unified weight allocation strategy needs to be adopted.

In order to verify the effectiveness of this weight selection method, a Hopkins statistic is introduced. The Hopkins statistic and the adaptive weighting method are used for verification. The Hopkins statistic explains the likelihood of dataset D following a uniform spatial distribution, and the process is as follows:

1. Randomly select n points from all samples. For each sampled point, find the nearest point in the sample and calculate the distance, represented by x_1, x_2, … x_n.
2. Randomly select n points in the sample space. For each point, find the nearest point in the sample and calculate the distance, represented by y_1, y_2, … y_n.
3. Calculate the Hopkins statistic H:

$$H = \frac{sum_{i=1}^{n} y_i}{sum_{i=1}^{n} x_i + sum_{i=1}^{n} y_i} \tag{6.2}$$

Table 6.8 Hopkins Statistics Values

Charge-discharge cycle number	Hopkins statistic value	
	Before weight adjustment	After weight adjustment
16	0.796352527409	0.908477997527
22	0.879751346778	0.862010899497
30	0.823770171744	0.913688086031
36	0.821897000873	0.889539068415
57	0.801573327229	0.939458597911
75	0.840932302602	0.881858467039
81	0.854802570108	0.899785496223
102	0.768091677497	0.840077358866
112	0.788865975425	0.855340986224
124	0.69586838624	0.882424664478

Before performing cluster analysis, the Hopkins statistic is used to verify the clustering trend of the organized data. In the above formula, if the clustering trend is not obvious, then H is close to 0.5; otherwise, H is close to 1.

After setting the dimensions and weights, we obtained the input data for clustering. Hopkins statistics were performed on the data for 10 cycles, and the results of the Hopkins statistics for the original three-dimensional data without weight adjustment were listed for comparison, as shown in Table 6.8.

From the characteristics of the Hopkins coefficient, it can be known that the closer the statistical result is to 1, the more the dataset tends to be convex data, and the more suitable it is for cluster analysis. From Tables 6.6, 6.7 and 6.8, it can be seen that after weight adjustment, the Hopkins statistics of each cycle have increased by 0.05 ~ 1 (except for cycle 22), and the coefficients of each cycle are all greater than 0.8, which proves that weight adjustment is effective for the preprocessing of cluster data, and the well-processed data can be used as the input of the clustering algorithm.

In addition, although the adjustment of weights does not cause a qualitative change from the perspective of the Hopkins coefficient, in fact, due to the order of magnitude difference between the three dimensions in this data model, the normalization of weight adjustment has a great impact on the effect of clustering, which can be determined by analyzing the data distribution. Taking the data of cycle 22, which has little change in the Hopkins coefficient, as an example, the three-dimensional data distribution before and after weight allocation is plotted, as shown in Fig. 6.13.

As can be seen from the figure, the data after weight normalization is more evenly distributed across the entire sample space. Although the data before weight adjustment may be more suitable for clustering from the Hopkins statistical results, it is clear that the clustering results will only be divided into large, medium, and small classes according to the parameter of "sum of voltage standard deviations". The more dispersed data in the right figure, after clustering analysis, can better reflect the distribution of the battery.

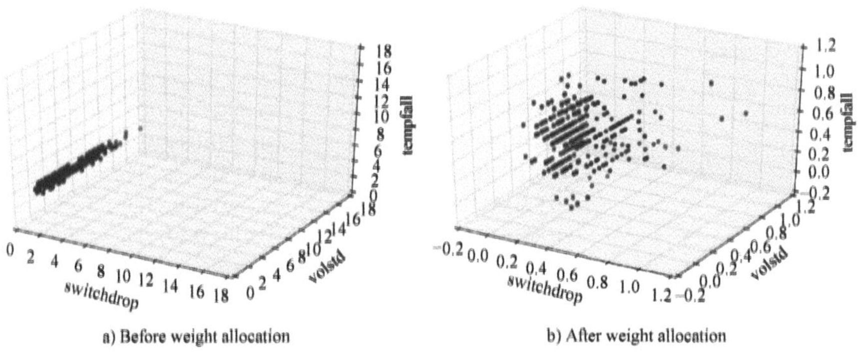

Fig. 6.13 Distribution of Cycle 22 data before and after weight allocation

In the absence of target values, it is difficult to perform parameter sensitivity analysis. The adaptive weighting method used in this section reflects the impact of the attributes of each dimension on the clustering results, to a certain extent, it has the effect of sensitivity analysis. The process of the adaptive weighting method is as follows:

1. Normalize the data set for clustering;
2. Perform clustering classification of number k on the normalized data;
3. For each dimension j, calculate the sum of the intra-class distances d_1:

$$d_1 = \sum_{k=1}^{K}\sum_{i=1}^{n_k}\left(x_{ij} - m_{kj}\right)^2 \tag{6.3}$$

m_{kj}*is* the mean of the k cluster on the j dimensional attribute;

4. Then calculate the sum of the distances between clusters d_2:

$$d_2 = \sum_{k=1}^{K}\left(m_{kj} - m_j\right)^2 \tag{6.4}$$

5. From the above results, we get the contribution of dimension j to the clustering $c_j = d_2 / d_1$;
6. Finally, we get the feature weight w_j:

$$w_j = c_j / \sum_{j=1}^{m} c_j \tag{6.5}$$

The adaptive weighting method can compare the contribution of data attributes in each dimension. For a certain dimension, if the clustering result is compact within the class and far away between classes, the distinguishing ability of the data

Table 6.9 Feature weights under the adaptive weighting method (22nd cycle)

k value	Weight 1 (voltage drop)	Weight 2 (discreteness)	Weight 3 (temperature difference)
3	0.14	0.41	0.45
3	0.14	0.41	0.45
4	0.08	0.43	0.49
5	0.07	0.18	0.75
6	0.21	0.35	0.44
6	0.21	0.49	0.3

in this dimension is stronger, and the contribution to clustering is greater, and vice versa.

The adaptive weighting method performs k-means clustering during calculation, so the clustering parameter k needs to be given at the input stage, which is the number of clusters. When the actual number of clusters in the dataset is unknown, to ensure the reliability of the results, several consecutive k values are selected for calculation, as shown in Table 6.9. It can be seen that due to the randomness of k-means clustering and the influence of different k values, the results of the algorithm show a certain degree of fluctuation. However, overall, the sensitivity of weight 1 (voltage drop) is slightly lower, the sensitivity of the other two weights is higher and relatively close, the weights of the three dimensions are basically in the same order of magnitude, and the distribution is relatively stable, indicating that this dataset can be used as the input for the clustering algorithm.

6.4.5 *Preprocessing of Clustering: Elbow Method to Determine the Number of Clusters*

After determining the weights, another important parameter to be determined before clustering is the number of categories k value.

At present, in the problem of determining the k value of the k-means clustering algorithm, the industry has proposed some clustering function indicators, such as the calinski-harabasz index, weighted inter-intra index, etc. These function indicators can be used to test the validity of the clustering results, and then calculate the optimal number of clustering centers. Based on the analysis of various indicators and optimization algorithms, this section uses the SSE elbow method to determine the number of clustering centers for the k-means algorithm.

The core indicator of the elbow method is the sum of squared errors (SSE):

$$SSE = \sum_{i=1}^{k} \sum_{p \in C_i} |p - m_i|^2 \tag{6.6}$$

Where, C_i is the i th cluster, p is a sample point in C_i, and m_i is the i centroid (the mean of all samples in C_i), SSE is the clustering error of all samples, representing

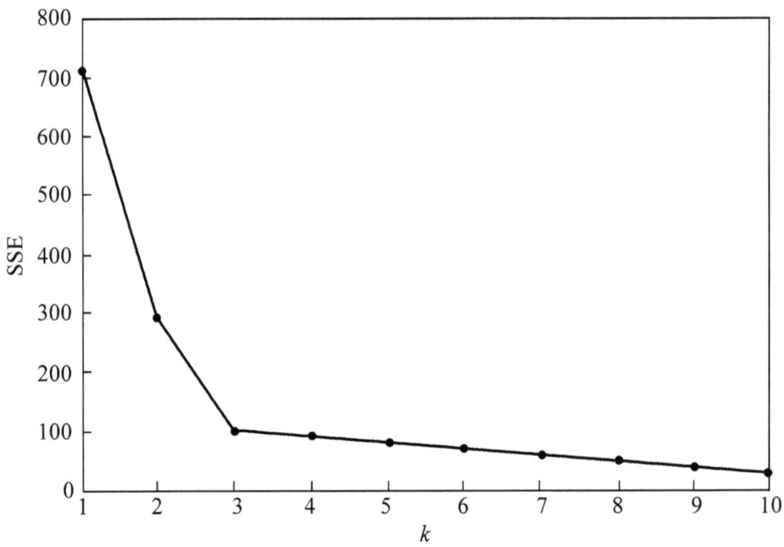

Fig. 6.14 Example of the Elbow Method

the quality of clustering. As the value of k increases, the clustering becomes more refined, and the SSE value gradually decreases (Fig. 6.14).

When the value of k is less than the actual number of clusters, the decrease in SSE will be significant; when the value of k exceeds the effective number of clusters, the return from increasing k will quickly diminish, so the decrease in SSE will sharply reduce, then tend to be flat. It can be seen that the graph of the relationship between SSE and k is an elbow shape, and the k value corresponding to this elbow position is the reasonable number of clusters.

In determining the value of k, in addition to considering the results of the elbow method analysis, it is also considered from the perspective of the battery bank. The properties of each battery in the battery bank, which is the source of data, will not differ too much. To avoid over-clustering, when selecting the number of k values, a smaller value is preferred. The elbow analysis results of 10 cycles of energy storage data are shown in Fig. 6.15.

It can be seen that in most cases, the decay curve of SSE is relatively smooth, and most of the SSE values have decreased to 35% of the maximum value when $k = 3$. The clustering is determined to be $k = 3$ for the following reasons:

1. The battery bank, which is the source of data, is installed in the same energy storage unit. When the batteries are manufactured, they are sorted and selected from the same batch with similar parameters to be configured in the same unit. Considering that the operation time of this energy storage system is not long, the consistency of the batteries will not deviate too much, so the degree of separation of battery health will not be too large, hence the selection of k value should be as small as possible.

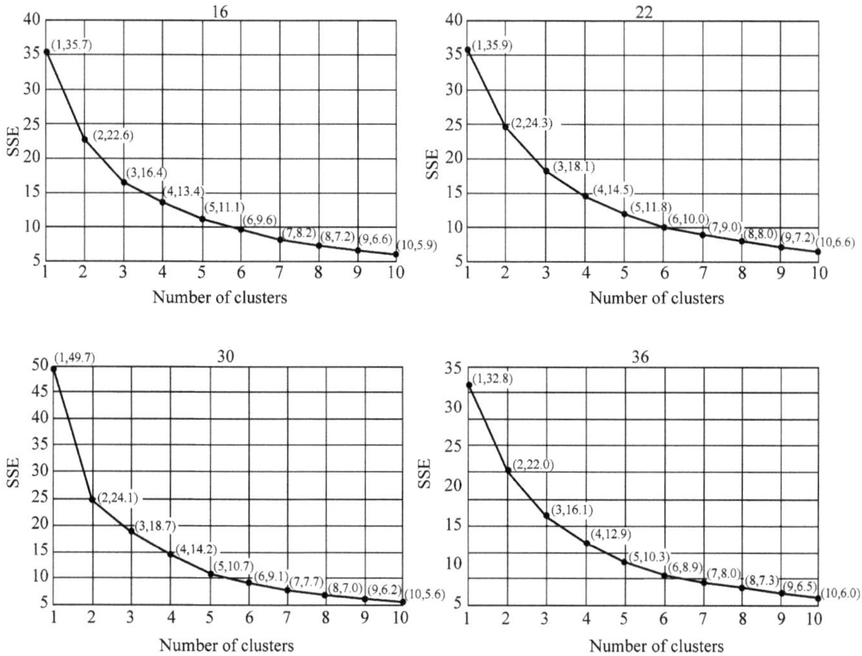

Fig. 6.15 Elbow Method Results for 10 Charging Cycles

2. In the case where the effective number of clusters in the battery data is unknown, blindly increasing the number of clusters will make the interpretation of the results more difficult.
3. As shown in the scatter plot of weight allocation in Fig. 6.13, the distribution of battery status is likely to be a central concentration-edge dispersion state. In this case, increasing the number of clusters will make the stability of the clustering results worse.

6.4.6 k-Means Clustering Results

According to the established model, the processed data is classified by the k-means clustering algorithm, and the three-dimensional scatter plot of some clustering results is shown in Fig. 6.16.

Overall, most of the single batteries are divided into two to three major categories based on different temperature changes, and some discrete data points are distributed at the edge. In most of the clustering results, these points are allocated to the same partition with a larger voltage drop value.

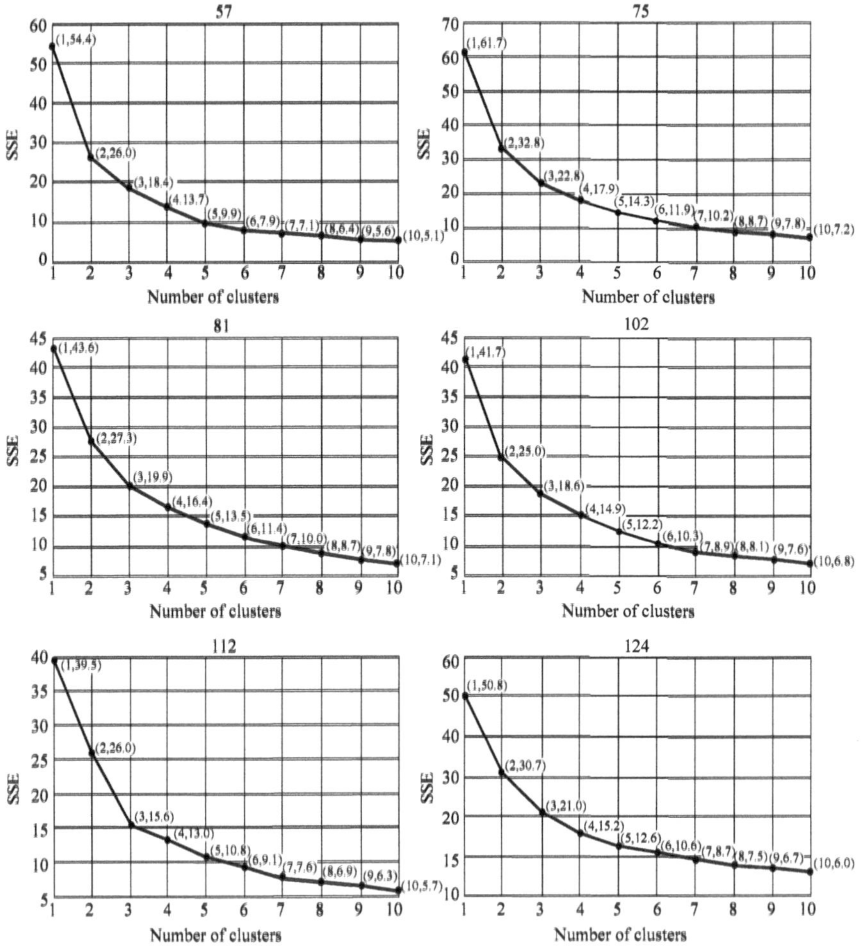

Fig. 6.15 (continued)

For battery classification, a single specific battery clustering result cannot serve as a strong basis, so statistical analysis of multiple clustering results is necessary. For 10 clustering results, the Euclidean distance from each point to its cluster center is calculated, and the results of multiple times are added. To locate the edge points, the number of times each battery was assigned to the group with the largest "voltage drop value" was counted, as shown in Table 6.10 and Fig. 6.17.

As shown in Fig. 6.17, batteries 160# and 328# have the highest degree of dispersion, and they are also more frequently classified into the group with the highest "voltage drop value". They are some of the more outlier points on the scatter plot, indicating that the health status of these batteries is relatively poor.

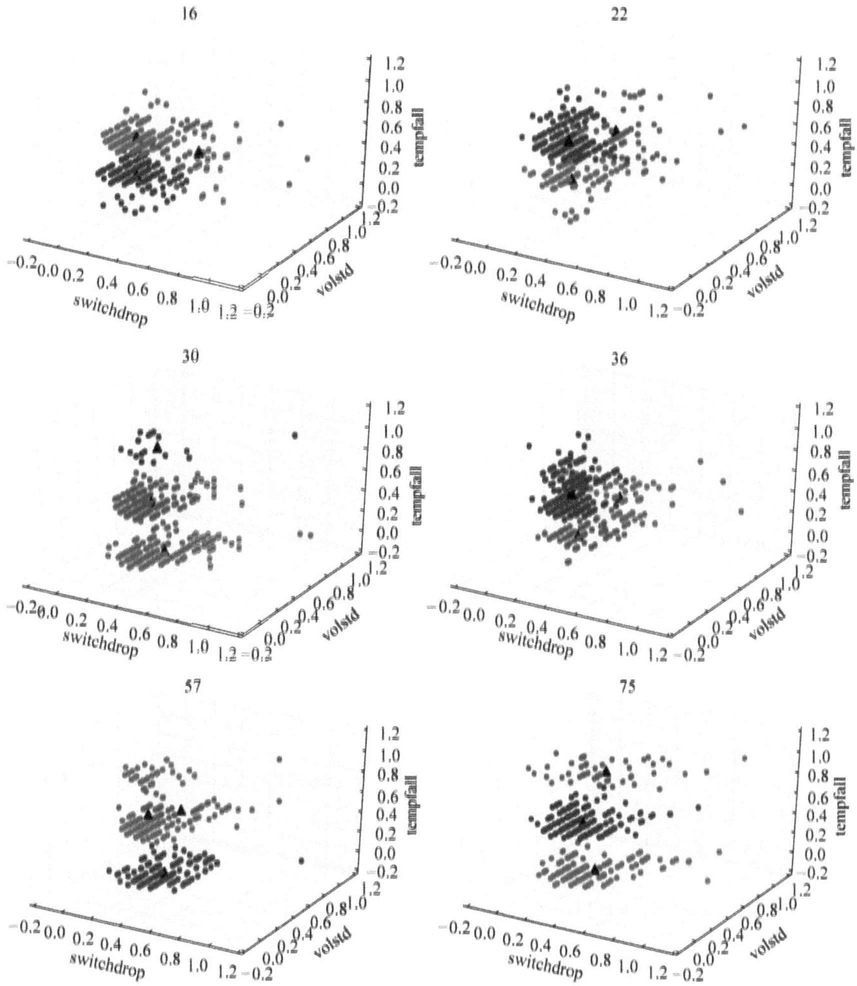

Fig. 6.16 Clustering Results for 10 Charging Cycles

6.4.6.1 DBSCAN Density Clustering Verification

The mechanism of the DBSCAN algorithm makes it very suitable for finding outliers in the dataset. In the output data of the DBSCAN algorithm, outliers will be assigned to a specific category. In this case, the input data for DBSCAN is the same as that for k-means above, with clustering parameters set to eps = 0.2, r = 5, and the clustering results are shown in Fig. 6.18.

In each clustering, the number of times each battery is marked as a noise point is separately counted and accumulated, as shown in Table 6.11.

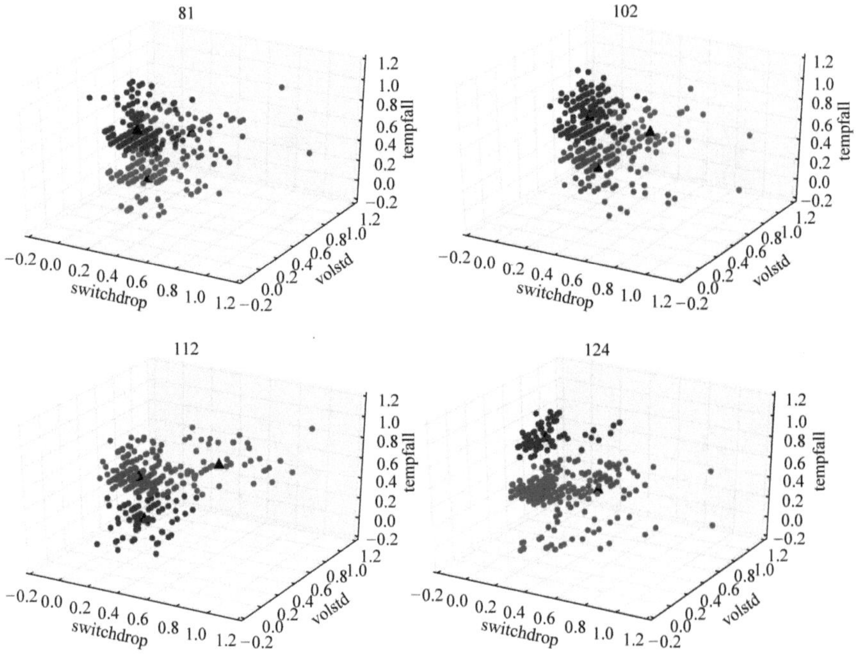

Fig. 6.16 (continued)

In the table, batteries 132#, 160#, and 328# are marked as outliers the most times, which is consistent with the statistical analysis of the k-means clustering results. It can be inferred that these batteries have a poorer health status, which reduces the consistency of the entire battery bank. Therefore, it can be seen that data analysis can locate a batch of batteries with poorer health status in the battery bank.

According to the statistical analysis of the clustering results, batteries 38# and 132# were identified as having poorer health status. The charging voltage change curves of these two batteries in the 22nd cycle were plotted and compared with the overall average voltage curve and the curve of battery 1#, as shown in Fig. 6.19.

In the figure, starting from the initial constant power charging, the voltage rise speed of battery 132#, which has the worst distribution in the clustering results, is significantly higher than the average voltage and the voltage of battery 1#. At the end of constant power charging, the voltage of battery 132# is already about 0.1 V higher than the average. When switching from constant power charging to constant voltage charging, the voltage drop of battery 132# is also significantly greater than the average level. In contrast, the voltage curve of battery 1# basically conforms to the average curve. It can be seen that compared with the overall average health status of the batteries, battery 132# has a poorer health status, while battery 38# is in between.

Table 6.10 k-means Statistics for Each Cycle

Battery number	Times	Distance sum	Battery number	Times	Distance sum
B_V_160	10	9.600892	B_V_171	7	4.848516
B_V_9	9	5.050595	B_V_172	7	4.522786
B_V_10	9	4.640823	B_V_174	7	4.303712
B_V_157	9	5.51755	B_V_175	7	4.72646
B_V_158	9	4.770785	B_V_176	7	4.286565
B_V_8	8	4.304551	B_V_177	7	3.849898
B_V_166	8	6.054962	B_V_185	7	3.220574
B_V_178	8	5.014531	B_V_191	7	3.606561
B_V_11	7	3.923634	B_V_244	7	4.685539
B_V_12	7	4.059658	B_V_319	7	5.207588
B_V_84	7	4.560755	B_V_325	7	5.420041
B_V_86	7	4.606139	B_V_326	7	4.262572
B_V_87	7	5.551331	B_V_327	7	4.423611
B_V_88	7	5.526712	B_V_328	7	9.075477
B_V_164	7	4.153094	B_V_329	7	4.145793
B_V_165	7	4.91528	B_V_6	6	3.114641
B_V_168	7	4.608586	B_V_7	6	3.118082
B_V_169	7	5.34698	B_V_31	6	4.516669
B_V_170	7	4.726273	B_V_72	6	4.140908

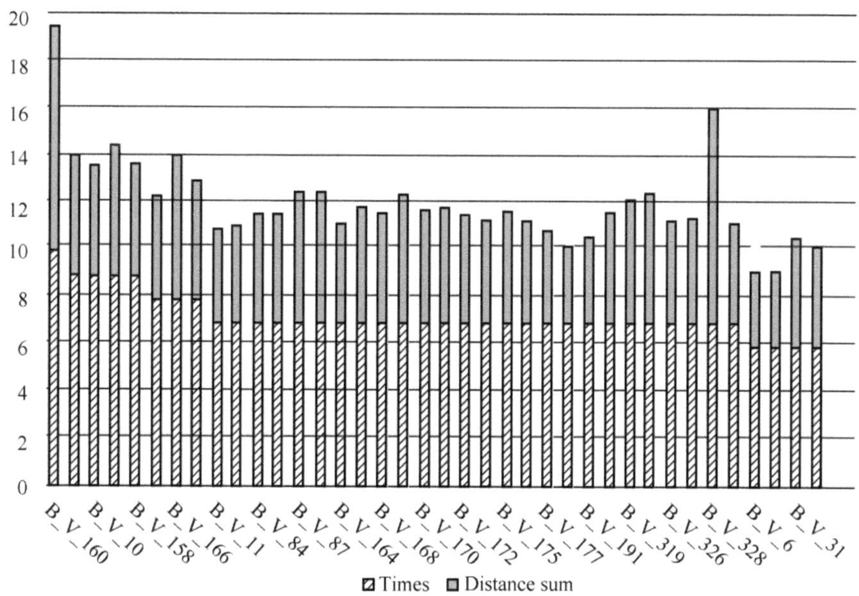

Fig. 6.17 k-means clustering result statistics

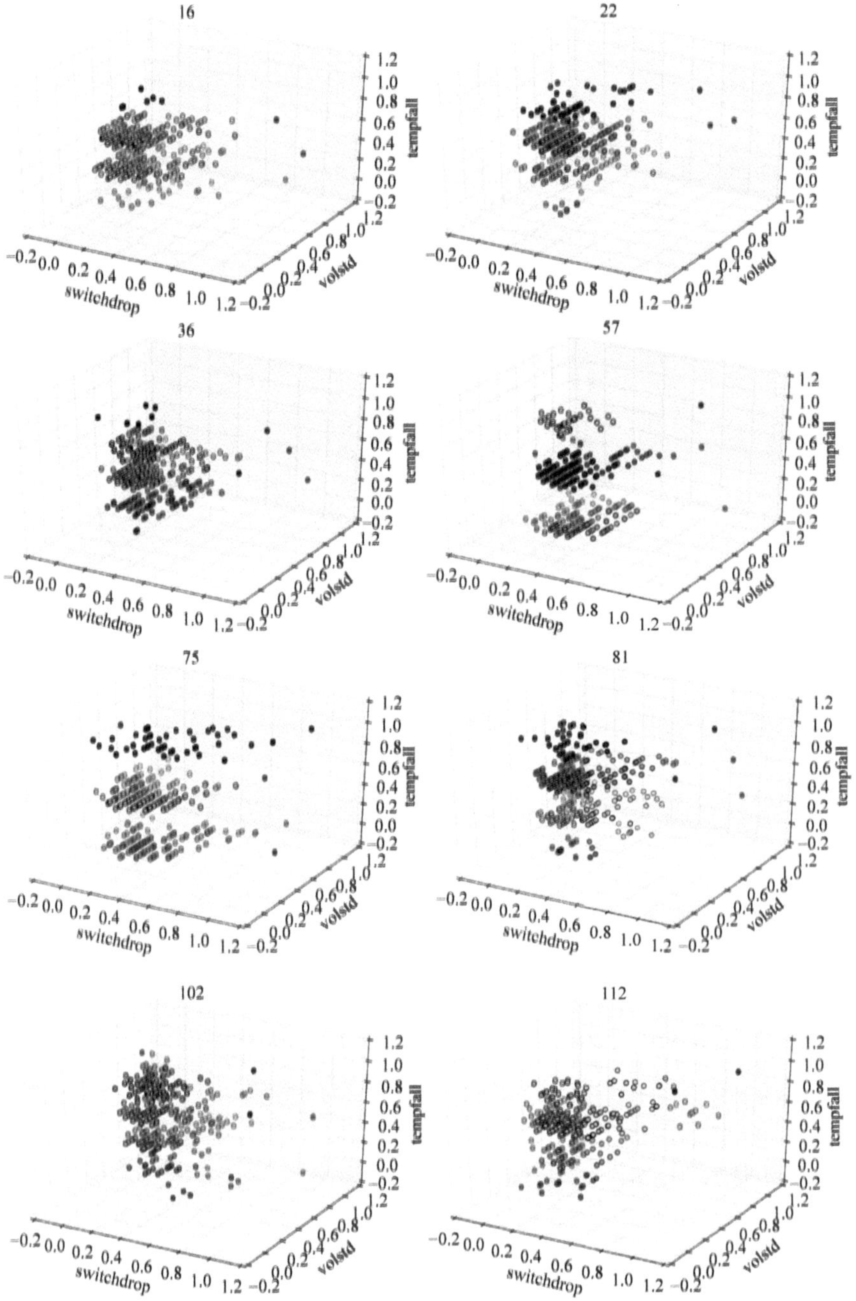

Fig. 6.18 DBSCAN clustering results

Table 6.11 Outlier marking count statistics

ID	Noise count	ID	Noise count
B_V_132	9	B_V_119	1
B_V_160	9	B_V_156	1
B_V_328	9	B_V_158	1
B_V_38	6	B_V_159	1
B_V_157	2	B_V_161	1
B_V_28	1	B_V_166	1
B_V_85	1	B_V_173	1
B_V_87	1	B_V_232	1
B_V_88	1	B_V_284	1
B_V_114	1	B_V_327	1
B_V_115	1	B_V_1	0
B_V_116	1	B_V_2	0
B_V_117	1	B_V_3	0
B_V_118	1	B_V_4	0

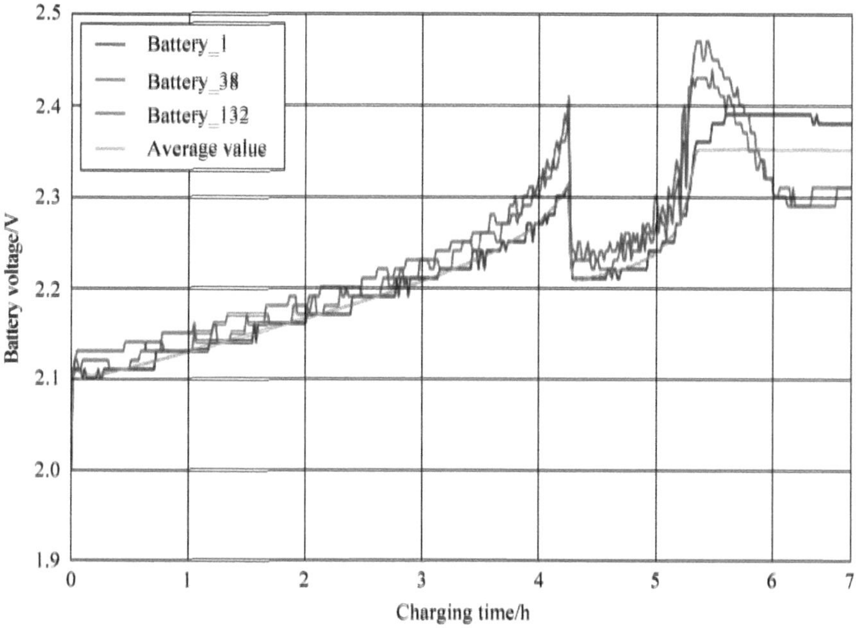

Fig. 6.19 Comparison of single cell battery charging voltage curves

6.4.7 Battery Health Differentiation

Figure 6.20 shows the temperature distribution at the beginning and end of charging of the energy storage system. The vertical axis represents temperature, the horizontal axis represents the temperature collector number, and a curve represents the temperature distribution of each battery in a certain cycle. It should be noted that in this case, there are 336 batteries, with 6 batteries sharing the same temperature sensor, and the temperature sensor numbers are 1–56.

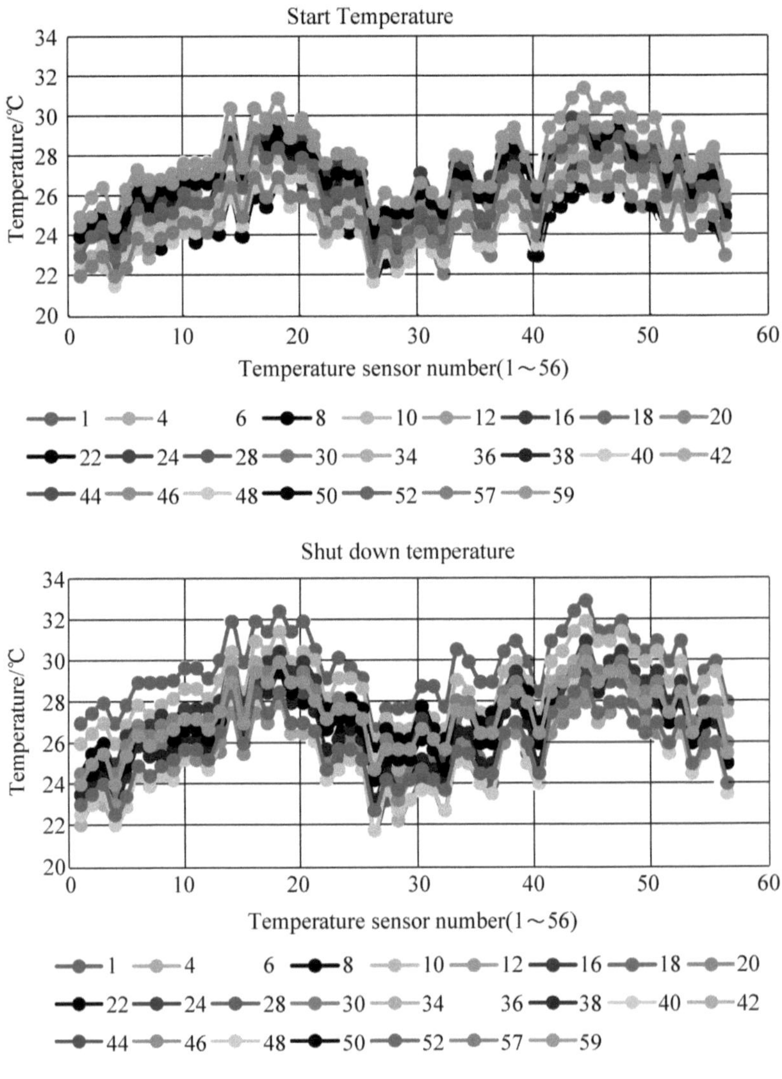

Fig. 6.20 Comparison of temperature distribution curves

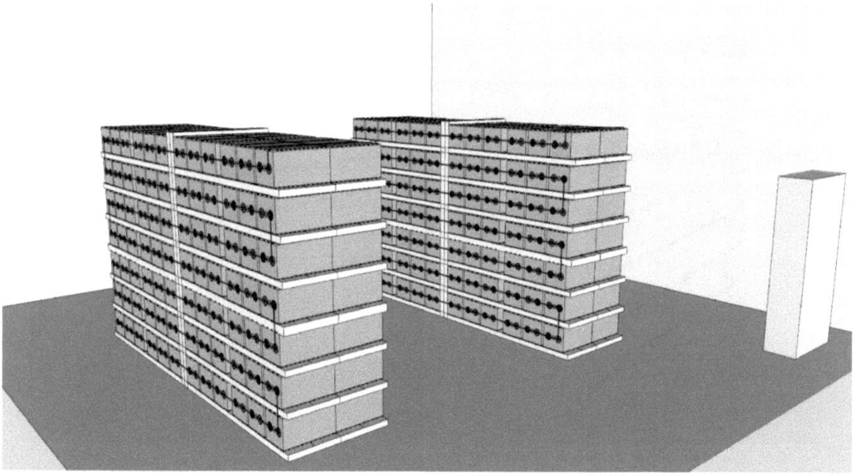

Fig. 6.21 Layout diagram of the battery room of the lead-carbon energy storage system

The temperature distribution shows that the temperature distribution patterns are similar across different cycles, and the overall temperature rises at the end of charging. However, each battery operates at different temperatures within the entire energy storage system, resulting in a certain temperature difference, which is the main reason for the differences in the operating status of individual batteries, leading to inconsistencies in the aging process. To further verify this point, an analysis was conducted in conjunction with the actual operating environment of this energy storage system, and its three-dimensional schematic diagram is shown in Fig. 6.21.

As shown in the figure, the battery energy storage system is arranged in the battery room, with 336 lead-carbon batteries assembled on four brackets, and the vertical air conditioner is placed in the right corner of the battery room. The batteries are numbered and wired according to their physical location. The external port of the entire battery bank is connected to the PCS, and the battery BMS system is also integrated on the battery shelf.

The batteries ranked at the top in the statistical results of the k-means clustering algorithm (Table 6.10) are marked in the physical distribution map, as shown by the green marked batteries in Fig. 6.22. The main feature of these batteries' statistical results is that the voltage drops sharply when switching from constant power to constant voltage, and they belong to the points leaning towards the edge in the clustering scatter plot.

At the same time, the discrete point batteries found by the DBSCAN algorithm are marked in the figure (orange marked), corresponding to the batteries with the highest degree of aging, #132, #160, #328, #38.

The above statistical analysis of the k-means clustering algorithm is based on the analysis of three data features related to the battery. Although there is no strict theoretical explanation for the relationship between these three parameters and the

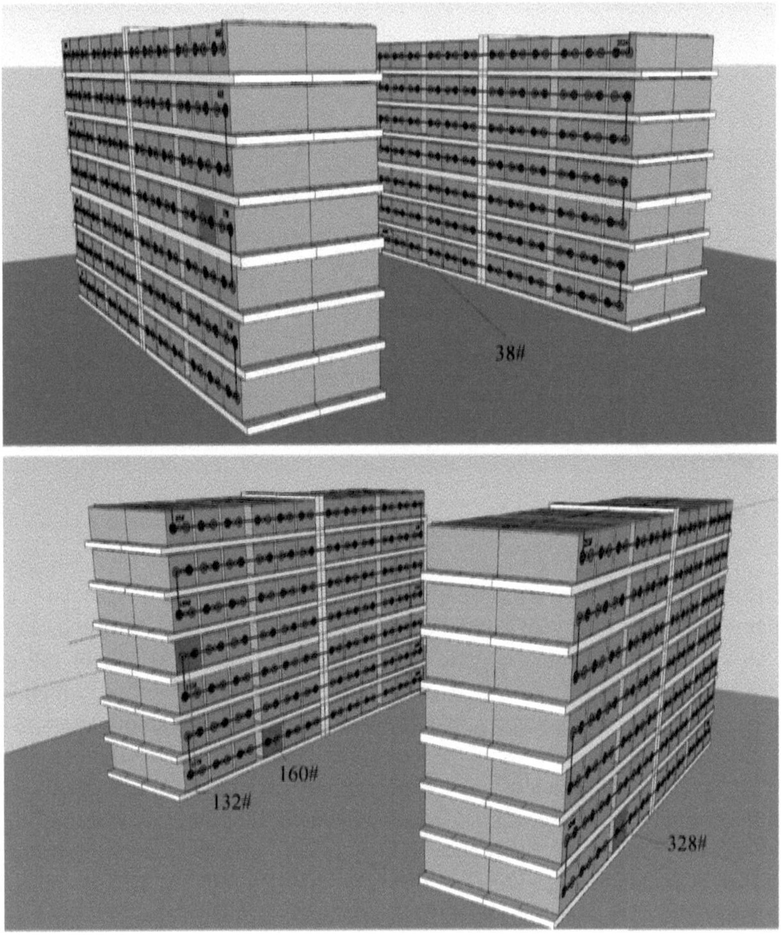

Fig. 6.22 Discrete batteries in the statistical analysis of clustering (the first and third rows in the upper figure, the second and fourth rows in the lower figure)

health status of the battery, the reasons for the increased aging of individual batteries can be further traced and analyzed based on the battery classification results, physical location distribution and connection relationship, physical environment, and battery batch. With sufficient cases and data, optimization solutions and references can be provided for battery material formulation and process flow, system configuration and dynamic environment design, operation and maintenance process, etc.

Chapter 3: Application of Energy Storage in Power Systems
On the one hand, the power system is a carrier for optimizing and adjusting the energy structure, which needs to absorb more clean energy. On the other hand, it needs to improve its own operation, and achieve efficient and economical

performance while ensuring safety and reliability. This section analyzes the role and control methods of energy storage from the development of microgrids, improvement of renewable energy generation performance, virtual power plants, etc., based on typical development modes of power systems.

Part III
Application of Energy Storage in Power Systems

Chapter 7
Application of Energy Storage in Microgrids

As an important form of efficient access to distributed generation and improving the reliability of power supply, microgrids will be widely present in power systems. As an important element in microgrids, energy storage is indispensable in the operation control and energy management of microgrids. This chapter analyzes the role of energy storage in microgrids and its operation control methods, etc.

7.1 The Role of Energy Storage in Microgrids

Microgrids are small, modular, decentralized energy supply networks formed based on distributed power generation technology, with small power stations close to distributed energy sources or users as the main energy sources, combined with end-user power quality management and energy cascade utilization technology [112]. Microgrids can operate in island mode, grid-connected mode, and seamlessly transits between the two modes. For the large grid, the microgrid can be seen as a "controllable unit" [113], with a certain predictability and dispatchability, and can quickly respond to system demands; for users, the microgrid can be seen as a customized power source, capable of meeting diversified electricity needs, such as enhancing local power supply reliability, reducing feeder lines losses, improving energy efficiency, correcting the lower limit of voltage, or uninterrupted power supply [114]. At present, microgrids have become an important solution to solve the safety and stability problems of power systems and achieve diversified and efficient use of energy.

Unlike traditional power grids, microsources in microgrids mostly generate power based on inverters or small-capacity asynchronous generators, with small system inertia and insufficient damping, lacking the disturbance resistance of traditional power grids. In microgrids, the intermittency and randomness of renewable energy generation such as wind power or photovoltaics, the random cut in/cut off of

X. Tang et al., *Electrical Energy Storage Technologies and Applications*,
https://doi.org/10.1007/978-981-96-5625-7_7

loads, and the process of grid connection/disconnection of microsources can have a significant impact on the stable operation of the system and the quality of electrical energy, causing voltage and frequency fluctuations, and even system instability [115–117].

Energy storage can achieve four-quadrant flexible operation of power through PCS, achieving instantaneous balance of active and reactive power in the microgrid, the effect of which is equivalent to enhancing system inertia and damping, and improving system stability [118–120]. Due to the role of energy storage, the microgrid can decouple the two unrelated random variables of microsources and loads, effectively reducing the negative impact of intermittent power sources such as wind power and photovoltaics on the microgrid and the large power grid, and improving the grid connection acceptance capacity of renewable energy generation. In addition, energy storage is the physical basis of customized power technology in microgrids, which can meet various user requirements for power quality, power supply reliability, safety, and economy.

This section analyzes the role of energy storage in microgrids from the perspective of operating mechanisms and basic functions, and introduces typical energy storage control strategies suitable for microgrids.

7.1.1 *Role of Energy Storage in Microgrids*

The primary goal of the microgrid is stable operation, which is the foundation of microgrid development; the next is to ensure the power quality and reliability of important loads, which is key to meeting users' high-quality electricity needs; finally, capacity credibility, which can achieve moderate dispatchability and predictability, is the guarantee for the microgrid to be connected to the large power grid on a large scale.

As an important functional unit of the microgrid, energy storage is the core and carrier of the microgrid to achieve stable control and energy management. The role of energy storage in the microgrid can be analyzed from several aspects such as system startup, stable control, power quality, and moderate capacity credibility [121], as shown in Fig. 7.1.

7.1.1.1 System Startup

The startup of the microgrid generally takes place in an independent operating mode, requiring a stable grid-connected power source to complete the system startup. Energy storage can achieve stable and controllable AC voltage output through PCS, possessing the technical conditions to serve as a grid-connected power source. Microgrids with energy storage grid connection are commonly seen in photovoltaic storage microgrids, wind-photovoltaic storage microgrids, etc.

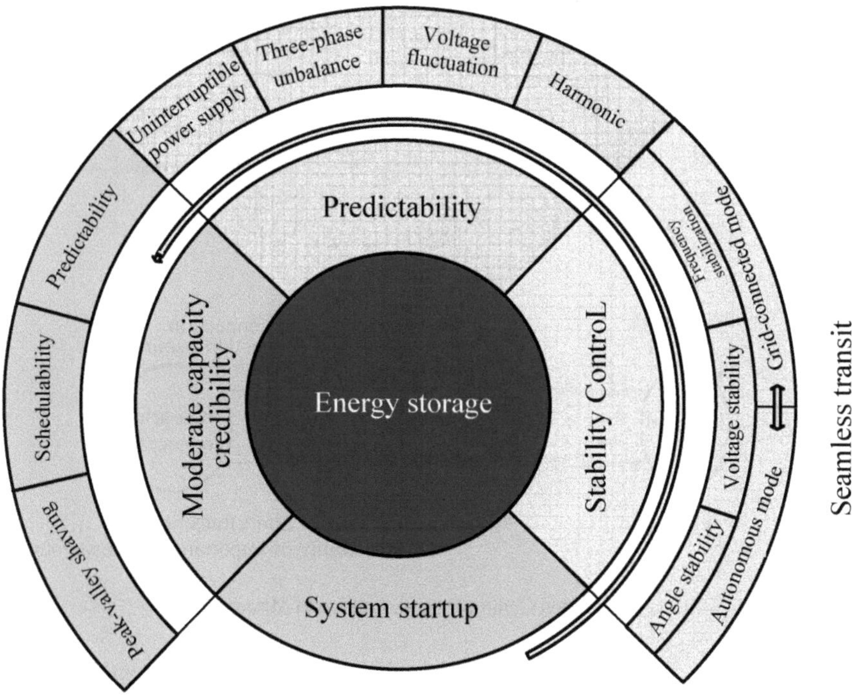

Fig. 7.1 Role of Energy Storage in Microgrids

After the microgrid startup is completed, in the independent working mode, energy storage can achieve or participate in the system's voltage and frequency control, and monitor the grid status in real time, adjusting its own output voltage amplitude and phase, and grid-connecting when conditions are met. In the grid-connected working mode, the energy storage unit can achieve its own energy optimization management, and can optimize the power flow at the point of common coupling (PCC) between the microgrid and the large power grid, to improve the controllability of the microgrid's grid-connected operation. When the microgrid transits between off-grid and grid-connected modes, energy storage needs to participate in achieving a smooth transition of the system. The operation process of the microgrid based on energy storage is shown in Fig. 7.2.

7.1.1.2 Stable Control

The microsources in the microgrid are mainly inverter-based, lacking the system inertia and disturbance resistance of traditional power grids. The intermittent changes of microsources and the random switching of loads can cause instantaneous imbalances of active or reactive power, thereby causing fluctuations in system voltage and frequency, affecting the stable operation of the system.

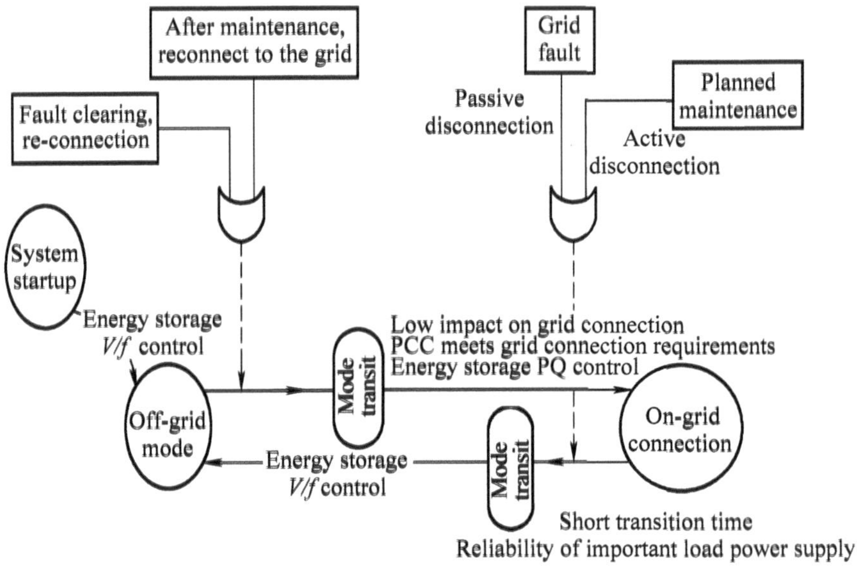

Fig. 7.2 Energy Storage Operation Control Process Applied to Microgrids

Furthermore, due to the large R/X parameter value of the microgrid line, the active and reactive power of the system cannot be fully decoupled, making traditional stability control methods ineffective.

Energy storage can quickly absorb and release active and reactive power, affecting the node voltage and current distribution within the microgrid, achieving regulation and control of the microgrid voltage and frequency, equivalent to the primary frequency regulation of traditional power systems. When the energy storage system performs stability control, the support time required is generally from seconds to minutes, and the required energy storage is less, which is technically and economically feasible.

With the scaled development of microgrids, the stability control role of energy storage systems is becoming increasingly important. Its significance is not only limited to the stable operation of the microgrid itself, but also can provide important technical support for the operation of the public grid through flexible control of the PCC point (Fig. 7.3).

7.1.1.3 Power Quality Improvement

The operation mechanism of the microgrid and the characteristics of distributed power sources determine that the microgrid is prone to power quality problems during operation. The switching process of distributed power sources and microgrids, the switching process of microgrids and public grids, and the random power changes of distributed power sources and loads can cause voltage quality and frequency

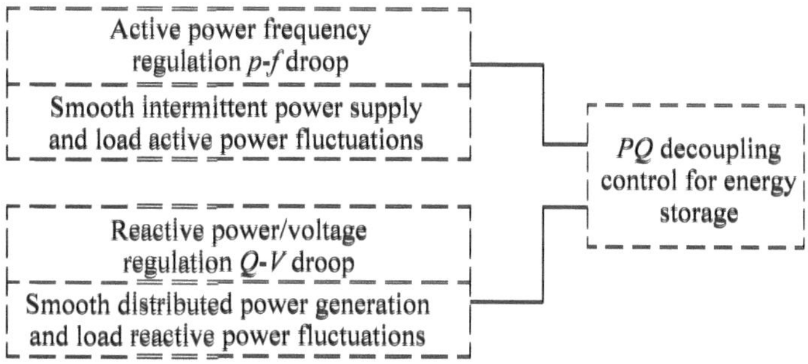

Fig. 7.3 Role of energy storage in microgrid stability control

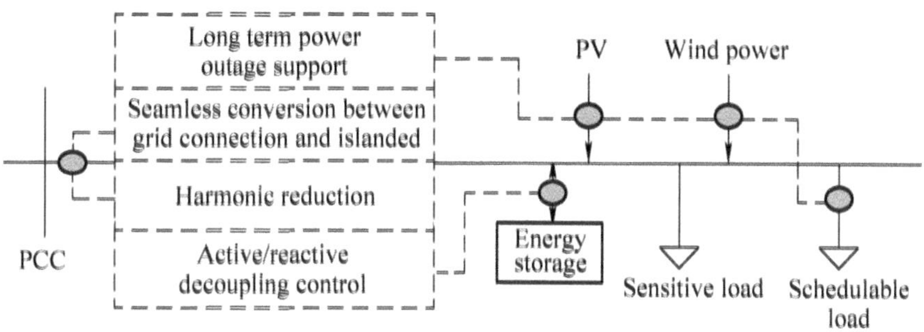

Fig. 7.4 Microgrid power quality improvement based on energy storage

quality problems. Especially in microgrids that include renewable energy generation such as wind power or photovoltaics, the intermittency and randomness of the micro-source output power and the power electronic device-based generation mode will further exacerbate the system's power quality problems.

The energy storage system can quickly adjust its power output according to the operating status of the microgrid, suppress system voltage and frequency fluctuations, reduce the system's main harmonic components and reactive components, and achieve three-phase balanced operation, improving the power quality of the microgrid.

In addition, the energy storage system can also ensure the uninterrupted power supply of important loads in the system when the micro-source power supply is insufficient or interrupted. The uninterruptible power supply function of energy storage is particularly important in independent microgrids with intermittent power sources (Fig. 7.4).

7.1.1.4 Moderate Capacity Credibility

In order for the microgrid to achieve scaled grid connection, in addition to ensuring its own stable operation, it also needs to have a moderate capacity credibility, that is, relative to the large grid, the microgrid can act as a controllable power source or

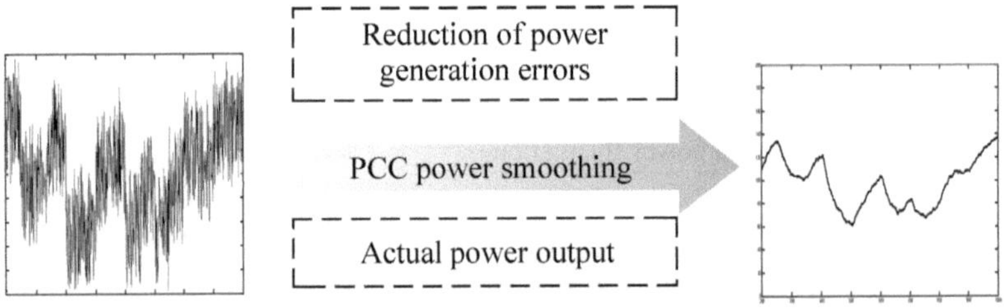

Fig. 7.5 Smoothing effect of energy storage on distributed power sources and PCC power

load, with a certain degree of dispatchability and predictability. Since there are two sets of unrelated random variables, micro-sources and loads, in the microgrid, configuring energy storage can achieve decoupling of the two, and can achieve system power leveling control on multiple time scales, achieving moderate dispatchability and predictability of the microgrid. In addition, the combination of energy storage and distributed power sources can also achieve peak and valley regulation of the large grid, alleviate the pressure of upgrading and rebuilding the distribution system, and improve the load rate.

Due to the high cost of energy storage, the economy is a key factor to consider when configuring energy storage. The role of energy storage is well combined with the energy management of the microgrid and the public grid, to achieve the greatest improvement in system function and performance with as little energy storage as possible, and achieve the best technical economy (Fig. 7.5).

7.1.2 Main Applications of Microgrids

Microgrids have very obvious localization characteristics and customization features. Different regions have large differences in natural conditions, resource endowments, and load characteristics. Therefore, there will be significant differences in the structure and operation control of microgrids, and the technical routes and implementation methods for developing microgrids in various places will also have their own characteristics. Overall, the main application scenarios of microgrids include:

7.1.2.1 Integration with Urban Industrial and Commercial Parks, and Public Enterprises

Large and medium-sized urban commercial areas and residential areas have very concentrated electricity, heat, and cooling loads. In accordance with the requirements of national energy structure adjustment, newly developed towns should not follow the old path of coal pollution or low-efficiency single burning of liquefied natural gas, nor is it suitable to use split air conditioning or window air conditioning. Therefore, it can be considered to use gas turbine and other combined heat and power units to replace traditional energy forms, and in areas conditions permitted, ground source heat pumps can be used, combined with building photovoltaics and energy storage, to form a microgrid at the level of a single building or multiple buildings, providing users with complete energy supply. On this basis, the overall consumption of electricity, heat, and cold in the microgrid can be controlled through energy management strategies and demand-side management, achieving energy cascade utilization, improving user energy flexibility, and alleviating the pressure of urban grid expansion. For important commercial and industrial users, the quality and reliability of power supply can be further improved through the seamless switching function of the microgrid.

7.1.2.2 Integration with Suburban Villas, Resorts, and Agricultural Ecological Parks

Suburban villas, resorts, agricultural ecological parks, and other areas usually have abundant renewable energy resources, with a large amount of rooftops and idle land that can be utilized, and some areas may even have some hydropower resources or biomass resources. Therefore, it can be considered to use small hydropower, photovoltaic power generation, biogas power stations, and other distributed power sources to form a feeder-level microgrid. On this basis, through the optimized management of the microgrid, renewable energy can be maximized, greenhouse gas emissions can be reduced, and reactive voltage support and power quality improvement and other auxiliary services can be provided for the feeder and neighboring areas.

7.1.2.3 Integration with Remote Areas, Such as Northern Mountainous Areas, Southwestern Arid River Valley Areas

Providing power to areas far from the grid, end-of-grid areas, or geographically isolated areas is a very important task. China's "three north" areas are rich in wind and light resources, while the southwest region has developed hydropower resources and abundant biomass resources. Therefore, it can be considered to reasonably allocate renewable energy and energy storage, such as wind-solar complementary power generation systems, hydro-solar complementary power generation systems, etc., to form a microgrid that can operate in isolation or connect with the large grid and serve as backup and support for each other, providing reliable power supply for these areas. While fully utilizing local resources and promoting the development and utilization of renewable energy in remote areas, it improves the power supply conditions in these areas, accelerates regional economic development, and achieves environmental protection.

7.1.2.4 Military Microgrid

Modern military, which is digital, informational, and networked, has an increasingly strong demand for electricity, and the requirements for the safety, reliability, confidentiality, economy, and efficiency of power supply are getting higher and higher. Through the integration function of the microgrid, distributed power generation is increased in military facilities or bases, reducing dependence on external energy and improving the safety of power supply; through energy storage, the self-healing and regeneration ability of the military microgrid is improved, and the reliability of power supply is improved; through energy storage, the adjustment and camouflage of power flow and harmonics are carried out, cutting off the enemy's way of judging our dynamics through power information, improving the confidentiality of military operations, and military applications have become an important direction of the microgrid.

7.2 Microgrid Control Based on Energy Storage

Generally, the microgrid needs to have two modes of operation: off-grid and on-grid. In order to ensure the reliability of power supply to important loads in the microgrid, the microgrid should also have the ability to smoothly switch between on-grid/off-grid modes, which is also an important supporting technology for the development of the microgrid.

For microgrids that use energy storage as a main power source, the energy storage system needs to switch between off-grid operation and grid-connected operation modes according to actual conditions. This is mainly reflected in the control method, where the PCS switches from the grid-connected operation's *PQ* control mode to the off-grid operation's *V/f* control mode, and from the off-grid operation's *V/f* control mode to the grid-connected operation's *PQ* control mode, allowing other distributed power sources and loads in the microgrid to continue operating, as shown in Fig. 7.6.

7.2.1 Grid-Connected Operation Control

When energy storage serves as the main power source of the microgrid, the control of the energy storage PCS is crucial. It is the main control body for the switch between the grid-connected/off-grid operation modes. The PCS generally uses a voltage source converter VSC and adopts an *LCL* type filter circuit, thus it can be controlled by three control loops, including the grid-connected inductor current loop, filter capacitor voltage loop, and filter inductor current loop.

As shown in Fig. 7.7, during grid-connected operation, the controller logic switch is set to the "grid-connected operation" mode. The grid-connected inductor

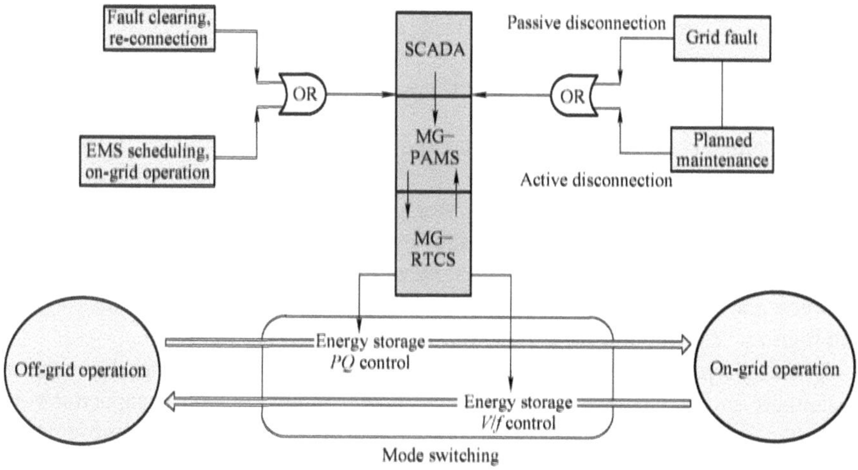

Fig. 7.6 Control process of microgrid operation based on energy storage as the main power

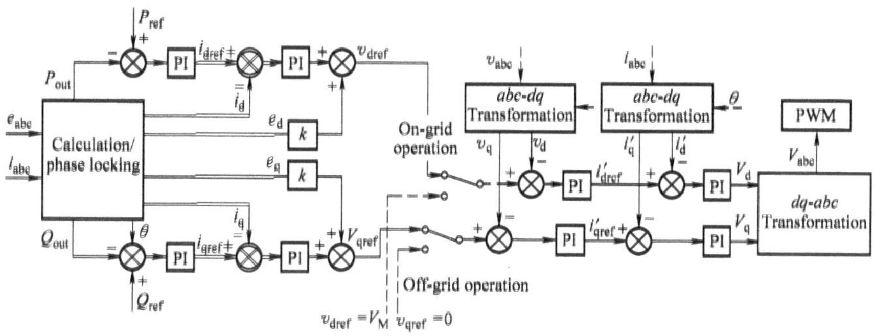

Fig. 7.7 Energy Storage PCS Grid-Connected Operation Control Diagram

current loop realizes the PCS's external power exchange adjustment, which can accept the dispatch of the microgrid EMS; the filter inductor current loop helps to improve the dynamic performance of the PCS and can implement overcurrent protection for the main circuit.

When the microgrid is operating in grid-connected mode, the energy storage PCS generally operates in *PQ* mode, accepting the power P_c, Q_c dispatch instructions issued by the microgrid energy management system. Using formula (7.1), we can solve for i_{dref}, i_{qref}, further obtain the reference values of filter capacitor voltage v_{dref} and v_{qref}, as well as the reference value of filter inductor current, to generate the PWM signal controlling the PCS. In addition, to improve the PCS controller's disturbance rejection ability to grid voltage changes, a grid voltage feedforward link is introduced.

$$\begin{cases} P_c = 1.5\left(e_d i_{dref} + e_q i_{qref}\right) \\ Q_c = 1.5\left(e_q i_{dref} - e_d i_{qref}\right) \end{cases} \quad (7.1)$$

In the formula, e_d, e_q are the dq axis components of the grid voltage e_{abc}, i_{dref}, i_{qref} are the dq axis components of the reference value of i_{abc}.

In this operation mode, the energy storage PCS can absorb or output a certain amount of active/reactive power from the public grid according to its own needs (or accept grid dispatch instructions), to achieve the power flow stability of the PCC between the microgrid and the public grid within a certain time, making the microgrid a "controllable unit" relative to the public grid.

The single-phase equivalent circuit and vector diagram of the PCS grid-connected system are shown in Fig. 7.8. In the figure, $\dot{V}$ represents the filtered capacitor voltage after filtering, which has a waveform similar to a sine wave, $\dot{E}$ represents the ideal grid voltage, $\dot{V}_L$ represents the voltage across the grid-connected inductor, $\dot{I}_g$ represents the grid-connected current [48].

By adjusting the amplitude of the filtered capacitor voltage $\dot{V}$ and its phase angle leading the grid voltage $\dot{E}$, the voltage across the grid-connected inductor θ can be changed. According to Kirchhoff's voltage law,

$$\dot{V}_L = \dot{V} - \dot{E} = j\omega L_g \cdot \dot{I}_g \quad (7.2)$$

Where is the ω grid angular frequency, and the grid-connected current is:

$$\dot{I}_g = \frac{\dot{V} - \dot{E}}{j\omega L_g} \quad (7.3)$$

Assuming the phase of the grid voltage is 0, and the amplitude is E, then $\dot{E} = E\angle 0$, the corresponding filtered capacitor voltage is $\dot{V} = V\angle\theta$, so the above formula can be written as:

$$\dot{I}_g = \frac{V\angle\theta - E\angle 0}{j\omega L_g} \quad (7.4)$$

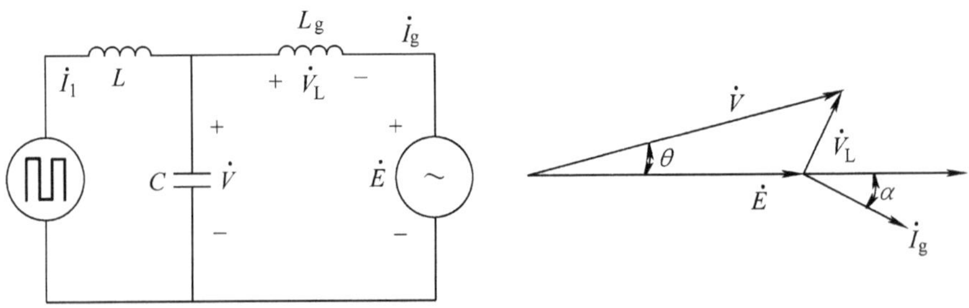

Fig. 7.8 Single-phase grid-connected equivalent circuit diagram and vector diagram

It can be seen that by adjusting the voltage amplitude and phase angle of the filter capacitor, the size and phase angle of the grid-connected current can be adjusted, thereby adjusting the active and reactive power injected into the grid. This adjustment control process is simple. The energy storage PCS, according to specific rules such as microgrid economic operation, load peak-valley regulation, participation in grid demand response, smoothing renewable energy fluctuations, load tracking, etc., adjusts the active and reactive power output to achieve specific functions during grid-connected operation.

7.2.2 *Off-Grid Operation Control*

When the microgrid operates off-grid, the energy storage PCS, which serves as the networked power source, generally adopts the V/f control method to establish and maintain the voltage and frequency of the microgrid's off-grid operation. The logic switch in Fig. 7.7 is set to the "off-grid operation" mode. v_{dref} and v_{qref} are taken from the system preset values, and after the filtered capacitor voltage loop and the filtered inductor current loop, they generate the PWM signal controlling the PCS, as shown in Fig. 7.9.

Microgrids often contain nonlinear loads, such as variable frequency drive devices or thyristor rectified DC devices, computers, UPS power supplies, etc. For this type of load, even if the supply voltage is a standard sine wave, the load current is severely distorted, containing a large amount of low-order harmonics. Due to the impedance of the PCS and the line, these harmonic currents will produce harmonic voltage drops at the output end of the PCS, causing distortion of the output voltage. Therefore, the PCS needs to add instantaneous waveform correction in control to maintain the output waveform as a standard sine wave. Otherwise, the harmonic voltage produced will generate harmonic circulating currents among various devices in the microgrid, affecting the normal operation of the system.

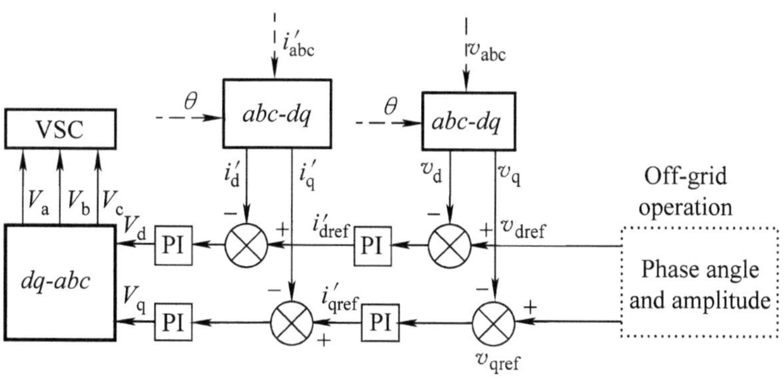

Fig. 7.9 Control Diagram of Off-grid Operation of Energy Storage PCS

At the same time, there are single-phase loads in the microgrid, and low-voltage microgrids often adopt a three-phase four-wire system structure. Interphase load imbalance will cause the microgrid to have zero-sequence and negative-sequence current components, which will then cause the three-phase voltage of the microgrid to be unbalanced.

Therefore, during the off-grid operation of the microgrid, the energy storage PCS, as the grid-connected power source, needs to solve the current harmonic and three-phase imbalance problems caused by the nonlinear load and three-phase unbalanced load in the microgrid, to ensure that the voltage quality of the microgrid meets the specified requirements without the support of the public grid. This is also the main task of control during the off-grid operation of the PCS.

7.2.2.1 Unbalanced Nonlinear Control Strategy

Since the three-phase unbalanced components can be decomposed into positive-sequence, negative-sequence, and zero-sequence of symmetrical components, the three-phase unbalance problem can be transformed into a compensation control problem for the negative-sequence component.

Adopting the negative-sequence compensation control strategy in the rotating coordinate system, the control structure is shown in Fig. 7.10. The sampled three-phase voltage values are passed through the positive-sequence and negative-sequence Park transformations, and the obtained negative-sequence component is compared with the component of the standard sine wave in the negative-sequence to eliminate the negative-sequence effect; the obtained positive-sequence voltage *dq* component is compared with the given value of the standard sine wave in the positive-sequence, and the deviation generated after the *PI* regulator is used as the inductive current inner loop control quantity [122, 123].

The inductive current control loop can improve the system's dynamic performance and steady-state performance, and facilitate overcurrent protection for the PCS.

Generally, the PCS needs to deal with voltage control problems under unbalanced and nonlinear loads at the same time. It can add the main harmonic

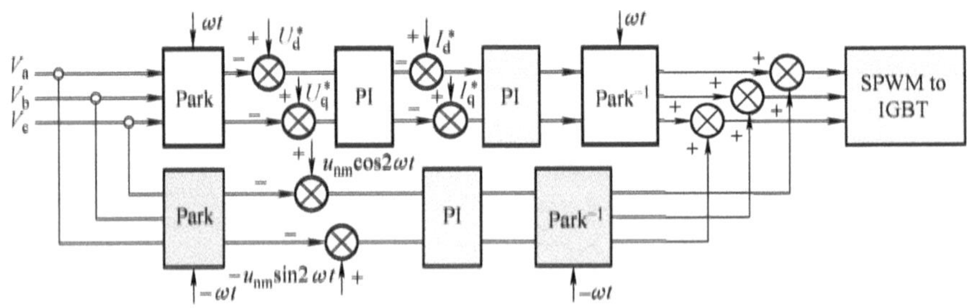

Fig. 7.10 Improved Negative-sequence Compensation Control Strategy

compensation control loop such as fifth and seventh times, and the negative-sequence control loop of unbalanced load based on the positive-sequence control. The control principle is shown in Fig. 7.11.

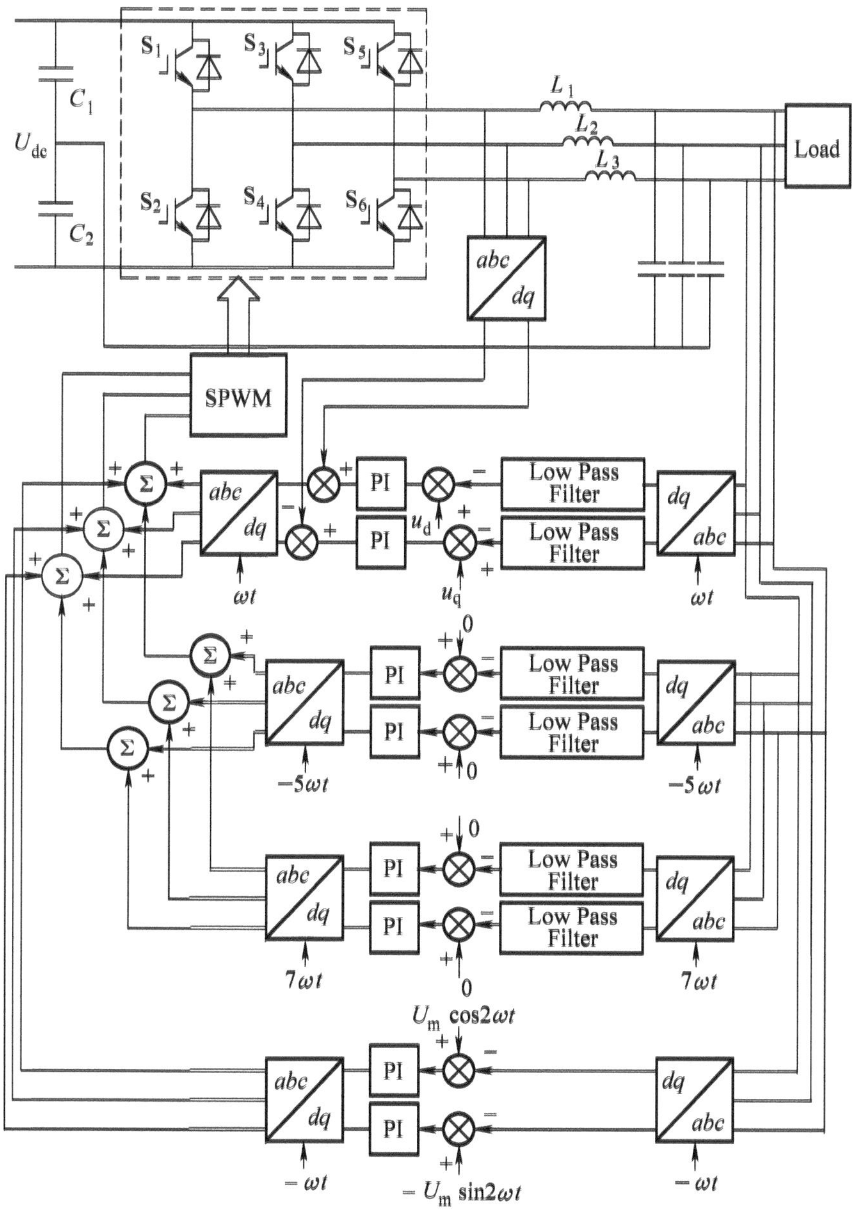

Fig. 7.11 Unbalanced and Nonlinear Load Control Strategy

Nonlinear Load Control Strategy

For most rectified power supply devices, such as variable frequency household appliances, energy-saving lamps, switch power supplies, etc., their harmonic content is mainly fifth and seventh times. The fifth and seventh harmonics can be transformed into *dq* components through the corresponding rotating coordinate transformation, the fifth harmonic becomes a DC quantity after the negative-sequence transformation of $-5\ \omega t$, and after low-pass filtering, it tends to zero through *PI* control, achieving the purpose of suppressing the fifth harmonic. Similarly, the control principle is similar for the seventh harmonic or other prominent nonlinear loads.

The fundamental three-phase voltage on the load side is:

$$\begin{cases} u_{a1} = U_1 \sin \omega t \\ u_{b1} = U_1 \sin\left(\omega t - \dfrac{2\pi}{3}\right) \\ u_{c1} = U_1 \sin\left(\omega t + \dfrac{2\pi}{3}\right) \end{cases} \tag{7.5}$$

Then the fifth harmonic voltage is

$$\begin{cases} u_{a5} = U_5 \sin 5\omega t \\ u_{b5} = U_5 \sin\left(5\omega t - 5\times\dfrac{2\pi}{3}\right) = U_5 \sin\left(5\omega t + \dfrac{2\pi}{3}\right) \\ u_{c5} = U_5 \sin\left(5\omega t + 5\times\dfrac{2\pi}{3}\right) = U_5 \sin\left(5\omega t - \dfrac{2\pi}{3}\right) \end{cases} \tag{7.6}$$

After performing a fifth order negative sequence transformation, the *d* axis rotates clockwise at an angular speed of 5 ω, under the "equal amplitude" transformation condition, the fifth harmonic is transformed in the *dq0* coordinate system.

$$\begin{bmatrix} u_{d5} \\ u_{q5} \\ u_{05} \end{bmatrix} = \mathbf{T}_{N5} \begin{bmatrix} u_{a5} \\ u_{b5} \\ u_{c5} \end{bmatrix} = \frac{2}{3} \cdot \begin{bmatrix} \sin 5\omega t & \sin\left(5\omega t + \dfrac{2\pi}{3}\right) & \sin\left(5\omega t - \dfrac{2\pi}{3}\right) \\ -\cos 5\omega t & -\cos\left(5\omega t + \dfrac{2\pi}{3}\right) & -\cos\left(5\omega t - \dfrac{2\pi}{3}\right) \\ 1/\sqrt{2} & 1/\sqrt{2} & 1/\sqrt{2} \end{bmatrix} \cdot$$

$$\begin{bmatrix} U_5 \sin 5\omega t \\ U_5 \sin\left(5\omega t + \frac{2\pi}{3}\right) \\ U_5 \sin\left(5\omega t - \frac{2\pi}{3}\right) \end{bmatrix} = \begin{bmatrix} U_5 \\ 0 \\ 0 \end{bmatrix} \tag{7.7}$$

Since we hope that the fifth harmonic is eliminated, we can set $U_5{}^* = 0$, after the system stabilizes, the fifth harmonic can tend to 0. When the load does not contain the fifth harmonic, Eq. (7.7) equals 0, that is, the fifth voltage loop does not work for balanced loads, and this algorithm can also run stably under balanced loads. The suppression principle of the seventh harmonic is the same, but the seventh harmonic uses the seventh order positive sequence transformation.

Unbalanced Load Control Strategy

When carrying an unbalanced load, in the positive sequence *dq* coordinate system, the fundamental positive sequence component of the load voltage is a DC quantity, and the negative sequence component is an AC quantity of 2ω. The negative sequence component is the main reason for generating asymmetric output voltage, therefore, as long as the negative sequence component is controlled to zero, a three-phase symmetrical load voltage can be output. To unify with the harmonic control method, unbalanced loads adopt negative sequence compensation control.

$$\begin{bmatrix} u_{d1\mathrm{N}} \\ u_{q1\mathrm{N}} \\ u_{01\mathrm{N}} \end{bmatrix} = \mathbf{T}_{N1} \begin{bmatrix} u_{\mathrm{a}1} \\ u_{\mathrm{b}1} \\ u_{\mathrm{c}1} \end{bmatrix} = \frac{2}{3} \begin{bmatrix} -\sin\omega t & -\sin\left(\omega t + \frac{2\pi}{3}\right) & -\sin\left(\omega t - \frac{2\pi}{3}\right) \\ \cos\omega t & \cos\left(\omega t + \frac{2\pi}{3}\right) & \cos\left(\omega t - \frac{2\pi}{3}\right) \\ 1/\sqrt{2} & 1/\sqrt{2} & 1/\sqrt{2} \end{bmatrix}$$

$$\cdot \begin{bmatrix} U_1 \sin\omega t \\ U_1 \sin\left(\omega t - \frac{2\pi}{3}\right) \\ U_1 \sin\left(\omega t + \frac{2\pi}{3}\right) \end{bmatrix} = \begin{bmatrix} U_1 \cos 2\omega t \\ -U_1 \sin 2\omega t \\ 0 \end{bmatrix} \tag{7.8}$$

Therefore, the given value of the negative sequence compensation amount is:

$$\begin{bmatrix} u_{dN}^* \\ u_{qN}^* \end{bmatrix} = \begin{bmatrix} U_1 \cos 2\omega t \\ -U_1 \sin 2\omega t \end{bmatrix} \tag{7.9}$$

From Eq. (7.9), it can be seen that as long as the given amplitude and frequency are considered, the system can be controlled for compensation.

After adding the negative sequence compensation control loop, the system control current command is:

$$\begin{bmatrix} i_{xcmd} \\ i_{ycmd} \end{bmatrix} = \mathbf{K}_p \begin{bmatrix} u_{xerr} \\ u_{yerr} \end{bmatrix} + \mathbf{T}_{P1} \begin{bmatrix} u_{dl} \\ u_{ql} \end{bmatrix} + \mathbf{T}_{N1} \begin{bmatrix} u_{ri} \\ u_{si} \end{bmatrix} \tag{7.10}$$

Where, $u_{x\,err}$, $u_{y\,err}$ are the voltage error components in the stationary coordinate system; is the voltage error proportional coefficient; $\boldsymbol{T}_{P1}$ is the positive sequence Park transformation rotation matrix; *N*1 *is the negative sequence Park transformation rotation matrix;* u_{dl} , u_{ql} *are the positive sequence error integral terms of the dq axis;* u_{ri}, u_{si} *are the negative sequence error integral terms of the dq axis.* $u_{x\,err}$, $u_{y\,err}$ *after the negative sequence Park transformation, we get:*

$$\begin{bmatrix} u_{rerr} \\ u_{serr} \end{bmatrix} = \mathbf{T}_{N1}^{T} \begin{bmatrix} u_{xerr} \\ u_{yerr} \end{bmatrix} \tag{7.11}$$

In the formula, u_{rerr}, u_{serr} are the negative sequence components of the voltage error. After the *PI* regulator, perform its integral operation:

$$\frac{1}{2\pi}\int_0^{2\pi} \begin{bmatrix} \cos\omega t & -\sin\omega t \\ -\sin\omega t & -\cos\omega t \end{bmatrix} \begin{bmatrix} U_{xerr}\cos(\omega t+\theta_x) \\ U_{yerr}\sin(\omega t+\theta_y) \end{bmatrix} d\theta = \frac{1}{2} \begin{bmatrix} U_{xerr}\cos\theta_x - U_{yerr}\cos\theta_y \\ U_{xerr}\sin\theta_x - U_{yerr}\sin\theta_y \end{bmatrix} \tag{7.12}$$

$u_{x\,err}$, $u_{y\,err}$ are represented by two arbitrary amplitude ($U_{x\,err}$, $U_{y\,err}$) and phase (θ_x, θ_y) sine functions. Because the integral operation can be seen as averaging the input, when the load is balanced, $U_{x\,err} = U_{y\,err}$, $\theta_x = \theta_y$, the result of formula (7.12) is zero. Therefore, it can be concluded that under balanced load, the integral part of the negative sequence component of voltage error has no effect on the command current, that is, this algorithm can also control the balanced load. Similarly, it can be inferred that under balanced load, the integral part of the positive sequence component of voltage error is the same as in conventional control [124].

Due to the disturbances that cause distortion of the voltage waveform during the off-grid operation of the microgrid, such as the harmonics of nonlinear loads, converter control dead zones, etc., all have periodic repetition characteristics, so the repetitive control algorithm can be introduced to suppress these disturbances.

At the same time, the 2ω harmonic corresponding to the negative sequence fundamental wave is alternating, but they repeat in exactly the same waveform in each fundamental wave period. Repetitive control can compensate for the problem that *PI* control cannot eliminate alternating dynamic errors, achieving zero-error regulation. It also has good suppression ability for the asymmetry of PCS output voltage caused by three-phase unbalanced loads, etc.

In the dual closed-loop control structure, the current loop uses a *PI* regulator, and the voltage loop uses a *PI* regulator and a repetitive controller in parallel, taking into account the dynamic and steady-state performance of the system [125], as shown in Fig. 7.12.

7.2.2.2 Unbalanced Nonlinear Load Experiment

Experiment with Single-Phase Rectifier Load

Using the *IPI* control strategy based on repetitive control, an experiment is conducted under the condition of a single-phase rectifier load in the microgrid. In the experimental platform, the *a* phase is connected to the rectifier load, the filter inductance is 0.5 mH, the filter capacitor is 3300 μF, and the load resistance is 11.2Ω. *b, c* phases are connected to the resistance 7.5Ω linear load, as shown in Fig. 7.13. In the repetitive controller, $N = 120$, 120 points are sampled per period, the feedback coefficient $Q(z)$ is 1, and the lead link z^k selects z^8 for phase compensation.

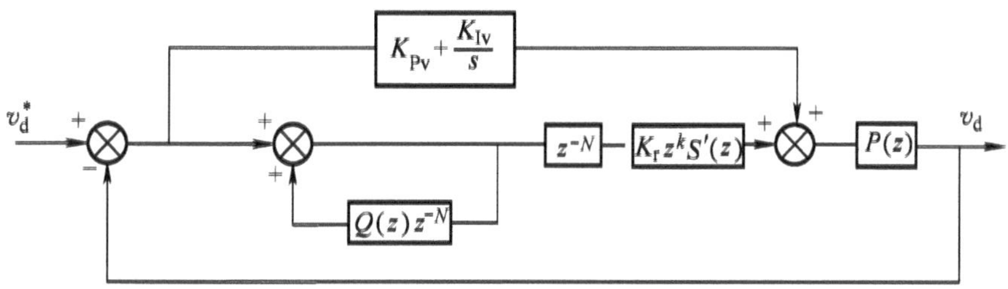

Fig. 7.12 *PI* control and repetitive control combined *IPI* control block diagram

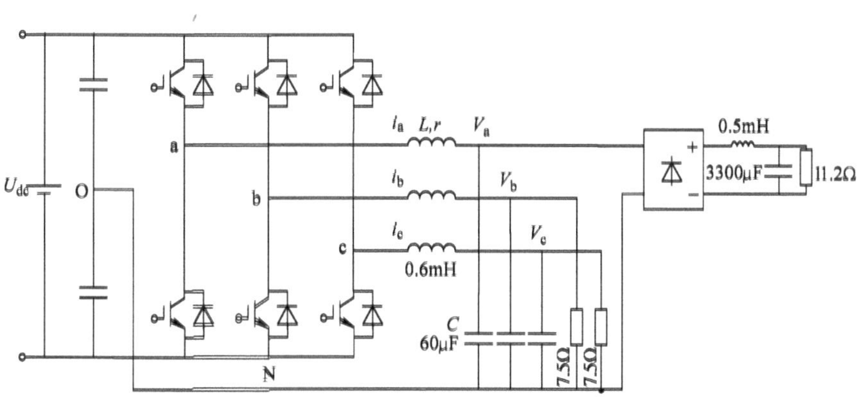

Fig. 7.13 Energy Storage PCS with Mixed Load Experiment

The experimental waveforms are shown in Figs. 7.14, 7.15, 7.16 and 7.17. It can be seen that when there is a single-phase nonlinear load, the PCS can output a well-balanced three-phase voltage waveform using the IPI control strategy based on

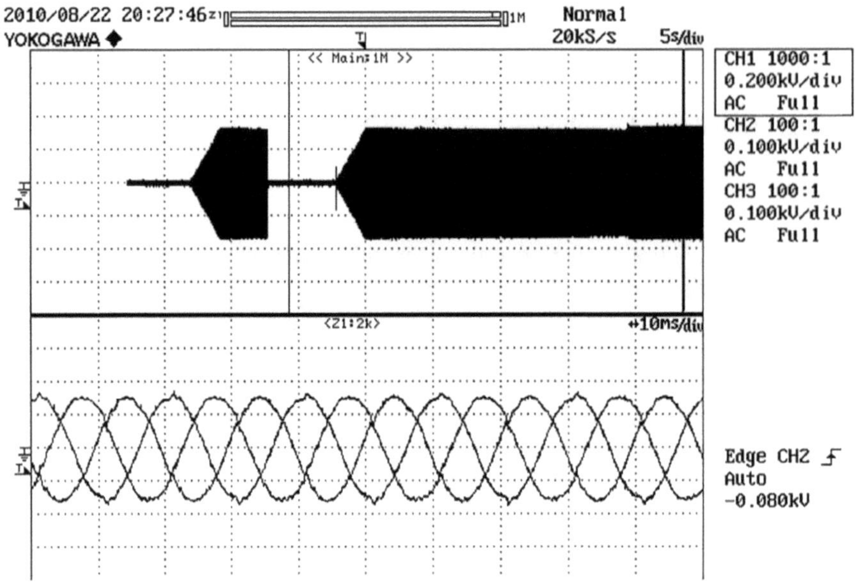

Fig. 7.14 PCS Output Voltage Waveform (CH2, CH3 are 200:1)

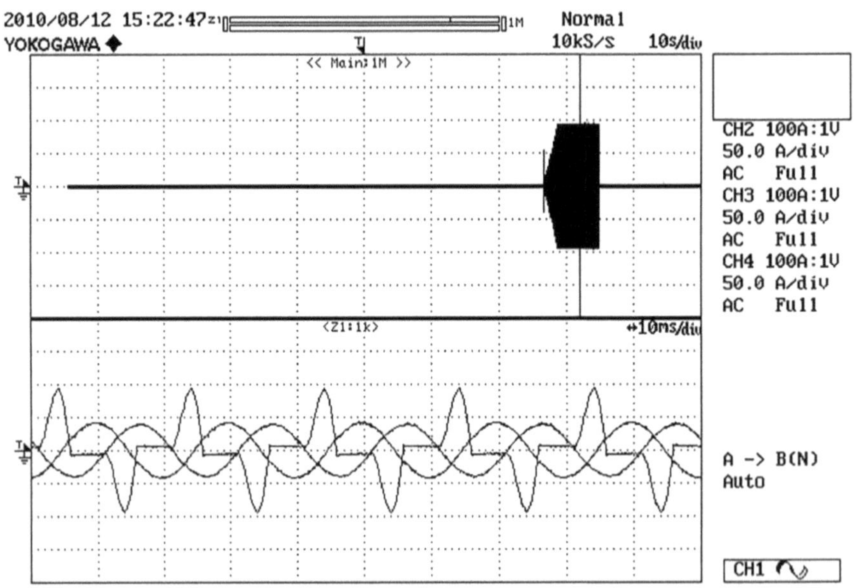

Fig. 7.15 PCS Output Current Waveform

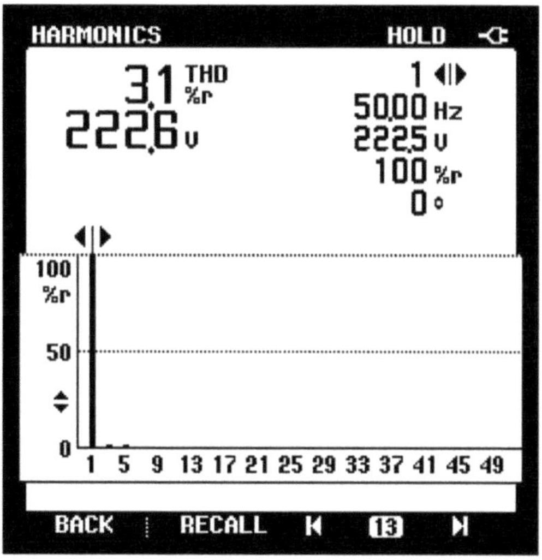

Fig. 7.16 PCS Output Voltage THD

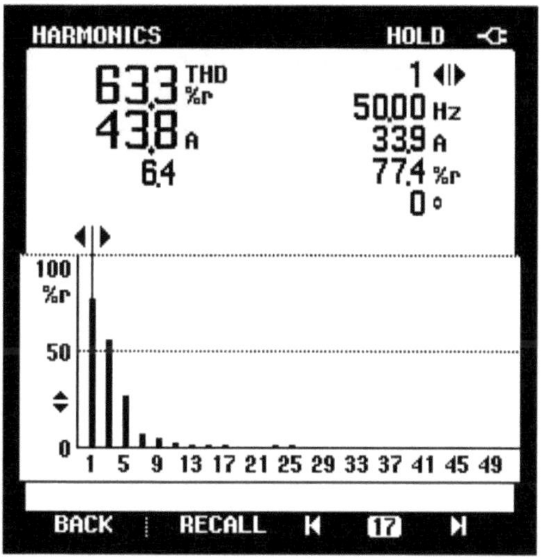

Fig. 7.17 Phase a Current THD with Nonlinear Load

repetitive control. The highest harmonic of the phase voltage is 3.1%, indicating good system control performance.

Experiment with Only Single-Phase Rectifier Load

To further verify the effectiveness of the strategy, an experiment was designed with a nonlinear load on phase a, while the other two phases were unloaded. The experimental platform and parameters are the same as above, and the main circuit is shown in Fig. 7.18.

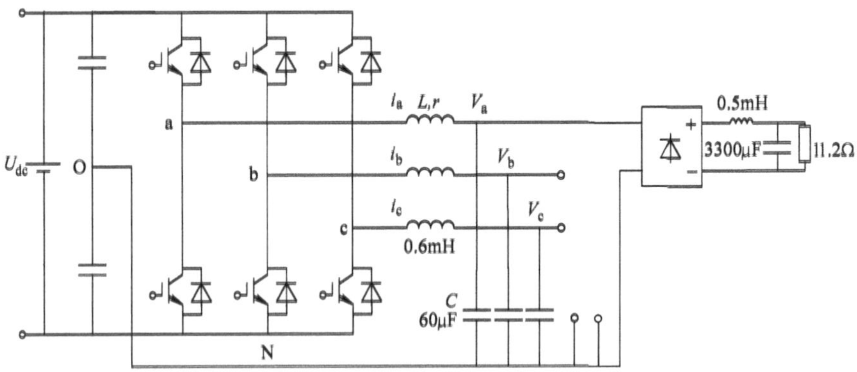

Fig. 7.18 Energy storage PCS with nonlinear load experiment

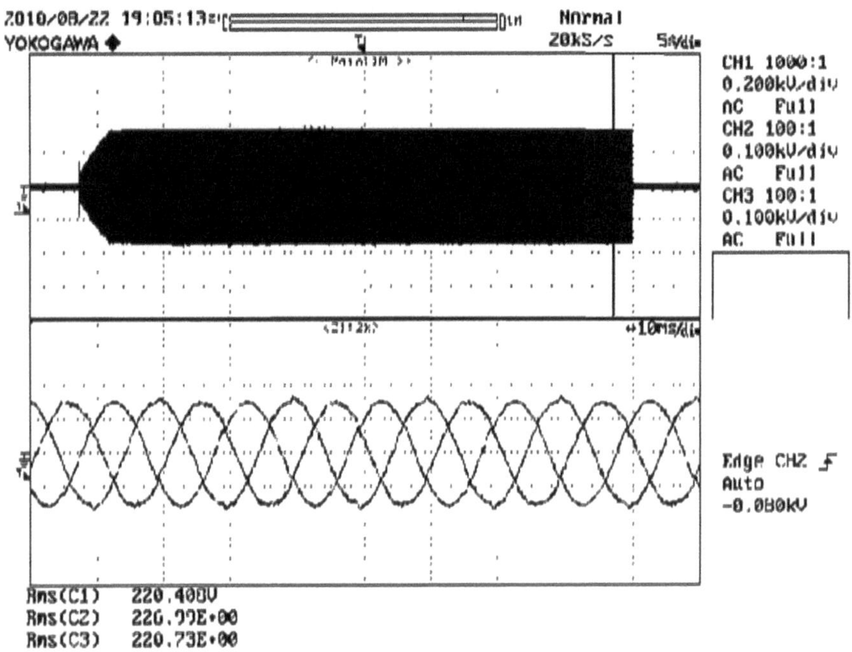

Fig. 7.19 Energy storage PCS output voltage waveform (CH2, CH3 are 200:1)

Through the experimental waveforms in Fig. 7.19, 7.20, 7.21 and 7.22 and Table 7.1, it can be seen that when the energy storage PCS carries a single-phase rectifier load, it can still output a well-balanced three-phase voltage waveform when the current harmonic reaches 63.5%. The highest phase voltage THD is 3.4%, and the overall system control performance is good.

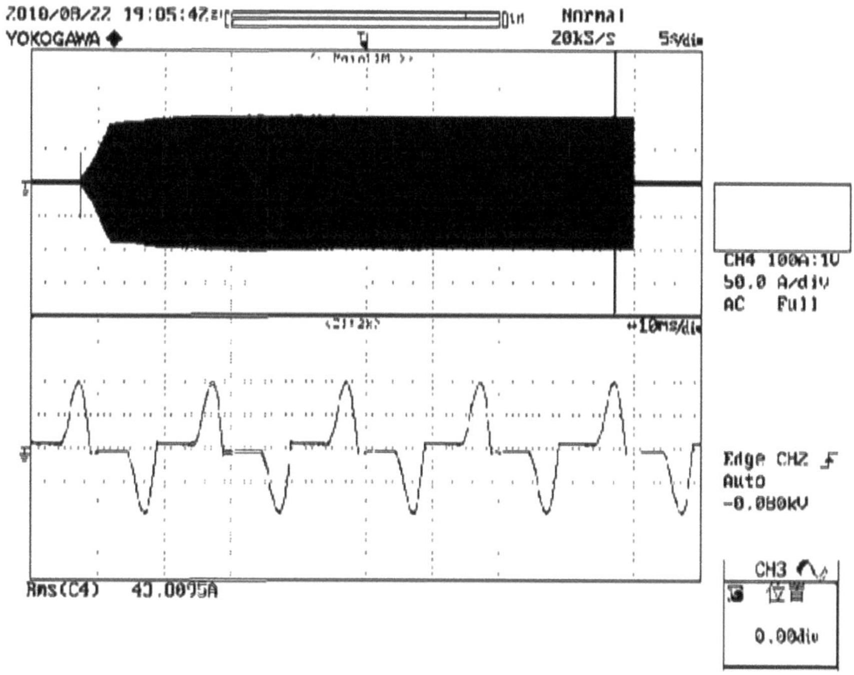

Fig. 7.20 Energy storage PCS output current waveform

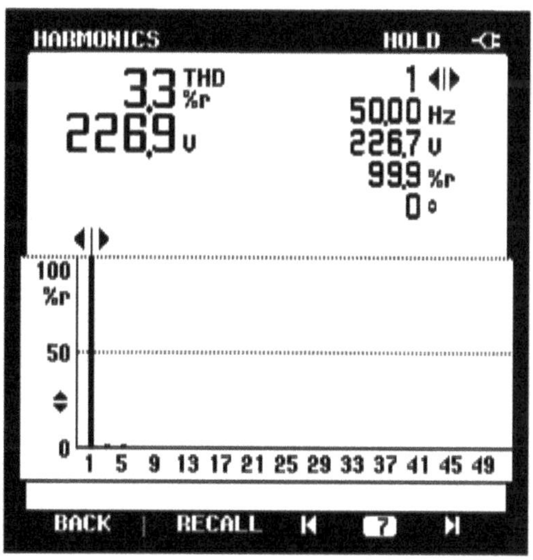

Fig. 7.21 Energy storage PCS output voltage THD

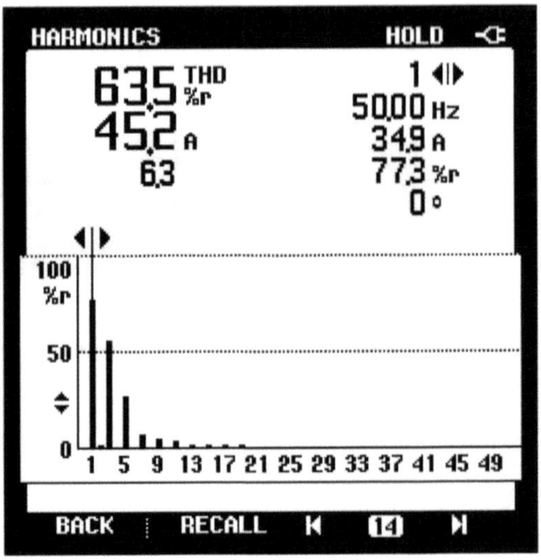

Fig. 7.22 Energy storage PCS output current THD

Table 7.1 Experimental output phase voltage and line-line voltage amplitude and THD

	Phase A	Phase B	Phase C	AB	BC	CA
Voltage value	220.41 V	226.99 V	220.73 V	386.13 V	381.87 V	386.77 V
THD	3.4%	3.3%	3.3%	0.2%	0.3%	0.2%

In summary, the IPI control strategy based on repetitive control has good off-grid operation effects when the energy storage PCS serves as a grid-connected power source. Although the design of the repetitive controller is complex and requires a large amount of computation, the current mainstream DSP can fully handle it, making it suitable for the control of energy storage PCS in microgrids.

7.2.3 Grid-Connected/Off-Grid Transition Control

The smooth transition of microgrid grid-connected/off-grid operation is key to ensuring the reliability of important load power supply. This includes the transition of the microgrid from grid-connected operation mode to off-grid operation mode, as well as the transition from off-grid operation mode to grid-connected operation mode. Because the microgrid often has sufficient time to synchronize and modes switch when transitioning from off-grid to grid-connected, the control difficulty is relatively small. Conversely, when the microgrid switches from grid-connected to off-grid, there are planned and unplanned scenarios. Especially in unplanned scenarios, the state flip of the main power sources such as energy storage is often large, the control mode switch requires high speed, and therefore the control difficulty is relatively large, even with the risk of switch failure.

7.2.3.1 Grid-Connected to Off-Grid Operation Mode Transition

When the public grid fails, the microgrid needs to quickly identify and swiftly switch to off-grid operation mode, which is unplanned off-grid. In this process, the switching process of the energy storage PCS as a networked power source needs to be fast enough to minimize the impact of grid failure on the load and distributed power sources within the microgrid. When the external grid needs to be powered off for planned maintenance, the microgrid EMS can actively switch to off-grid operation mode after receiving the power-off notice, ensuring the continuity of power supply to the load within the microgrid and maintaining the normal operation of distributed power sources.

When energy storage serves as the main power source of the microgrid, the energy storage PCS often adopts the indirect current control method of three-loop control during grid-connected operation of the microgrid. At the moment of off-grid, when it is confirmed that the grid-connected switch has been disconnected, the PCS switches to the dual-loop working mode. Keeping the voltage loop of the filter capacitor and the current loop of the filter inductor basically unchanged under the two operating modes at the moment of off-grid, thus ensuring the smooth and fast process of mode conversion (Fig. 7.23).

It should be noted that the microgrid is connected to the public grid through the public connection point PCC, and the grid-connected and off-grid operation is realized by controlling the grid-connected switch. In order to improve the success rate of the microgrid's switch from grid-connected to off-grid, the switch of the energy storage PCS mode needs to be coordinated with the grid-connected switch in logic, ensuring that the PCS operates in the *V/f* mode when the grid-connected switch has reliably disconnected.

Common grid-connected switches can be mechanical contactors or solid-state switches. Because the action time of solid-state switches is shorter than that of contactors, they are more commonly used. The solid-state switch is generally an AC switch composed of two SCRs connected in reverse parallel, and its closing and opening are

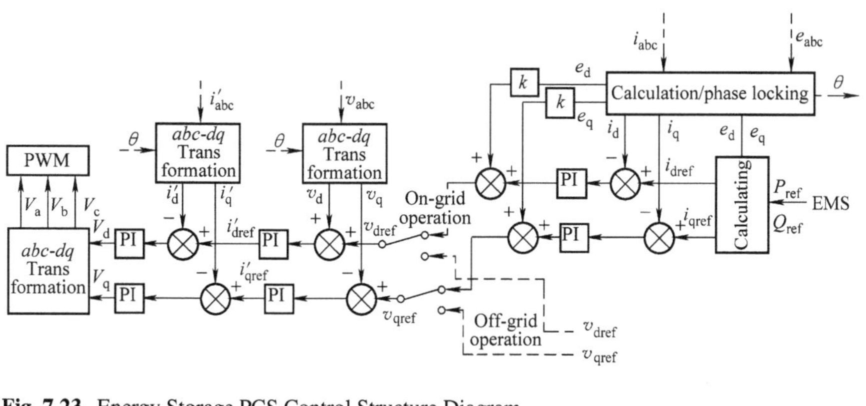

Fig. 7.23 Energy Storage PCS Control Structure Diagram

realized by the logic controller. Since the prerequisite for SCR to achieve free turn-off is that the anode voltage is lower than the cathode voltage, theoretically, the longest turn-off time of SCR is half a cycle, which is 10 ms for a 50 Hz power system.

In order to shorten the turn-off time of the solid-state switch, a forced turn-off strategy of SCR can be adopted. The basic idea is to change the amplitude of the PCS filter capacitor voltage to make it higher or lower than the grid voltage, thereby forming a reverse voltage at both ends of the grid inductor. This reverse voltage forces the grid current to drop rapidly. When it drops below the holding current of the SCR, the SCR changes from the conduction state to the blocking state, thereby disconnecting the connection with the grid [48].

7.2.3.2 Off-Grid to Grid-Connected Operation Mode Switch

When the microgrid is in off-grid operation mode, it monitors the status of the public grid in real time. When it determines that the public grid has resumed power supply and the microgrid has obtained grid-connected permission, the microgrid can gradually adjust the voltage status (frequency, amplitude, and phase) at the public connection point. At the moment when it reaches the synchronous state with the public grid, it closes the grid-connected switch, and the microgrid is connected to the public grid.

The energy storage PCS, as the main power supply, uses the real-time detected voltage amplitude and phase information of the public grid as reference control quantities. It adjusts the voltage amplitude and phase at the common connection point of the microgrid. When the conditions for parallel grid connection are met and the grid connection point switch is confirmed to be closed, the energy storage PCS adds a grid inductance current outer loop based on the dual-loop operation of the filter capacitor voltage loop and the filter inductor current inner loop, switching to indirect current control mode. Fast and accurate grid status detection and phase-locked control can reduce grid impact and achieve smooth mode switching.

By setting strict grid synchronization conditions, the impact current of the microgrid at the moment of grid connection can be reduced, which is beneficial to the stable operation of the microgrid and the public grid, but it will cause the grid connection time to be correspondingly extended. Considering that there is generally no strict time requirement when the microgrid switches from off-grid operation mode to grid-connected operation mode, and considering the future large-scale development of the microgrid, the grid connection conditions of the microgrid can be set stricter, which can refer to or be higher than the IEEE 1547 standard for distributed power grid connection conditions [126].

7.2.3.3 Microgrid Grid-Connected/Off-Grid Transition Experiment

A microgrid experimental platform is set up, using a three-phase four-wire system, with a line-line voltage of 380 V, a frequency of 50 Hz, an impedance of 0.0774 Ω for line 1, an impedance of 0.018 Ω for line 2, and an impedance of 0.1548 Ω for line 3. The microgrid and the distribution network are controlled by a solid-state switch. The photovoltaic unit capacity is 30 kW, the wind power unit capacity is 11 kW, the energy storage serves as the main power supply, using lead-acid batteries, and the PCS capacity is 50kVA [127].

As shown in Fig. 7.24, the microgrid operation experiment process is as follows (Fig. 7.25):

1. System startup. The microgrid starts and enters off-grid operation, with energy storage working in the *V/f* control mode, establishing and maintaining the system's voltage and frequency. After the system stabilizes, the photovoltaic unit and wind power unit are put into operation in succession, and the microgrid achieves off-grid operation.
2. Switching from off-grid to grid-connected mode. The energy storage accepts the grid connection command from the microgrid EMS, adjusts the voltage amplitude and phase of the microgrid, and at the moment of meeting the grid connection requirements, the static switch closes, and the microgrid switches to the grid-connected operation mode, with the energy storage switching to the *PQ* control mode. In the grid-connected mode, the energy storage can absorb power from the distribution network, or it can output power to maintain the stability of the PCC current when the load, photovoltaic/wind power fluctuates.

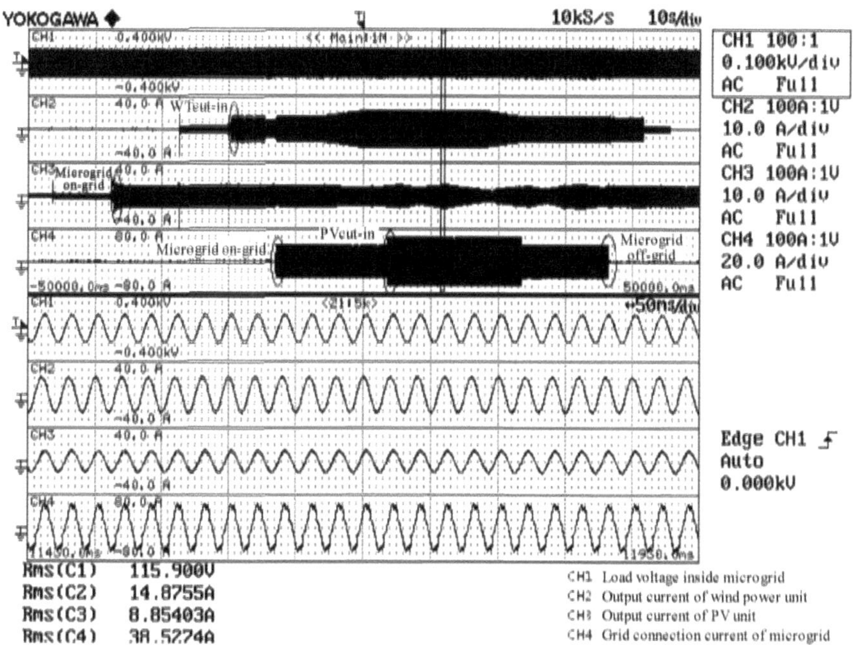

Fig. 7.24 Microgrid operation experiment process (CH1: Load voltage, CH2: Wind power output current, CH3: Photovoltaic output current, CH4: Current at PCC)

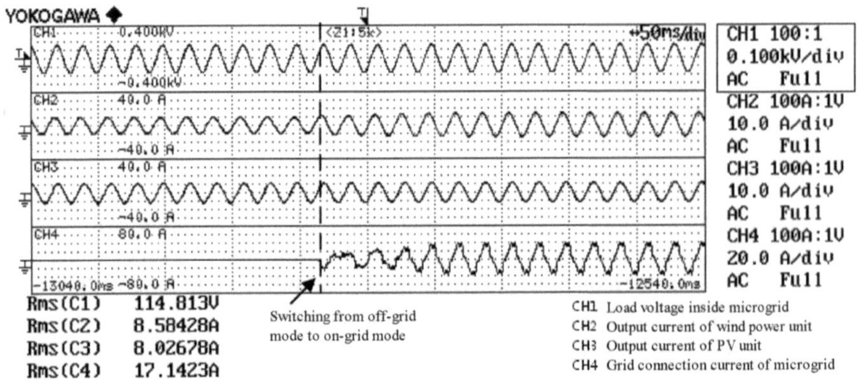

Fig. 7.25 Microgrid off-grid/grid-connected transition process

3. Microgrid grid-connected power dispatching. The microgrid accepts the dispatching command from the EMS, adjusts the current fed into the public grid (mainly realized by energy storage or controllable load), making it a controllable unit for the public grid.
4. Switching from grid-connected to off-grid mode. The energy storage detects an external distribution network fault, or accepts the off-grid command from the microgrid EMS, turns off the drive signal of the solid-state switch, adjusts the output voltage according to the current situation at the PCC using the *V/f* control mode, forcibly turns off the solid-state switch, and when it is judged to be completely disconnected, it switches to the off-grid operation mode.

Figure 7.25 shows the process of a microgrid transitioning from off-grid operation to grid-connected operation. Due to strict control of the synchronization link and grid-connection conditions, the current at the PCC of the microgrid and the public grid is smooth, without causing grid-connection problem. The voltage waveform within the microgrid hardly fluctuates or flickers during the transition, maintaining good power supply continuity and quality.

Figure 7.26 shows the process of a microgrid transition from grid-connected operation to off-grid operation. Due to the use of solid-state switch forced shutdown technology, the microgrid can complete the off-grid transition process within 3 to 5 ms, and the voltage waveform within the microgrid is hardly affected.

7.3 Energy Storage-Based Microgrid Peer Control

Various types of battery energy storage systems have good characteristics as microgrid main power sources. However, due to the complexity and uncertainty of the microgrid scale and structure, multiple storage units often need to run in parallel, while also undertaking the task of main power sources.

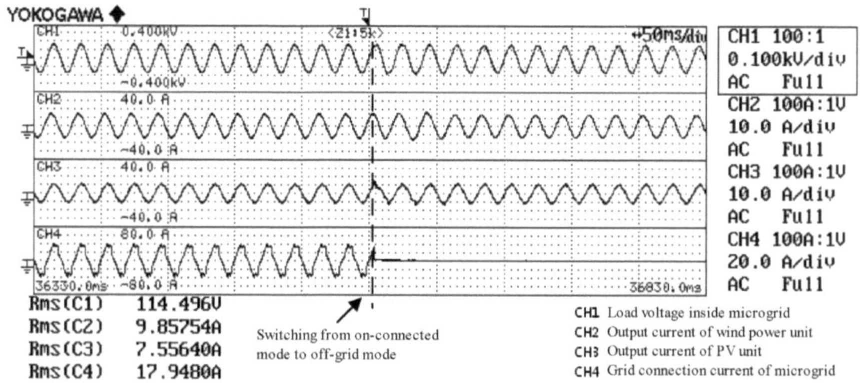

Fig. 7.26 Microgrid grid-connected/off-grid transition process

The parallel coordinated operation of multiple energy storage units aims to evenly distribute the load among the parallel units to share the load stress, and to ensure that the voltage and frequency of the microgrid system are maintained within the specified range. According to whether there is a communication line interconnection between the parallel units, the parallel scheme can be divided into two types: the interconnected line scheme and the non-interconnected line scheme [128, 129]. In the interconnected line scheme, the PCS communicates with the host controller or other units to obtain information such as current, voltage, frequency, or the dispatching instructions. The non-interconnected line scheme is represented by droop control, which originates from the static power-frequency characteristics of synchronous generator sets in traditional power systems. Each PCS unit only relies on local frequency and voltage deviation information to adjust its own output voltage amplitude and phase, achieving adaptive coordination between units without communication. The non-interconnected line scheme is a peer control idea, each PCS is equivalent, and the system will not collapse due to the failure of a PCS.

7.3.1 Peer to Peer Control

The advantage of peer to peer control is that it does not depend on communication. Each energy storage PCS acts as an independent unit, which is easy to expand, and each unit can be flexibly distributed at different nodes of the microgrid, bringing convenience to local voltage control. In control, the energy storage PCS is controlled as an equivalent voltage source, and the droop characteristics similar to synchronous generator sets are used to coordinate each unit [112].

Under peer to peer control, the energy storage unit can be equivalent to a structure in which a voltage source is connected in series with its output impedance. To simplify the analysis, take a system with two energy storage units in parallel as an example for analysis, as shown in Fig. 7.27. In the figure, $U_1 \angle \varphi_1$, $U_2 \angle \varphi_2$ are the

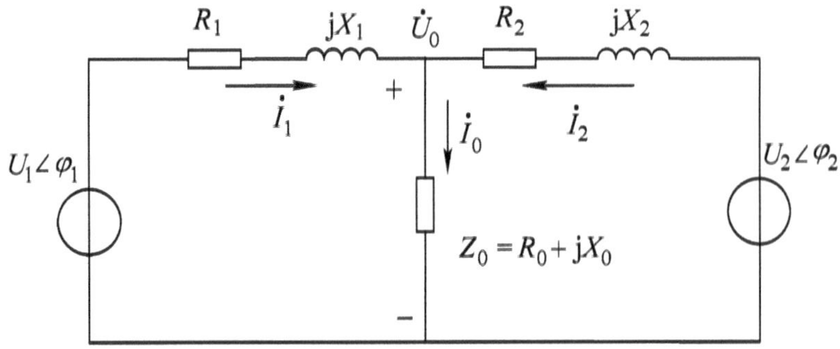

Fig. 7.27 Schematic diagram of equivalent peer to peer control system

equivalent output voltages of each energy storage unit, R_1, R_2 are the sum of the output resistance of each energy storage unit and the line resistance, X_1, X_2 are the sum of the output reactance of each energy storage unit and the line reactance, Z_0 For load impedance.

It can be obtained:

$$\dot{U}_0=\frac{\frac{\dot{U}_1}{Z_1}+\frac{\dot{U}_2}{Z_2}}{\frac{1}{Z_0}+\frac{1}{Z_1}+\frac{1}{Z_2}} \tag{7.13}$$

Where, $Z_1 = R_1 + jX_1Z_2 = R_2 + jX_2$, so the two branch currents are:

$$\begin{aligned}\dot{I}_1&=\frac{\dot{U}_1-\dot{U}_0}{Z_1}=\frac{(Z_2+Z_0)\dot{U}_1-Z_0\dot{U}_2}{Z_1Z_2+Z_0Z_2+Z_1Z_0}\\ \dot{I}_2&=\frac{\dot{U}_2-\dot{U}_0}{Z_2}=\frac{(Z_1+Z_0)\dot{U}_2-Z_0\dot{U}_1}{Z_1Z_2+Z_0Z_2+Z_1Z_0}\end{aligned} \tag{7.14}$$

The apparent power of the two energy storage units are:

$$\begin{aligned}\dot{S}_1&=\dot{U}_1\times\dot{I}_1^*=\dot{U}_1\times\frac{(Z_2^*+Z_0^*)\dot{U}_1^*-Z_0^*\dot{U}_2^*}{Z_1^*Z_2^*+Z_0^*Z_2^*+Z_1^*Z_0^*}\\ \dot{S}_2&=\dot{U}_2\times\dot{I}_2^*=\dot{U}_2\times\frac{(Z_1^*+Z_0^*)\dot{U}_2^*-Z_0^*\dot{U}_1^*}{Z_1^*Z_2^*+Z_0^*Z_2^*+Z_1^*Z_0^*}\end{aligned} \tag{7.15}$$

It can be seen that the currents of the two energy storage units $\dot{I}_1$, $\dot{I}_2$ and the apparent power $\dot{S}_1$, $\dot{S}_2$ are all simultaneously affected by $\dot{U}_1$, $\dot{U}_2$ and line parameters Z_0, Z_1, Z_2.

When the line parameters are certain, if equal current control is realized, that is, $\dot{I}_1 = \dot{I}_2$, then it is necessary to adjust $\dot{U}_1$, $\dot{U}_2$ simultaneously. Let $\dot{I}_1 = \dot{I}_2$, then,

$$\frac{(Z_2+Z_0)\dot{U}_1 - Z_0\dot{U}_2}{Z_1Z_2+Z_0Z_2+Z_1Z_0} = \frac{(Z_1+Z_0)\dot{U}_2 - Z_0\dot{U}_1}{Z_1Z_2+Z_0Z_2+Z_1Z_0} \tag{7.16}$$

Multiply both sides by $\dot{U}_2^*$ and rearrange to get:

$$\angle(\varphi_2-\varphi_1) = \frac{U_2}{U_1}\frac{2Z_0+Z_1}{2Z_0+Z_2} \tag{7.17}$$

In the formula, φ_1, φ_2 are the phase angles of $\dot{U}_1$, $\dot{U}_2$ respectively, U_1, U_2 are respectively the amplitudes of $\dot{U}_1$, $\dot{U}_2$. By expanding the complex components in the formula, making the real and imaginary parts on both sides equal, we can get:

$$\begin{aligned}\cos(\varphi_1-\varphi_2) &= \frac{U_2}{U_1}\frac{(2R_0+R_1)(2R_0+R_2)+(2X_0+X_1)(2X_0+X_2)}{(2R_0+R_2)^2+(2X_0+X_2)^2}\\ \sin(\varphi_1-\varphi_2) &= \frac{U_2}{U_1}\frac{(2X_0+X_1)(2R_0+R_2)-(2R_0+R_1)(2X_0+X_2)}{(2R_0+R_2)^2+(2X_0+X_2)^2}\end{aligned} \tag{7.18}$$

This indicates that it is necessary to adjust the four quantities of U_1, U_2, φ_1, φ_2 to make the above two formulas hold, in order to make $\dot{I}_1=\dot{I}_2$. Any deviation in any quantity will lead to the appearance of circulating current. It can be seen that in complex microgrid structures, especially when there are differences in output impedance and line parameters among various energy storage PCS, it is difficult to accurately control power distribution, which can easily lead to circulating current phenomena, resulting in increased electrical and thermal stress on individual PCS units.

7.3.2 *Improved Droop Control*

7.3.2.1 Analysis of Droop Control Circulating Current

The premise of droop control is that the power line between the PCS and the PCC is inductive, that is, the reactance is much greater than the impedance, $X > > R$, at this time the power transmission expression of the PCS is:

$$\begin{aligned}P &= \frac{EV}{X}\sin\phi \approx \frac{EV}{X}\phi\\ Q &= \frac{EV\cos\phi - V^2}{X} \approx \frac{V}{X}(E-V)\end{aligned} \tag{7.19}$$

It can be seen that active/reactive power can be decoupled through frequency/voltage control, thus determining the droop relationship as:

$$\begin{aligned} \omega &= \omega^* + m(P_0 - P) \\ E &= E^* + n(Q_0 - Q) \end{aligned} \tag{7.20}$$

Under droop control, the PCS has the ability to "self-synchronize". Even if the initial amplitudes of each PCS are not the same, they can be made consistent through dynamic adjustment, effectively suppressing parallel circulating currents.

However, in low-voltage microgrids, the resistance of the power line is significantly enhanced, often R and X cannot be ignored, at this time the power transmission expression becomes:

$$\begin{aligned} P &= \left(\frac{EV}{Z}\cos\phi - \frac{V^2}{Z}\right)\cos\theta + \frac{EV}{Z}\sin\phi\sin\theta \\ Q &= \left(\frac{EV}{Z}\cos\phi - \frac{V^2}{Z}\right)\sin\theta - \frac{EV}{Z}\sin\phi\cos\theta \end{aligned} \tag{7.21}$$

As shown in Fig. 7.28, because active/reactive control cannot be decoupled, if $P - f/Q - v$ droop is still used, it will lead to the appearance of reactive circulating current, and in severe cases, it may even cause the PCS operating state to flip [130].

Figure 7.29 shows the circulating current of parallel PCS units under different line impedance characteristics. It can be seen that when X> > R, i_1 and i_2 basically coincide, the circulating current i_h is very small, and the system works within a reasonable range. As the resistance in the line R increases, i_1 and i_2 differences gradually increase. When X << R, the circulating current is very high, increasing the electrical stress and thermal stress of the PCS, affecting its service life and safety.

7.3.2.2 Principle of Virtual Inductance and Suppression of Circulating Current

By adjusting the control strategy, the impedance characteristics of the low-voltage line are corrected to make it reappear inductive, thereby suppressing the parallel circulating current [131]. Figure 7.30 shows the main circuit of a single-phase PCS,

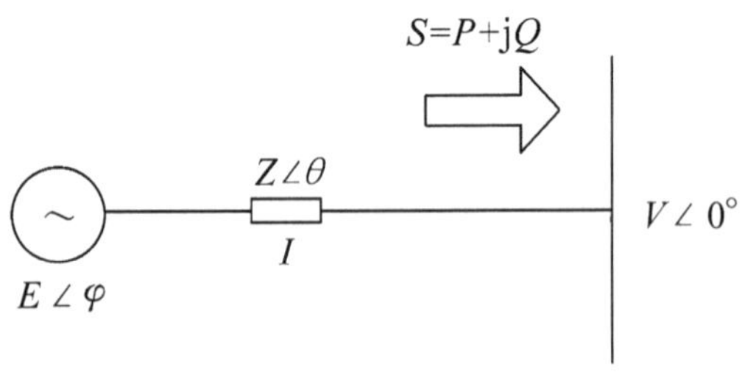

Fig. 7.28 Equivalent circuit of PCS access to microgrid

Fig. 7.29 Current and circulating current output of PCSs under different impedance characteristics. (**a**) Line X >> R. (**b**) Line X = R. (**c**) Line X << R

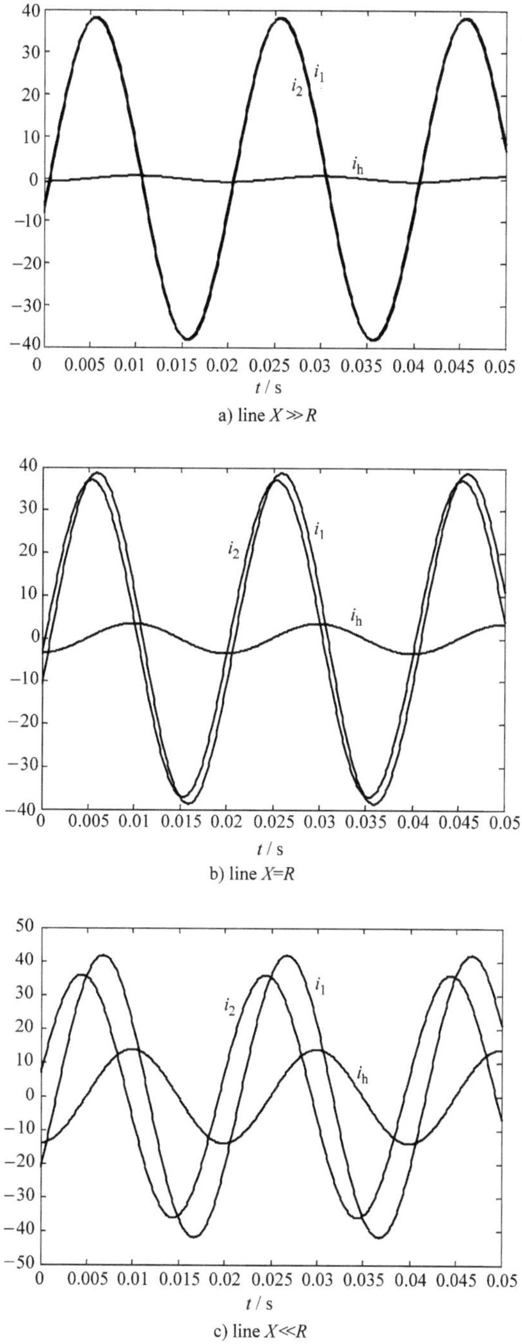

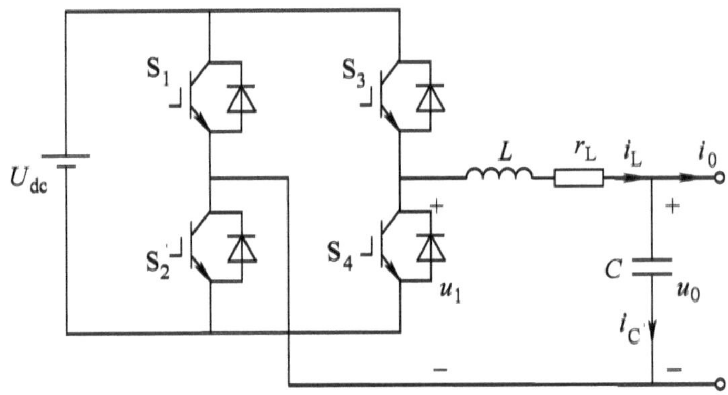

Fig. 7.30 PCS single-phase main circuit topology with LC filtering

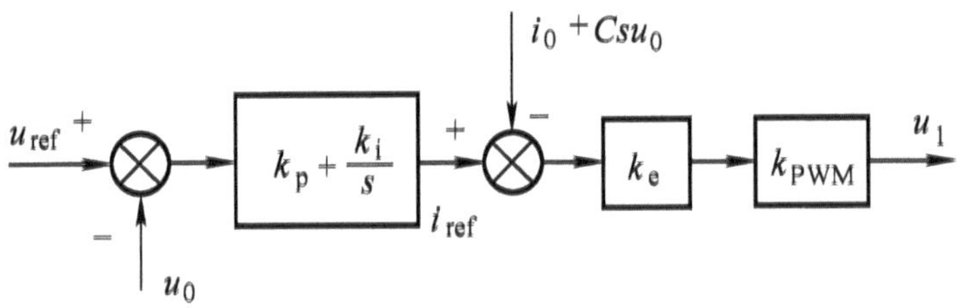

Fig. 7.31 Voltage and current dual closed-loop control diagram

Fig. 7.32 Equivalent circuit model of PCS under voltage and current dual closed-loop control

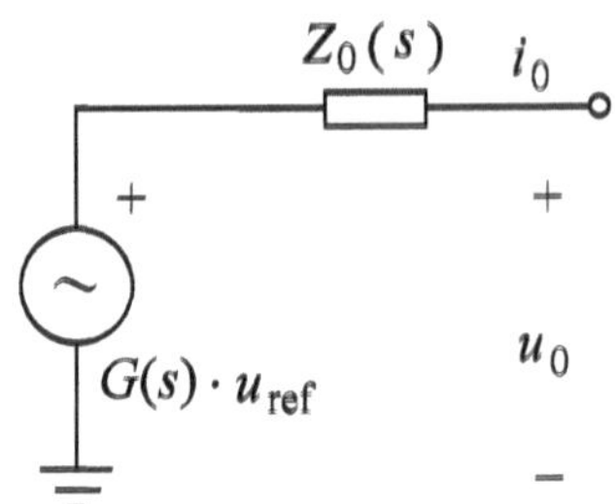

where L and r_L are the filter inductance and parasitic resistance, respectively, C is the filter capacitor, u_0 and i_0 representing the output voltage and current respectively, U_{dc} represents the DC bus voltage.

Under the voltage and current dual closed-loop control shown in Fig. 7.31, PCS can be equivalent to the form of a voltage source in series with the output impedance as shown in Fig. 7.32, and its mathematical expression is

$$u_0 = G(s)u_{ref} - Z_0(s)i_0 \tag{7.22}$$

The introduction method of virtual inductance is shown in Fig. 7.33. After introducing the virtual inductance, the impedance between PCS and the bus is composed of three parts: PCS output impedance $Z_0(s)$, virtual reactance $Z_D(s)$ and actual line impedance $Z_L(s)$. Its total equivalent impedance is shown as the form of the sum of the three in series, as shown in Fig. 7.34.

$$Z(s) = Z_0(s) + Z_D(s) + Z_L(s) \tag{7.23}$$

Figure 7.35a, b respectively show the amplitude frequency characteristics and phase frequency characteristics of the total equivalent impedance of the line $Z(s)$ before and after the introduction of virtual inductance. It can be seen that before the introduction of virtual inductance, its phase value is close to 0°(5.62°), so its characteristics are close to pure resistance. After introducing virtual inductance, its phase value becomes 87.1°, close to 90°, indicating that after the correction of virtual inductance, the line impedance has changed from resistance to inductance.

Figure 7.36a, b respectively show the simulation results of the output current and circulating current of the two PCSs before and after the introduction of virtual inductance. The line impedances of the two sections are set to $0.0642 + j0.0083\,(\Omega)$ and $0.1284 + j0.0166\,(\Omega)$ respectively, and the load is set to: initially $1.92 + j0.55\,(\Omega)$, it suddenly changes to $4.36 + j1.45\,(\Omega)$ at 0.6 s, and then changes to $2.79 + j0.56\,(\Omega)$ at 0.8 s. From the results, it can be seen that before the introduction of virtual inductance, the circulating current i_H is very large in amplitude. After introducing virtual inductance, the circulating current is significantly suppressed, and the two PCSs basically achieve balanced operation.

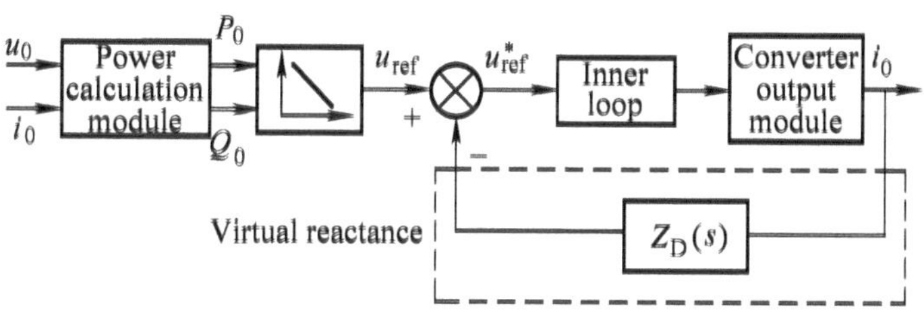

Fig. 7.33 Schematic diagram of PCS control after introducing virtual inductance

Fig. 7.34 Equivalent model of PCS connected to the bus after introducing virtual inductance

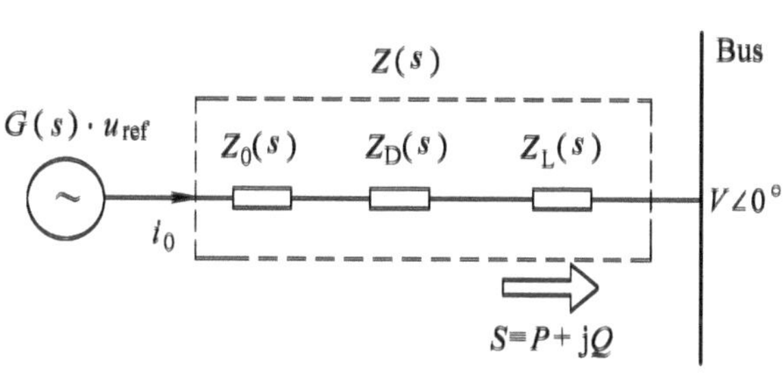

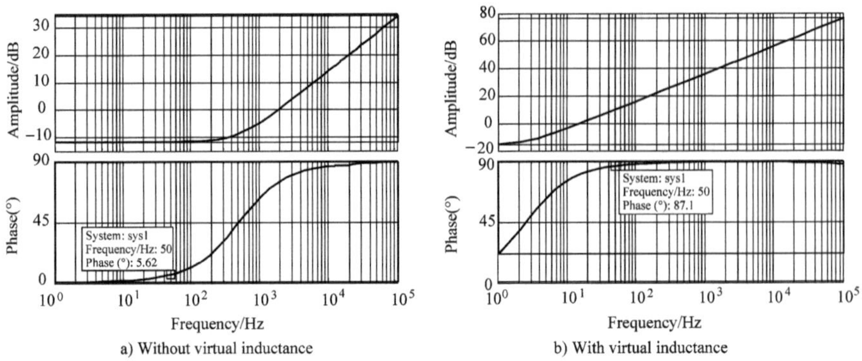

Fig. 7.35 Total line impedance Bode plot before and after introducing virtual inductance. (**a**) Before introducing virtual inductance. (**b**) After introducing virtual inductance

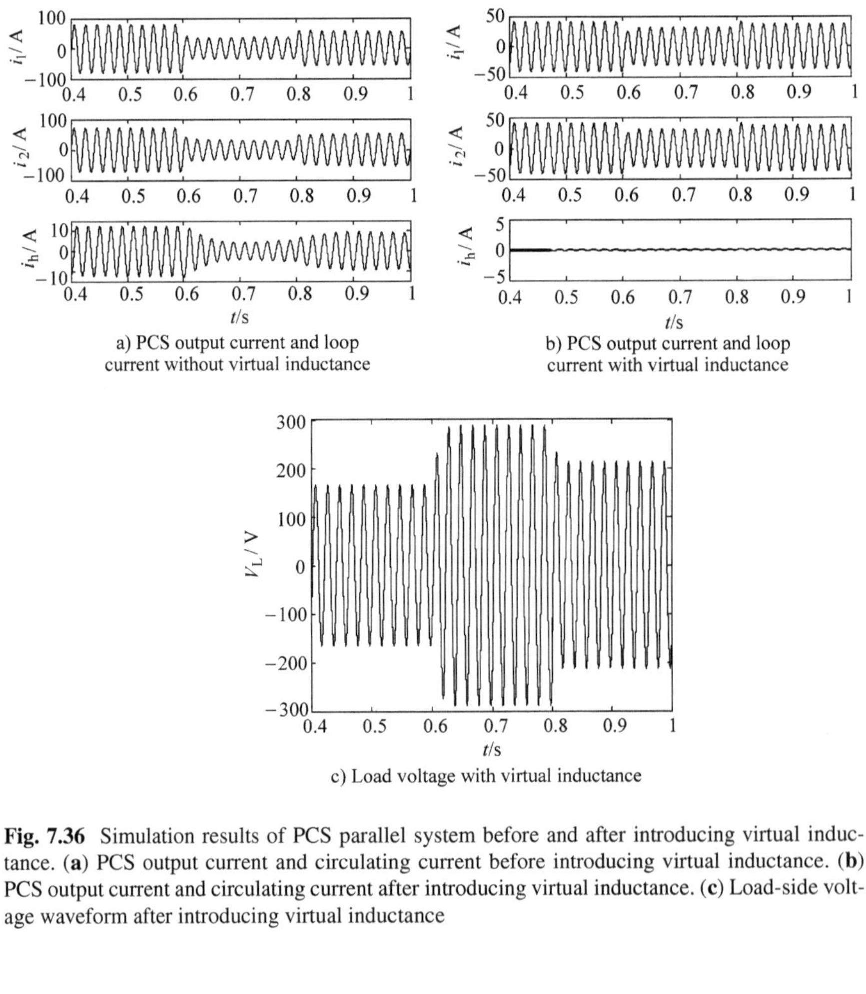

Fig. 7.36 Simulation results of PCS parallel system before and after introducing virtual inductance. (**a**) PCS output current and circulating current before introducing virtual inductance. (**b**) PCS output current and circulating current after introducing virtual inductance. (**c**) Load-side voltage waveform after introducing virtual inductance

However, as can be seen from Fig. 7.36c, after introducing virtual inductance, the load-side voltage has significantly deviated from the rated value. This is because introducing virtual inductance is equivalent to inserting a large inductance in the line, which affects the voltage at the load end. Therefore, to ensure the quality of the microgrid voltage, after introducing virtual inductance, it is necessary to correct the amplitude to keep the load-side voltage within a reasonable range.

7.3.2.3 Analysis and Correction of the Impact of Virtual Inductance on Load-Side Voltage

Since the mechanism of the impact of virtual inductance on each phase voltage is the same and there is no interaction, to simplify the analysis process, a single-phase circuit is used to replace the three-phase circuit for analysis, as shown in Fig. 7.37. Where, L_D is the virtual inductance, $R + jX$ is the equivalent load of the system.

The load is divided into pure resistance, resistive, and capacitive types for separate analysis.

1. **Voltage analysis under pure resistive load**

Line current:

$$\dot{I} = \frac{\dot{U}}{R + j\omega L_D} = \frac{U_{ref}\angle 0^\circ}{R + jX_D} \tag{7.24}$$

In the formula, U_{ref} is the effective value of the PCS output voltage u_{ref}, assuming its phase is 0°, ω is the angular frequency, then the load-side voltage is:

$$\dot{U}^*_{ref} = \dot{I} \cdot R = U_{ref} / \left(1 + j\frac{X_D}{R}\right) \tag{7.25}$$

Its effective value is:

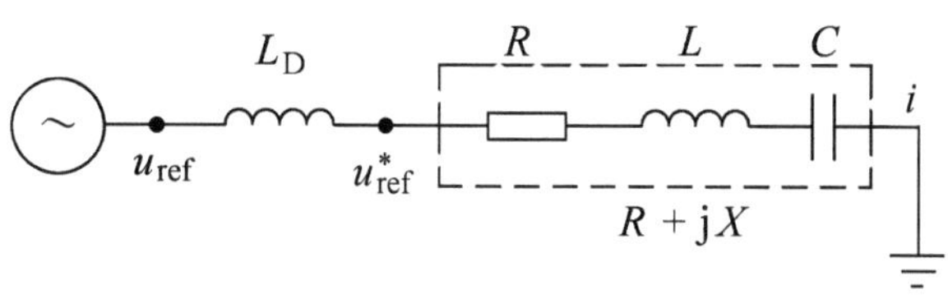

Fig. 7.37 Equivalent structure diagram of the converter with virtual inductance

$$U_{ref}^{*} = \frac{1}{\sqrt{1+\left(\frac{X_D}{R}\right)^2}} U_{ref} < U_{ref} \tag{7.26}$$

It is known that when the load is purely resistive, the amplitude of the load voltage will inevitably be lower than the rated output voltage, and the larger the virtual reactance X_D, the lower the load voltage, as shown in Fig. 7.38.

In the figure, i_0 is the line current, v_L is the voltage drop on the virtual inductance, v_L and i_0 have an angle of 90° between them, v_{ref} and v_{ref}^* are the PCS output voltage and the load voltage, respectively.

2. **Analysis of end voltage under resistive-inductive load**

Line current:

$$\dot{I} = \frac{\dot{U}}{R + j(\omega L_D + \omega L)} = \frac{U_{ref}\angle 0°}{R + j(X_D + X)} \tag{7.27}$$

Load voltage:

$$\dot{U}_{ref}^{*} = \dot{I}\cdot(R + jX) = \frac{U_{ref}(R + jX)}{R + j(X_D + X)} \tag{7.28}$$

Effective value of terminal voltage:

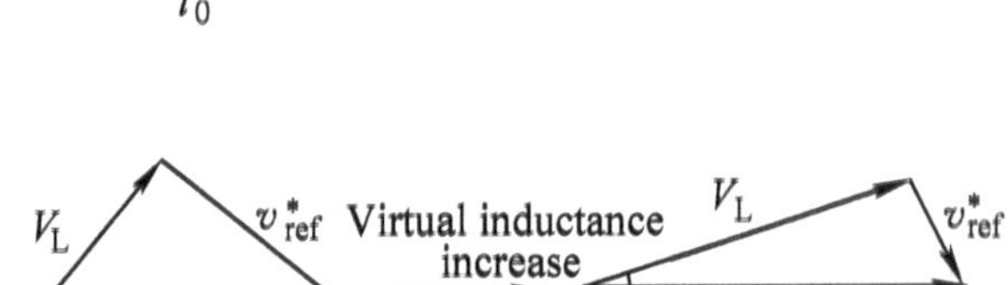

Fig. 7.38 Vector diagram of end voltage variation with virtual inductance under pure resistive load

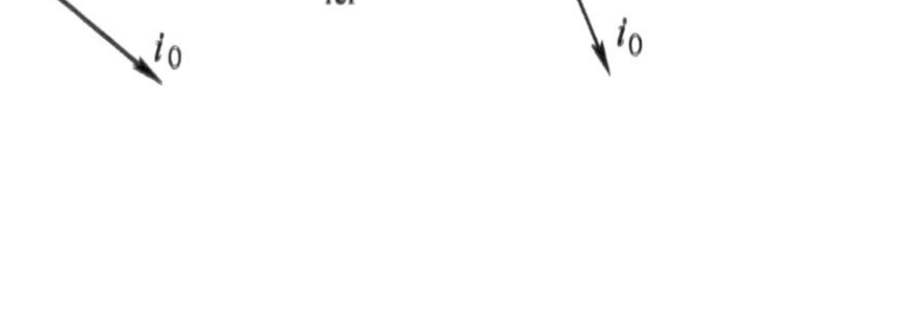

Fig. 7.39 Vector diagram of terminal voltage variation with virtual inductance under resistive-inductive load

$$U_{ref}^{*} = U_{ref} \frac{\sqrt{\left(R^2 + X^2 + XX_D\right)^2 + \left(RX_D\right)^2}}{R^2 + \left(X_D + X\right)^2} < U_{ref} \tag{7.29}$$

It can be seen that when the load is resistive-inductive, the amplitude of the load terminal voltage will inevitably be lower than the rated output voltage. Similarly, the larger the virtual reactance X_D, the lower the load terminal voltage, as shown in Fig. 7.39.

3. **Terminal voltage analysis under resistive-capacitive load**

Line current:

$$\dot{I} = \frac{\dot{U}}{R + j\left(\omega L_D - \frac{1}{\omega C}\right)} = \frac{U_{ref}\angle 0^\circ}{R + j\left(X_D - X\right)} \tag{7.30}$$

Load terminal voltage:

$$\dot{U}_{ref}^{*} = \dot{I}\cdot\left(R - jX\right) = \frac{U_{ref}\left(R - jX\right)}{R + j\left(X_D - X\right)} \tag{7.31}$$

Effective value of terminal voltage:

$$U_{ref}^{*} = U_{ref}\sqrt{\frac{\left(R^2 + X^2 - XX_D\right)^2 + \left(RX_D\right)^2}{\left[R^2 + \left(X_D - X\right)^2\right]^2}} \tag{7.32}$$

When $X_D > 2X$,

$$\frac{\left(R^2 + X^2 - XX_D\right)^2 + \left(RX_D\right)^2}{\left[R^2 + \left(X_D - X\right)^2\right]^2} < 1 \tag{7.33}$$

When $X_D = 2X$,

$$\frac{\left(R^2 + X^2 - XX_D\right)^2 + \left(RX_D\right)^2}{\left[R^2 + \left(X_D - X\right)^2\right]^2} = 1 \tag{7.34}$$

When $X_D < 2X$,

$$\frac{\left(R^2 + X^2 - XX_D\right)^2 + \left(RX_D\right)^2}{\left[R^2 + \left(X_D - X\right)^2\right]^2} > 1 \tag{7.35}$$

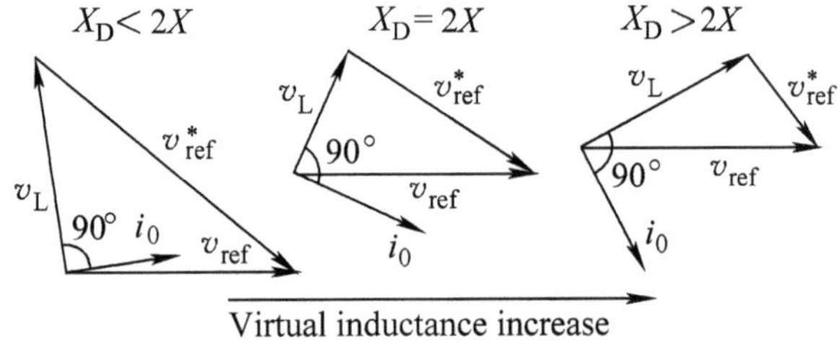

Fig. 7.40 Vector diagram of voltage changes at the resistive-capacitive load end with virtual inductance

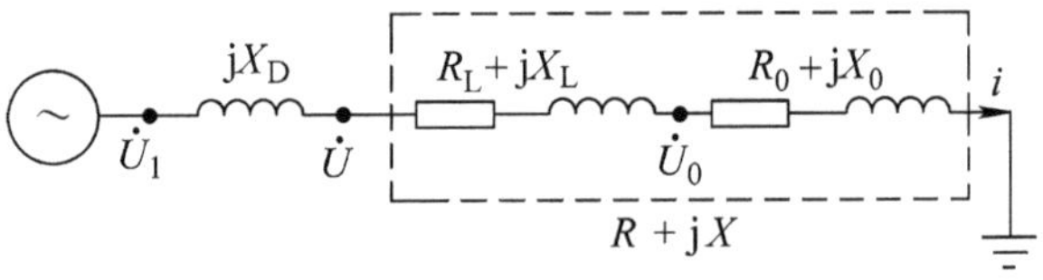

Fig. 7.41 Equivalent structure diagram of the converter with virtual inductance

It can be seen that when the load is resistive-capacitive, the situation is more complex. The voltage at the load end may decrease, increase, or remain unchanged, as shown in Fig. 7.40.

4. **Amplitude correction scheme**

To correct the voltage offset at the load end caused by the introduction of virtual inductance, this section proposes an amplitude correction scheme based on real-time estimation of the external system state of PCS. The basic idea is: first, estimate the equivalent impedance of the external system of PCS based on local voltage and current sampling R + jX, then calculate the amplitude of the PCS output voltage U_0 when the voltage at the load end U_0 equals the rated value U_N, and use U_1 as the reference value for the inner loop to track and control.

Figure 7.41 is an equivalent structure model of a single-phase system, where $\dot{U}_1$ is the PCS output voltage, U is the voltage after virtual inductance, U_0 is the voltage at the load end, jX_D is the virtual reactance, $R_L + jX_L$ is the line impedance, $R_0 + jX_0$is the equivalent load. The implementation method is as follows:

1. According to the real-time sampled output voltage u_{abc} and output current i_{abc}, with the phase voltage a u_a as the reference, the i_{abc} undergoes a $dq0$ transformation. Then, the angle θ between i_a and u_a is

$$\theta = \arctan \frac{i_q}{i_d} \tag{7.36}$$

2. Estimation of the Equivalent Impedance of the External System of PCS

$$\frac{U\angle 0}{I\angle \theta} = R + jX \tag{7.37}$$

Therefore, we can get,

$$R = \frac{U}{I}\cos\theta$$
$$X = -\frac{U}{I}\sin\theta \tag{7.38}$$

Thus, the equivalent load $R_0 + jX_0$ is

$$R_0 = R - R_L$$
$$X_0 = X - X_L \tag{7.39}$$

In the formula, R_L and X_L represent the resistance and reactance of the line, respectively.

3. Let $U_0 = U_N$, Then the Line Current Is

$$\dot{I} = \frac{U_N}{R_0 + jX_0} \tag{7.40}$$

The output voltage of the PCS is:

$$\begin{aligned}\dot{U}_1 &= \dot{I}\cdot\left[R + j\left(X + X_D\right)\right]\\ &= \frac{U_N}{R_0^2 + X_0^2}\left[\left(RR_0 + XX_0 + X_D X_0\right) + j\left(R_0 X + R_0 X_D - RX_0\right)\right]\end{aligned} \tag{7.41}$$

Its effective value is:

$$U_1 = \sqrt{A^2 + B^2} \tag{7.42}$$

Where,

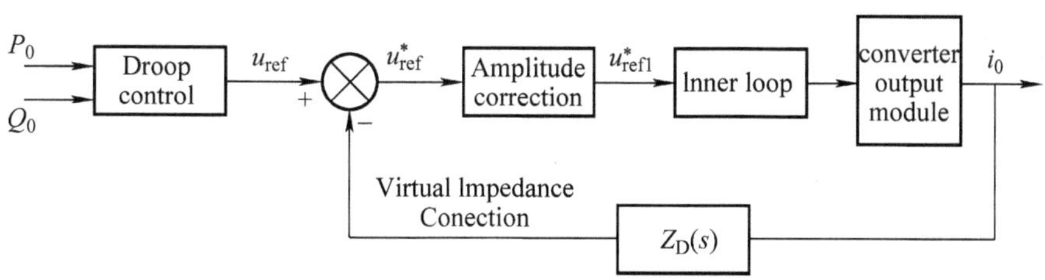

Fig. 7.42 PCS control diagram with amplitude correction

$$A = \frac{U_N}{R_0^2 + X_0^2}\left(RR_0 + XX_0 + X_D X_0\right)$$
$$B = \frac{U_N}{R_0^2 + X_0^2}\left(R_0 X + R_0 X_D - RX_0\right) \tag{7.43}$$

By taking U_1 as the amplitude reference, the reference value of the PCS output voltage can be corrected, which can compensate for the voltage drop at the load end caused by the introduction of virtual inductance. The PCS control diagram with amplitude correction added is shown in Fig. 7.42. The reference value u_{ref} obtained from droop control, first goes through virtual reactance correction to get u_{ref}^{*} and then send it to the inner control loop.

Figure 7.43 shows the simulation results of the load voltage with amplitude correction for the same system as in Fig. 7.36. It can be known that without the amplitude correction the load voltage deviation from the rated value, and the voltage fluctuates with the load power changes. When the amplitude correction is applied, the load voltage maintains at rated value and can soon recover after load power change.

To verify the effectiveness of the proposed strategy, a parallel energy storage system experimental platform consisting of two 100kVA PCS was constructed, as shown in Fig. 7.44. Set the switching frequency to 6 kHz, the filtering inductance value to 0.75 mH, and the filtering capacitance value to 60 μ F; The impedance of

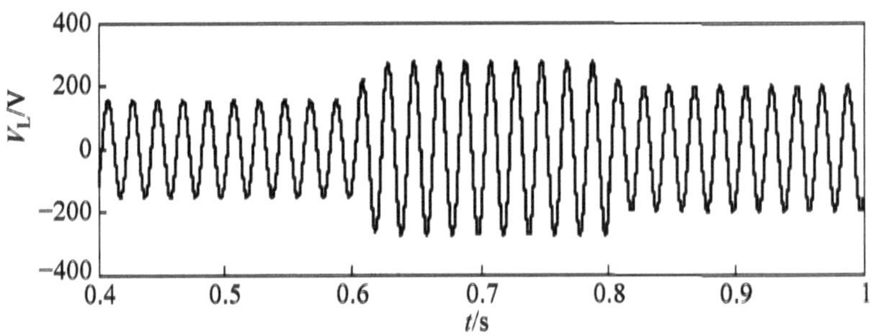

a) Without amplitude correction

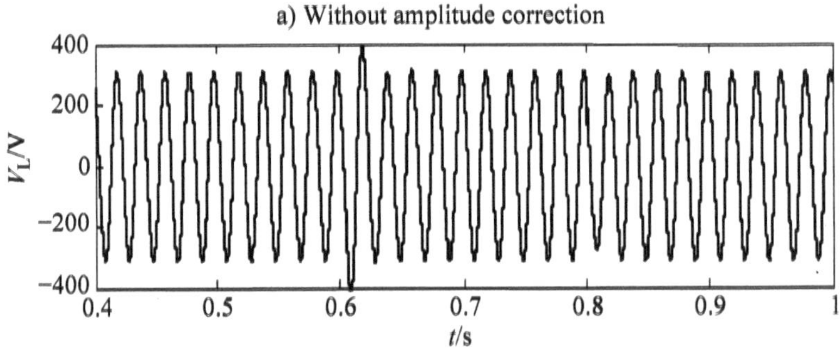

b) With amplitude correction

Fig. 7.43 Load voltage with/without amplitude correction

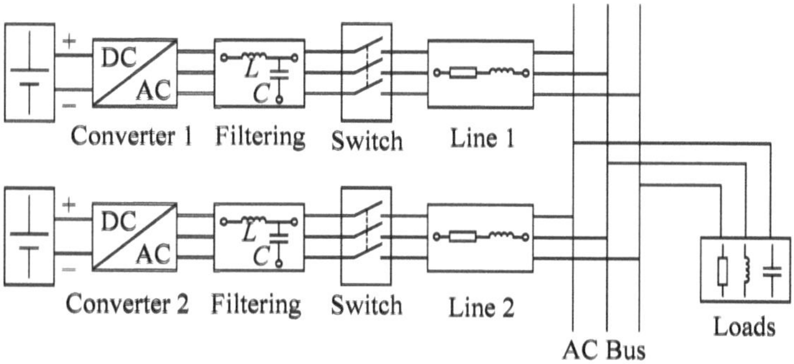

Fig. 7.44 Paralleled Energy System Experimental Platform

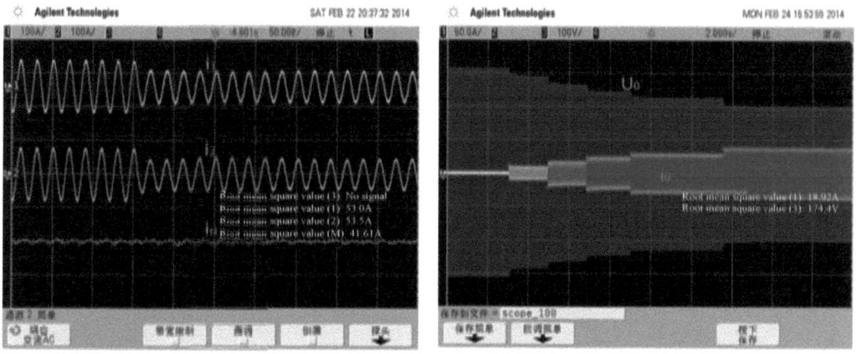

a) PCS output current i_1, i_2 and loop current i_h b) Load voltage U_0 and load current i_0

Fig. 7.45 Operating conditions when virtual inductance is introduced without amplitude correction. (**a**) Output currents i_1 and i_2 and circulating current i_H waveform. (**b**) Load voltage U_0 and load current i_0 waveform

the two power lines is 0.4 Ω/0.17 mH and 0.24 Ω/0.1 mH, respectively; The rated line-line voltage is 380 V and the rated frequency is 50 Hz.

Figure 7.45 shows the operation of the parallel system when the virtual inductance is introduced but no amplitude correction is performed. Figure 7.45a shows the output current of the two PCSs and the circulating current before and after the load suddenly decreases by 40 kW. The parallel system can transition smoothly, and the circulating current is very small. Figure 7.45b shows the waveform of the load voltage when the load is increased from 5kVar to 30kVar. Since no amplitude correction is performed, the load voltage deviates significantly from the rated value, and the deviation becomes more serious as the load increases.

Figure 7.46 shows the operating conditions of the system after amplitude correction is added, with an experimental setting of a sudden increase of 30 kVar inductive load. Comparing Fig. 7.46a with Fig. 7.45a, it can be seen that adding amplitude correction does not affect the equal current effect of the parallel system, the system transitions smoothly, and the circulating current is very small. As can be seen from Fig. 7.46b, after adding amplitude correction, the load voltage can be maintained

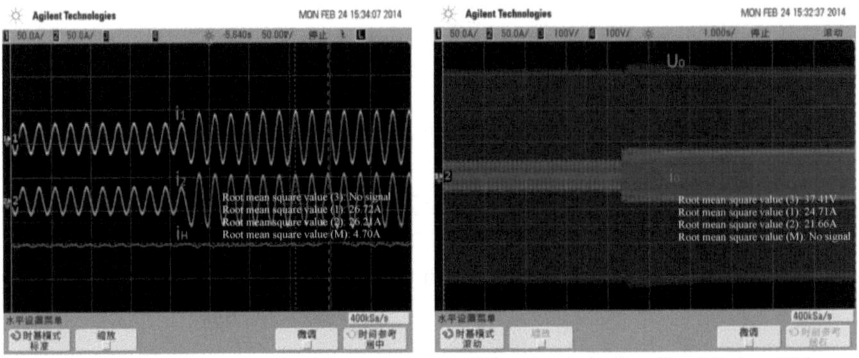

a) Two PCS output current i_1, $i_2$and circulating current i_h b) Load terminal voltage U_0 and load current i_0

Fig. 7.46 Operating conditions when virtual inductance is introduced with amplitude correction. (**a**) Two PCSs output currents i_1, i_2 and circulating current i_H waveform (**b**) Load voltage U_0 and load current i_0 waveform

near the rated value. After a sudden change in load, it returns to the rated voltage after a brief adjustment, without a large overshoot, and the voltage fluctuation is very small.

7.3.3 Master-Slave Droop Control

This section proposes a voltage source/current source master-slave droop control mode. Through the voltage/frequency droop control (Voltage and Frequency Droop Control, VFDC) in the previous section, some energy storage units are controlled as voltage-type power sources and serve as the main grid power source. Through active/reactive droop control (PQ Droop Control, PQDC) controls another part of the energy storage unit as a current-type power source, serving as an auxiliary grid-connected power source. By utilizing the control ability of the current-type power source over the power flow, it reduces the control difficulty brought by the differences in output impedance and line impedance, and shields the problem of circulating current [132].

The master-slave droop control multi-energy storage parallel system is equivalent to a voltage source/current source parallel system. Figure 7.47 describes the equivalent diagram of the master-slave parallel system composed of a voltage source and a current source.

According to circuit theory, we get:

$$\dot{I}_1 = \frac{\dot{U}_1 - \dot{I}_2 Z_0}{Z_0 + Z_1} \tag{7.44}$$

Therefore, the apparent power output of the voltage source is:

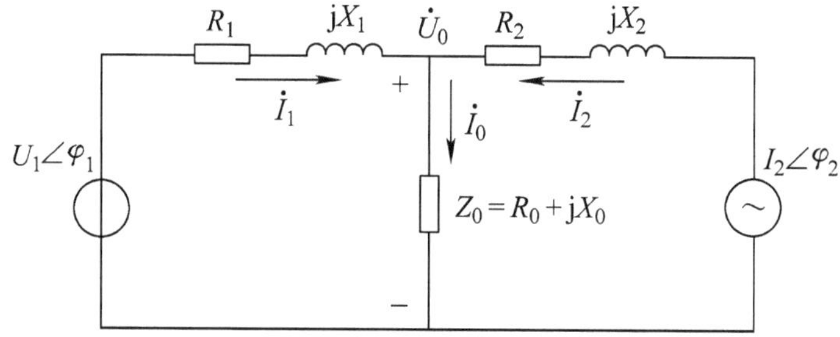

Fig. 7.47 Equivalent schematic diagram of the master-slave control system

$$\dot{S}_1 = \dot{U}_1 \dot{I}_1^* = \dot{U}_1 \left(\frac{\dot{U}_1 - \dot{I}_2 Z_0}{Z_0 + Z_1} \right)^* = \frac{U_1^2 - \dot{U}_1 \dot{I}_2^* Z_0^*}{Z_0^* + Z_1^*} \tag{7.45}$$

It can be seen that, $\dot{S}_1$ is independent of Z_2, assuming $\dot{U}_1$ and line parameters are given, then S_1 is only determined by I_2, therefore, adjusting I_2 can effectively control S_1. This also shields the difficulty brought to the power flow control in the voltage source parallel system due to $Z_1 \neq Z_2$.

Load voltage U_0:

$$\dot{U}_0 = \dot{U}_1 - \dot{I}_1 Z_1 = \frac{\dot{U}_1 Z_0 - \dot{I}_2 Z_0 Z_1}{Z_0 + Z_1} \tag{7.46}$$

This formula shows that the load voltage $\dot{U}_0$ is also independent of Z_2, and can be controlled by adjusting I_2.

From the perspective of line loss,

$$\left|\dot{I}_1\right|^2 Z_1 + \left|\dot{I}_2\right|^2 Z_2 = \left|\frac{\dot{U}_1 - \dot{I}_2 Z_0}{Z_0 + Z_1}\right|^2 Z_1 + \left|\dot{I}_2\right|^2 Z_2 \qquad 7.47$$

This formula shows that the system grid loss is also controlled.

The above analysis shows that in the master-slave droop control system, due to the strong power control ability of the current source, the system power flow is easy to control, and it can indirectly affect the output of the voltage source. Because it shields the influence of the current source branch impedance on the control, it brings convenience to the control.

7.3.3.1 VFDC Control

The voltage/frequency droop VFDC control strategy is shown in Fig. 7.48.

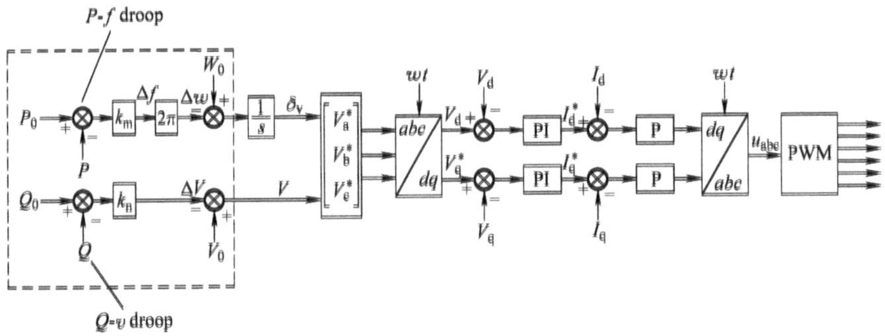

Fig. 7.48 VFDC Control Strategy Diagram

VFDC control is a voltage-type control scheme with droop characteristics. It calculates the frequency reference value f_{ref} and voltage reference value U_{ref} based on the active/reactive power output detected at the PCS port according to formulas (7.48) and (7.49). It controls the PCS output through a voltage/current dual-loop and establishes the frequency and voltage of the microgrid as a grid-connected power source.

$$f_{ref} = f_0 - k_m (P - P_0) \tag{7.48}$$

$$U_{ref} = U_0 - k_n (Q - Q_0) \tag{7.49}$$

Where, k_m is the P-f droop coefficient, k_n is the Q-V droop coefficient, P_0, Q_0, f_0, U_0 are the set active, reactive, frequency, and voltage base points respectively. By setting the droop coefficient and base points according to the capacity of the microgrid and the requirements for frequency and voltage deviation, the frequency and voltage deviation of the microgrid during operation can be controlled within a reasonable range.

7.3.3.2 PQDC Control

The active/reactive power droop PQDC control strategy is shown in Fig. 7.49.

PQDC is a current-type control with droop characteristics. Unlike VFDC, PQDC works by detecting system frequency and node voltage deviation, calculating the active and reactive reference values through formulas (7.50) and (7.51), and controlling the output power of the converter through a current loop. In terms of external characteristics, both PQDC and VFDC have droop characteristics and can share load power. In terms of working principle, the VFDC unit provides frequency and

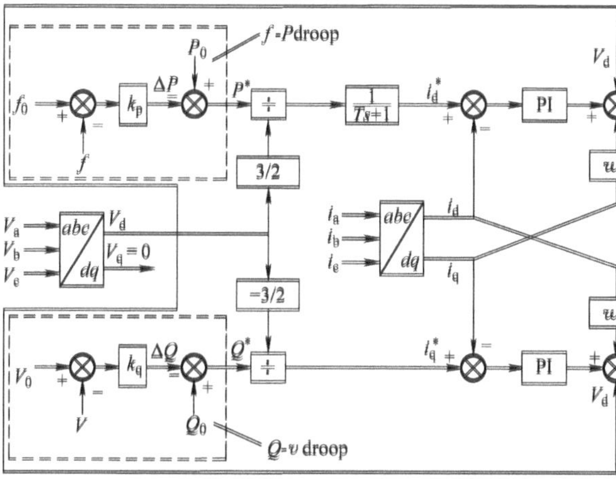

Fig. 7.49 PQDC Control Strategy Diagram

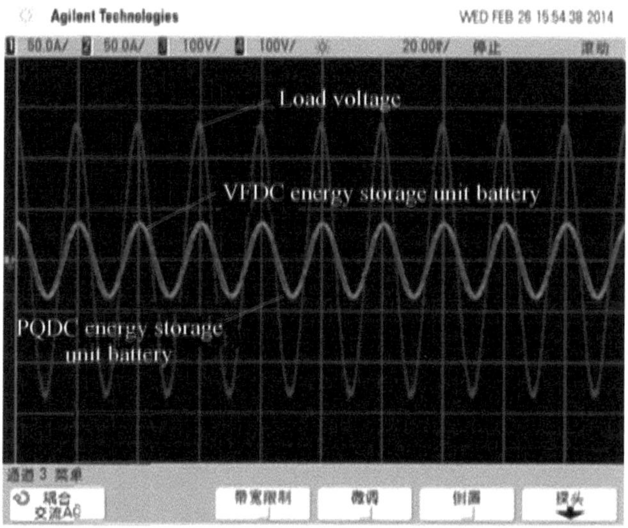

Fig. 7.50 Steady-state operation process of the parallel connected system (load 40 kW + 5kVar)

voltage support for the PQDC unit, while the PQDC unit indirectly adjusts the output of the VFDC unit by controlling its own output.

$$P_{ref} = P_0 - k_p\left(f - f_0\right) \tag{7.50}$$

$$Q_{ref} = Q_0 - k_q\left(U - U_0\right) \tag{7.51}$$

Where, k_p is the f-P droop coefficient, k_q is the V-Q droop coefficient, P_0, Q_0, f_0, U_0 are respectively set for active power, reactive power, frequency, and voltage base points.

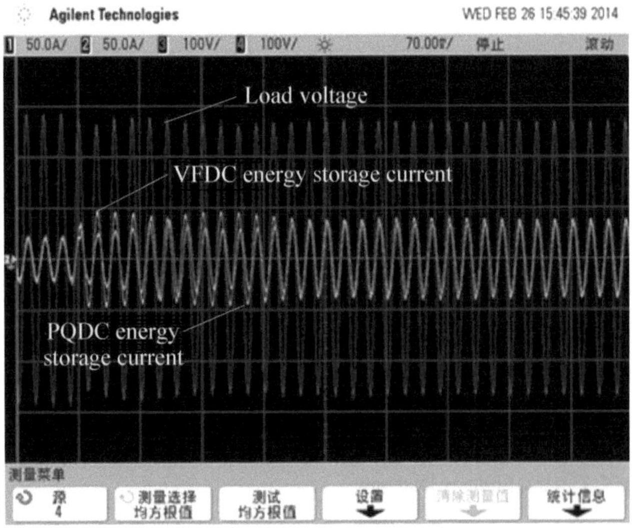

Fig. 7.51 Dynamic process of sudden load increase (20 kW + 5kVar to 40 kW + 5kVar)

7.3.3.3 Droop Coefficient and Base Point Configuration

An analysis of the output distribution of multiple energy storage systems under the coordinated control of VFDC and PQDC is conducted. The output power of the VFDC energy storage unit is set as P_1/Q_1, and the droop coefficient is k_{p1}/k_{q1} (where $k_{p1} = 1/k_m$, $k_{q1} = 1/k_n$), and the power base point is P_{10}/Q_{10}. The output power of the two PQDC energy storage units are respectively P_2/Q_2 and P_3/Q_3, the droop coefficients are k_{p2}/k_{q2} and k_{p3}/k_{q3}, and the power base points are P_{20}/Q_{20} and P_{30}/Q_{30}.

The output power of other non-dispatchable distributed power sources in the microgrid is set as P_{DG}, the total load is P_{load}, and the frequency base point is f_0. Therefore, the active power of each energy storage unit is:

$$\begin{aligned} P_1 &= P_{10} + \Delta P_1 = P_{10} + \Delta f \times k_{p1} \\ P_2 &= P_{20} + \Delta P_2 = P_{20} + \Delta f \times k_{p2} \\ P_3 &= P_{30} + \Delta P_3 = P_{30} + \Delta f \times k_{p3} \end{aligned} \tag{7.52}$$

According to the power balance of the system, we can get:

$$P_1 + P_2 + P_3 + P_{DG} = P_{load} \tag{7.53}$$

By combining the above equations, we can get:

$$
\begin{aligned}
P_1 &= P_{10} + \frac{P_{load} - P_{DG} - P_{10} - P_{20} - P_{30}}{k_{p1} + k_{p2} + k_{p3}} \times k_{p1} \\
P_2 &= P_{20} + \frac{P_{load} - P_{DG} - P_{10} - P_{20} - P_{30}}{k_{p1} + k_{p2} + k_{p3}} \times k_{p2} \\
P_3 &= P_{30} + \frac{P_{load} - P_{DG} - P_{10} - P_{20} - P_{30}}{k_{p1} + k_{p2} + k_{p3}} \times k_{p3} \\
f &= f_0 + \Delta f = f_0 + \frac{P_{load} - P_{DG} - P_{10} - P_{20} - P_{30}}{k_{p1} + k_{p2} + k_{p3}}
\end{aligned}
\tag{7.54}
$$

As can be seen, the reasonable setting of droop coefficients k_{p1}, k_{p2}, k_{p3} and active power base points P_{10}, P_{20}, P_{30}, can change the output ratio of each energy storage unit and control the system frequency deviation within the allowable range.

The reactive power output of each energy storage unit is:

$$
\begin{aligned}
Q_1 &= Q_{10} + \Delta Q_1 = Q_{10} + \Delta U_1 \times k_{q1} \\
Q_2 &= Q_{20} + \Delta Q_2 = Q_{20} + \Delta U_2 \times k_{q2} \\
Q_3 &= Q_{30} + \Delta Q_3 = Q_{30} + \Delta U_3 \times k_{q3}
\end{aligned}
\tag{7.55}
$$

This formula describes the reactive power output expression of each energy storage unit. Each energy storage unit adjusts its reactive power output by detecting the voltage deviation at its node, inversely adjusting the voltage deviation, and controlling the node voltage within the allowable range.

7.3.3.4 Experimental Verification

To verify the effectiveness of the strategy proposed in this paper, experimental verification is carried out on the experimental platform in the previous section. The platform consists of two 100 kVA PCSs connected to the system, one of which uses the VFDC control mode and the other uses the PQDC control mode. The PCS switch frequency is 6 kHz, the filter inductance value is 0.75 mH, the filter capacitance value is 60 μF; the line impedance at both ends is 0.4 Ω/0.17 mH and 0.24 Ω/0.1 mH respectively; the rated line-line voltage is 380 V, and the rated frequency is 50 Hz.

Through the change in system load power, the transient performance and dynamic performance of the adopted control strategy and method are verified. Figure 7.50 shows the steady-state operation process of the two energy storage units under resistive-inductive load. It can be seen that the load sharing of the VFDC energy storage unit and the PQDC energy storage unit is relatively good. Of course, there is also a certain imbalance, mainly caused by sampling errors and control accuracy.

Next, a sudden load change experiment is conducted, where the load suddenly increases from 20 kW + 5kVar to 40 kW + 5kVar. Figure 7.51 shows the dynamic

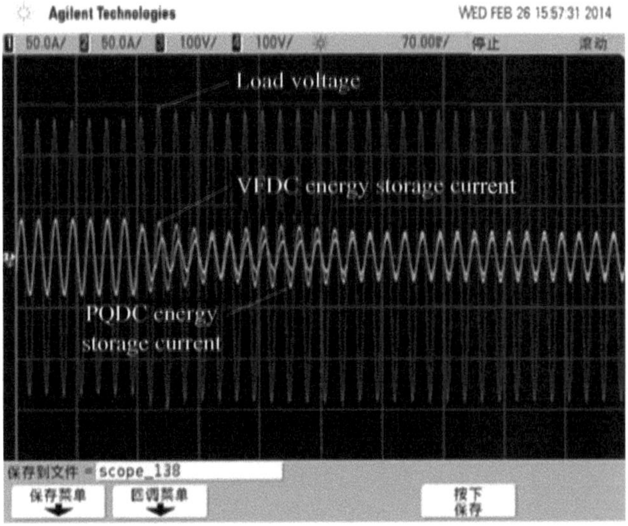

Fig. 7.52 Process of sudden load reduction (40 kW + 5kVar to 20 kW + 5kVar)

process of the output current of the two energy storage units. It can be seen that after a transient process of 200–300 ms, the VFDC energy storage unit and the PQDC energy storage unit achieved good load sharing. At the same time, because the droop control has differential regulation, as the load increases, the output voltage amplitude of the energy storage unit decreases correspondingly.

Figure 7.52 shows the process of sudden load reduction, and its dynamic process is similar to Fig. 7.51.

Compared with peer control, the operation of the master-slave control system depends on the reliability of the main power source. The auxiliary power source can share the stress of the main power source, and theoretically can be expanded to any number; while in the peer control system, each unit is equal in status, and the system will not collapse due to the failure of a unit, which is also the advantage of peer control compared to master-slave control.

Of course, in the peer control system, the electrical quantities are strongly coupled, and the output of each unit is affected by line parameters and control parameters. For complex systems with multiple energy storage units, the control is difficult. The master-slave control system is easy to implement decoupling control, shields the influence of different line impedances, and has strong control capabilities for system power flow and node voltage, making it suitable for large-capacity complex systems.

For some larger-scale practical application systems, a combination of peer control and master-slave control strategies can be used, i.e., some units are controlled by VFDC as the main power source, and some units are controlled by PQDC as the auxiliary power source. Moreover, their roles can be interchanged during operation through a certain mechanism.

Fig. 7.53 Main equipment of the intelligent microgrid demonstration system

7.4 Microgrid Application Case

The intelligent microgrid demonstration system is located in a research park. The distributed power sources include photovoltaics, direct-drive small wind power, micro gas turbine CHP, and battery energy storage systems. The part of the power load of the electrical research building is the critical load, and the lighting load is the normal load. Various distributed power generation units, energy storage, and loads are connected to the microgrid through an energy router, and intelligent electricity use transformation is carried out for some offices and conference rooms. The demonstration system fully demonstrates the concepts and effects of intelligent microgrids in renewable energy distributed power generation access, high-reliability power supply, and intelligent electricity use (Figs. 7.53 and 7.54).

A solid-state switch based on a bidirectional thyristor is used as the interconnection switch between the microgrid and the external power grid to achieve seamless switching of its grid-connected/off-grid mode; each branch integrates short-circuit and overcurrent protection, which can ensure the safety and stability of multi-route operation; it is equipped with a distributed power grid-connected interface and power monitoring equipment, forming a data communication network through multiple types of communication interfaces, uploading real-time operation data and status information, and receiving control instructions issued by the management center; it is also equipped with a maintenance circuit for system maintenance and repair.

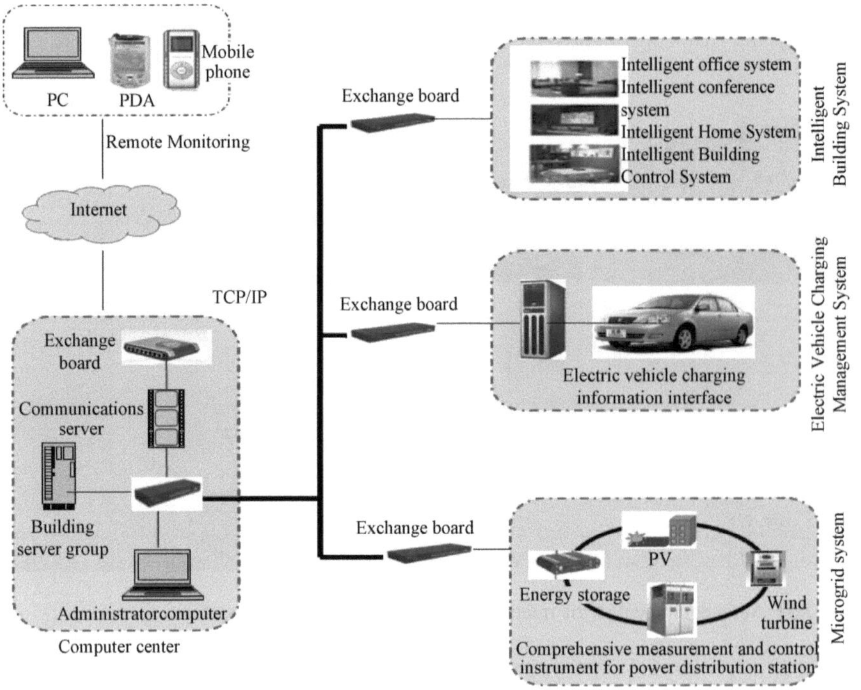

Fig. 7.54 Intelligent microgrid communication architecture diagram

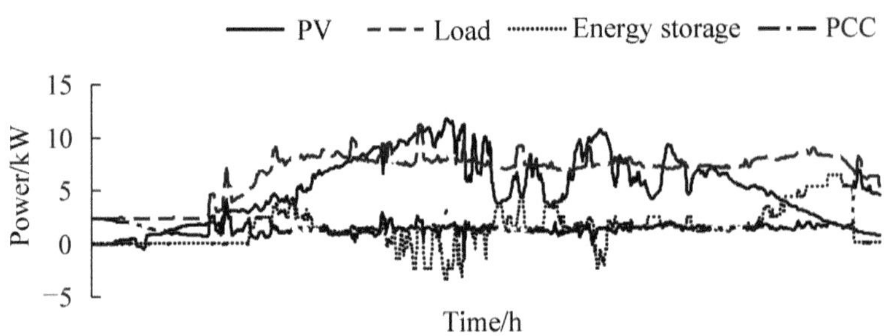

Fig. 7.55 Energy storage stabilizes the PCC exchange power of the microgrid

Based on energy storage, the operation control and management of the microgrid fully consider the factors of maximizing the use of renewable energy and the reliability of power supply to important loads. It can smooth intermittent power fluctuations, adjust load peaks and valleys, track loads, and control PCC point power flow, as shown in Figs. 7.55 and 7.56.

When the microgrid is grid-connected, the energy storage detects the status of the public grid in real time. When the public grid power supply is abnormal or fails, the solid-state switch is disconnected, and the energy storage operates in the V/f

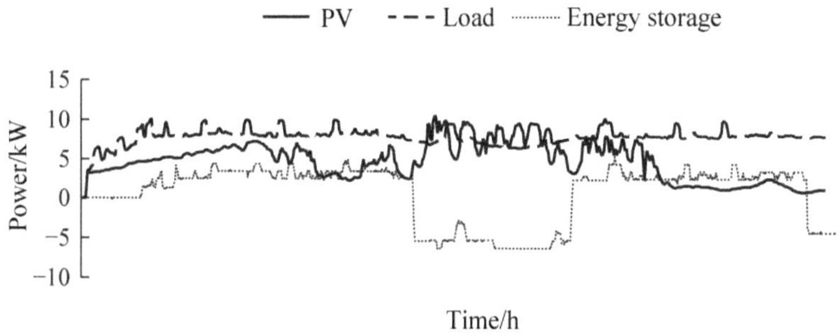

Fig. 7.56 Energy storage adjusts the peak and valley of the microgrid load

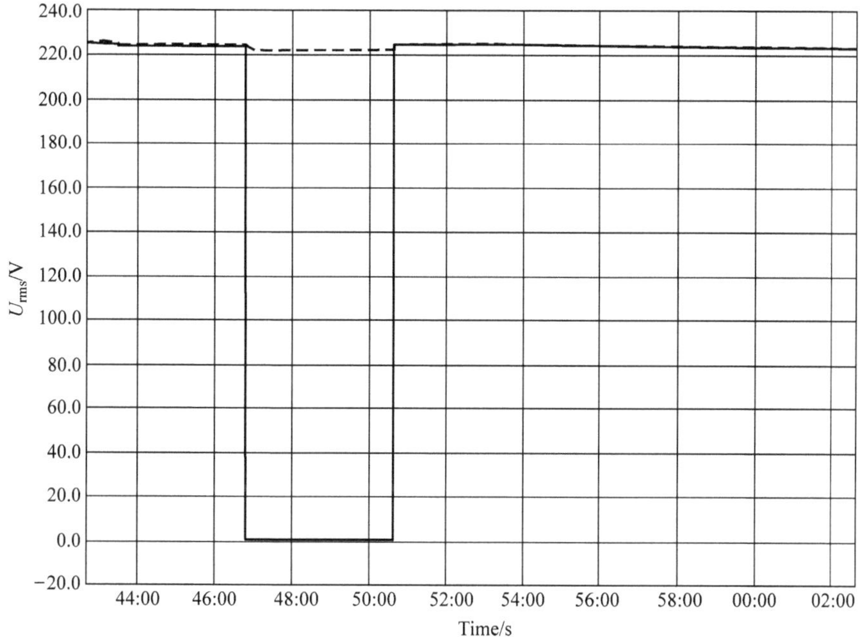

Fig. 7.57 Microgrid internal and external voltage (red curve) at PCC point during unplanned off-grid

mode to provide stable voltage and frequency support for the microgrid. When operating off-grid, if the public grid resumes power supply, the energy storage adjusts its operating status, synchronizes with the public grid after closing the solid-state switch, and the microgrid switches to the grid-connected operation mode.

During the routine maintenance of the park's power distribution system, two transformers were switched, and a power outage occurred during the period. Figure 7.57 shows the process of the microgrid switching from grid-connected operation to off-grid operation and then back to grid-connected operation. It can be seen that the internal voltage of the microgrid (red) remains stable during the two conversions.

Chapter 8
Energy Storage for Smoothing Renewable Energy Fluctuations

The fluctuation and unpredictability of renewable energy generation pose challenges to the safety, stability, and economic operation of power systems. Energy storage can effectively smooth the fluctuations of renewable energy generation and track the power generation output plan, eliminating the impact of prediction errors. This chapter mainly analyzes the impact of renewable energy generation fluctuations on the operation of power systems, and the main control methods of energy storage to smooth wind power fluctuations, with wind power as the main focus.

8.1 Characteristics and Impact Analysis of Wind Power Fluctuation

8.1.1 Analysis and Modeling of Wind Power Fluctuation Characteristics

Processing the time-series power signal of wind power, extracting the distribution and fluctuation characteristics of wind power, is an important basis for studying wind power and its grid absorption capability. It is also an important reference for considering the operation scheduling and stable control of the wind power grid.

The output power of the wind farm, after acquisition, is a series of discrete time-series signals, which can be analyzed for power fluctuation characteristics using digital signal processing methods. Generally, time-domain analysis, frequency-domain analysis, and time-frequency domain analysis can be used, as shown in Fig. 8.1.

X. Tang et al., *Electrical Energy Storage Technologies and Applications*, https://doi.org/10.1007/978-981-96-5625-7_8

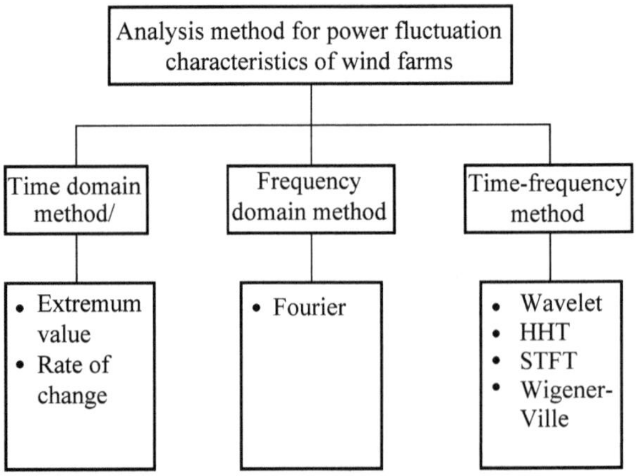

Fig. 8.1 Analysis methods of wind farm power characteristics

8.1.1.1 Time-Domain Analysis Method

The time-domain analysis of the output power of the wind farm mainly conducts statistical analysis on the distribution and change rate of the wind farm's output from a statistical perspective. The time-domain characteristic quantity is the basis for determining the maximum capacity of the wind farm, planning the transmission channel capacity when the wind farm is concentrated and sent out, improving the load rate of the transmission system, reducing investment, and improving economic benefits. The power change rate of the wind farm is used to evaluate the impact of wind power changes on different time scales on grid voltage, frequency, and safe operation, to determine the limit of wind power penetration, the selection of reserve capacity, and the arrangement of frequency regulation units.

The change in output power of the wind farm:

$$\Delta P_n = P_{n+1} - P_n \tag{8.1}$$

Where P_n represents the output power of the wind farm at the n sampling point.

The time-domain method can statistically analyze the distribution and change rate of the wind farm's output, but it is difficult to obtain quantitative characteristic quantities of the wind power output change rate.

Taking the output data of a wind farm with an installed capacity of 14 MW (14 wind turbines, single machine capacity 1 MW, wind power sampling period is 30 s) on a certain day as an example. The power time-series curve of the wind farm is shown in Fig. 8.2. It can be seen that the wind power output is large at night, and the output is small during the daytime peak load, showing obvious "counter-peak" characteristics.

Using formula (8.1) to process the wind power fluctuation, the wind power change time-series graph as shown in Fig. 8.3 and the probability distribution as shown in Fig. 8.4 are obtained.

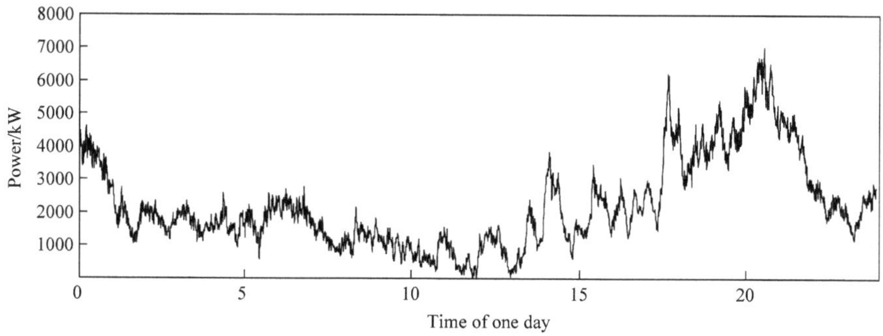

Fig. 8.2 Actual output power graph of a wind farm (24 h)

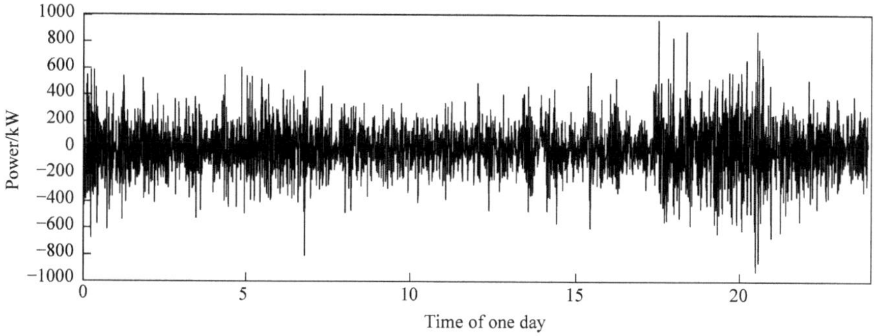

Fig. 8.3 Wind power change time-series graph (24 h)

It can be seen that most of the wind power fluctuations are distributed below 400 kW, accounting for 2.85% of the total installed capacity of 14 MW of wind power. During steady-state operation, it will not affect the safe operation of the grid; but a very small number of power fluctuations are close to 1 MW, and it is necessary to combine the grid structure at the PCC, system frequency regulation capability, and reserve capacity, etc., to analyze the impact of such changes in wind farm output on system voltage and frequency and countermeasures, including the configuration of faster start-up frequency regulation units, such as gas units and energy storage stations, etc.

8.1.1.2 Frequency-Domain Analysis Method

The frequency-domain analysis method usually uses Fourier transform, and in practice, it mostly uses the Discrete Fourier Transform (DFT) method for discrete time series $\{f_n\}$'s DFT is defined as:

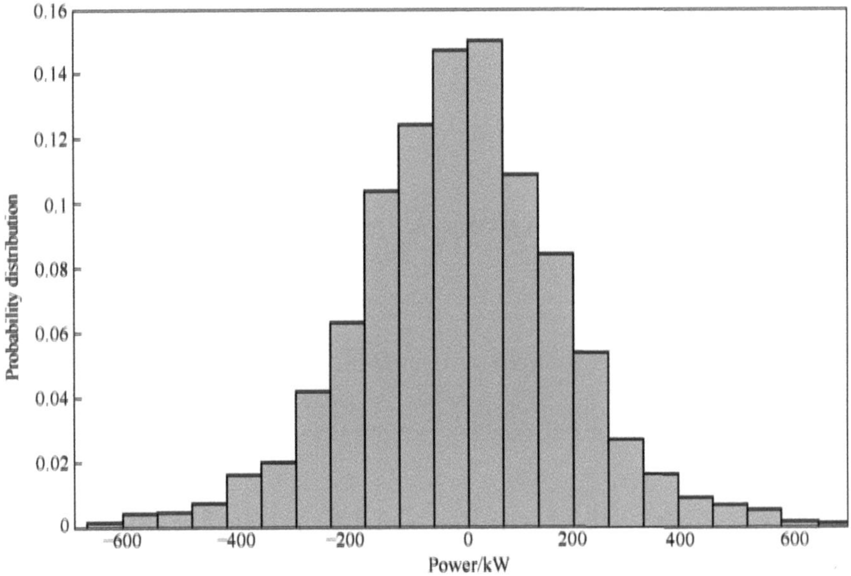

Fig. 8.4 Probability distribution of wind power changes

$$X(k) = F(f_n) = \sum_{n=0}^{N-1} f_n e^{-i\frac{2nk}{N}n} \tag{8.2}$$

where $k = 0, 1, \cdots, N - 1$

Fast Fourier Transform (FFT) is a fast processing method for DFT

$$X(k) = \sum_{j=1}^{N} x(j) e^{-\frac{i2\pi}{N}(k-1)(j-1)} \tag{8.3}$$

where $k = 0, 1, 2 \cdots, N - 1$

FFT can reflect the overall frequency domain characteristics of the entire signal at all times, and quantitatively give the power fluctuation characteristics at different frequencies. The FFT spectrum analysis of the above wind power is shown in Fig. 8.5. It can be seen that although the wind power is randomly changing, it can be decomposed into low, medium, and high different frequency bands, but most of the wind power is distributed near DC and low frequency.

As a typical frequency domain analysis method, FFT cannot perform time domain positioning, cannot combine time domain and frequency domain, and make the signal have time domain and frequency domain characteristics to form the time-frequency spectrum of the signal. At the same time, FFT requires the data to have strict periodicity and stationarity, the system has linear characteristics, and is suitable for steady-state signals. FFT, as an analysis method that focuses on the whole, lacks local time domain and is difficult to provide frequency information in local

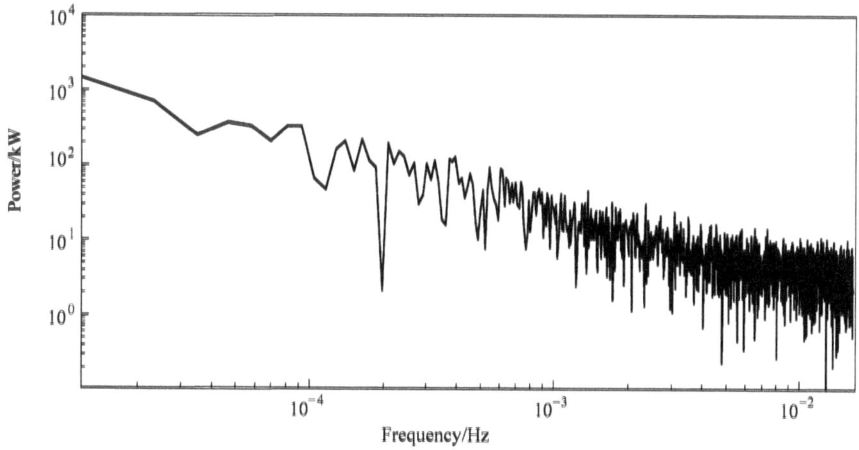

Fig. 8.5 Frequency domain distribution of wind power

time periods, so it has certain limitations in analyzing non-stationary, non-linear wind power signals.

8.1.1.3 Time-Frequency Analysis Method

The time-frequency analysis method combines the advantages of time domain and frequency domain analysis methods. Common time-frequency analysis methods include Short Time Fourier Transform (STFT), Wavelet Transform, Wingner-Ville Transform, and Hilbert-Huang Transform. This section focuses on Wavelet Transform and Hilbert-Huang Transform.

Wavelet Transform Method

Wavelet Transform (WT) is a common tool for time-frequency analysis of non-stationary, non-linear signals [133]. As a new branch of applied mathematics, it has been widely used in image processing, speech analysis, pattern recognition and other engineering fields.

Wavelet analysis compares a known basic function after translation and scaling with the signal to be analyzed, to analyze the characteristics of the signal at each moment and in various local ranges. In practical applications, Discrete Wavelet Transform (DWT) is usually used, and the wavelet mother function $\varphi(t)$ is scaled and translated to obtain a symmetric wavelet basis function with tight support in the time domain, rapid decay in the frequency domain, and a certain order of vanishing moments.

$$\varphi_{a,b}(t) = \frac{1}{\sqrt{a}} \varphi\left(\frac{t-b}{a}\right) \tag{8.4}$$

Where, a and b are called the scaling factor and translation factor, respectively, $a = 2^j$, $b = 2^j k$. That is:

$$\varphi_{j,k}(t) = 2^{-j/2} \varphi\left(2^{-j} t - k\right) \tag{8.5}$$

The discrete wavelet transform of the corresponding function x(t) can be represented as:

$$W_x(j,k) = \left(x(t), \varphi_{j,k}(t)\right) \tag{8.6}$$

DWT overcomes the disadvantage of information redundancy in Continuous Wavelet Transform (CWT); it has the property of frequency analysis and can represent the occurrence time, which is beneficial for analyzing phenomena that occur at a certain time; DWT computation time is an order of magnitude faster than FFT, making it more suitable for situations where high signal analysis processing time is required, such as real-time analysis and safety control, etc.

Wavelet transform is essentially a window-adjustable Fourier transform. The signal within the wavelet must be stationary, so it does not fundamentally escape the limitations of Fourier analysis; the priority determination of the wavelet basis can cause signal energy leakage, and the energy-frequency-time distribution of the signal is quite broad, making it difficult to perform quantitative analysis.

The selection of the mother wavelet is crucial to the reliability of wavelet analysis. Tight support in the time domain makes the wavelet have good computational performance; tight support in the frequency domain ensures strict division of the wavelet frequency domain; the use of Meyer wavelets with good local time-frequency characteristics, infinitely differentiable, and tight frequency domain support, is very smooth and has good fluctuation characteristics.

Wavelet analysis of the wind power in Fig. 8.3 is shown in Fig. 8.6. Extracting the fluctuation characteristics of wind power at different time scales, d_1, d_2, d_3 is the high-frequency component of wind power fluctuation, with a small change in amplitude, usually caused by wind farm terrain, turbulence effects, data collection and statistical errors; d_1, d_2 component fluctuation frequency distribution is in 0.01 ~ 1 Hz, its impact on the power grid needs further analysis. d_4, d_5, d_6 represents the mid-frequency component of wind power, with a larger fluctuation amplitude, which can be used as a reference for primary and secondary frequency regulation of the power grid. a_6 as the residual component of the wavelet transform, represents the trend of wind farm output power fluctuation, which can be used as a reference for wind farm power prediction and power generation plan.

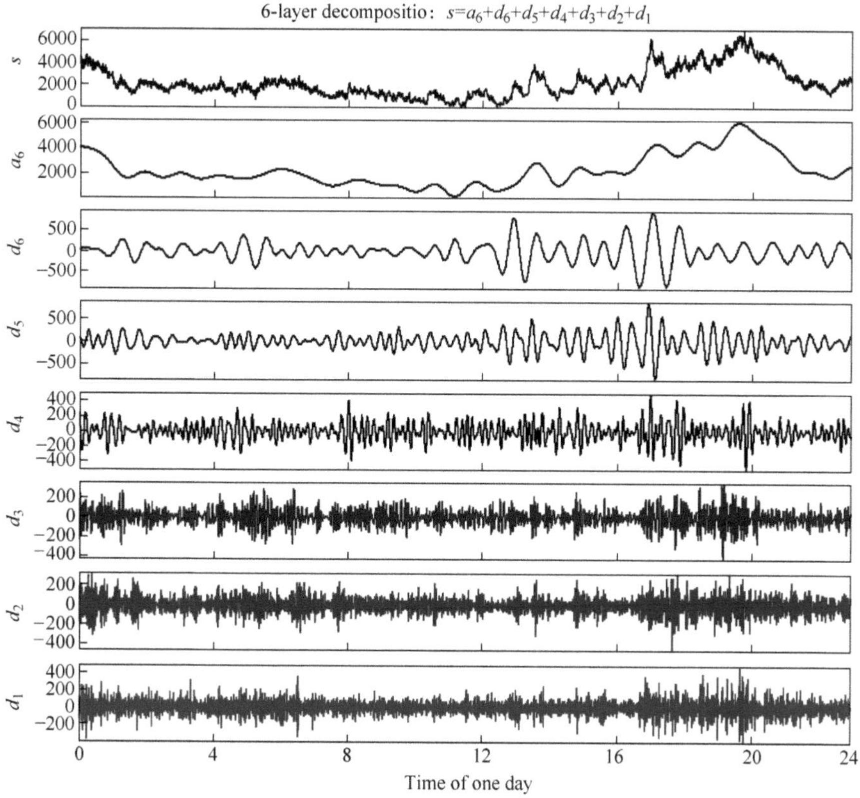

Fig. 8.6 Wavelet analysis of wind power

Hilbert Analysis Method

In 1998, Norden E. Huang, a Chinese-American at NASA, proposed the Hilbert Huang Transform (HHT) signal processing method for non-stationary, non-linear, and non-periodic signals. It differs from signal processing methods based on Fourier transform, as it does not require pre-definition of function bases and has adaptive characteristics; HHT has strong local characteristics and has unique advantages in dealing with non-stationary random signals.

The HHT method first uses Empirical Mode Decomposition (EMD) to decompose the signal into a series of Intrinsic Mode Function (IMF) components with specific physical meanings; then performs Hilbert Transform (HT) on each IMF component to obtain instantaneous frequency and instantaneous amplitude, thereby obtaining the Hilbert spectrum of the signal, which represents the complete amplitude and frequency distribution of the signal [134].

Empirical Mode Decomposition (EMD) is a process of stabilizing time series, decomposing the fluctuations or trends of different scales in the signal step by step, generating a series of data sequences with different scales, each of which is an Intrinsic Mode Function (IMF). IMF must meet the following two conditions:

1. Throughout the entire data sequence, the number of extreme points and the number of zero-crossings must be equal or differ by at most one;
2. The data sequence is locally symmetric about the time axis, that is, the local mean is zero at any time point.

According to the meaning of IMF, the algorithm flow of EMD is as follows (Fig. 8.7):

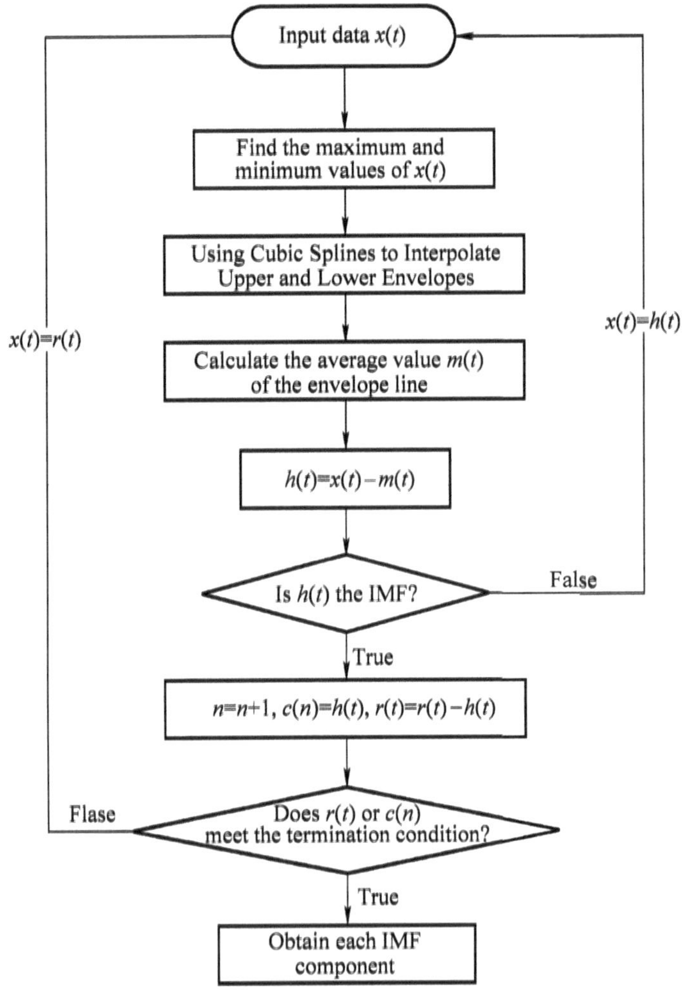

Fig. 8.7 EMD algorithm flowchart

1. Let $x(t) = r(t)$;
2. Determine whether $x(t)$ is a monotonic function or whether its amplitude difference is less than a predetermined threshold. If it is, the algorithm stops, otherwise, proceed to step (3);
3. Let $h(t) = r(t)$;
4. Determine whether $h(t)$ is an IMF. If it is, proceed to step (7);
5. Find the low-frequency trend $h(t)$ of the signal $m(t)$;
6. $h(t) = h(t) - m(t)$, go back to step (4);
7. $c(t) = h(t)$;
8. $r(t) = r(t) - c(t)$, go back to step (2)

After the above process, we finally get:

$$x(t) = \sum_{i=1}^{n} c_i(t) + r_n \tag{8.7}$$

Where $c_i(t)$ represents the i-th IMF component, which represents the signal components of different feature scales in the original signal $x(t)$; $r_n(t)$ is the residual component, reflecting the trend of the original signal $x(t)$. Therefore, EMD can decompose the signal $x(t)$ into n different frequency stationary components IMF and a trend term.

For the signal $x(t)$, its HT is:

$$y(t) = \frac{1}{\pi}\int_{\infty}^{+\infty} \frac{x(\tau)}{t-\tau} d\tau \tag{8.8}$$

Its inverse transform is:

$$x(t) = \frac{1}{\pi}\int_{\infty}^{+\infty} \frac{y(\tau)}{\tau - t} d\tau \tag{8.9}$$

The analytic signal obtained is:

$$z(t) = x(t) + iy(t) = a(t)e^{i\theta(t)} \tag{8.10}$$

Where, a(t)—Instantaneous amplitude, θ(t)—Instantaneous phase.

The instantaneous frequency is:

$$f(t) = \frac{1}{2\pi}\frac{d\theta(t)}{dt} \tag{8.11}$$

For each IMF component, perform HT to get:

$$s(t) = \mathrm{Re}\sum_{i=1}^{n} a_i(t)e^{j\phi_i(t)} = \mathrm{Re}\sum_{i=1}^{n} a_i(t)e^{j\int \omega_i(t)dt} \tag{8.12}$$

Where, Re represents the real part. This equation is called the Hilbert amplitude spectrum, also known as the Hilbert spectrum, denoted as:

$$H(\omega,t) = \mathrm{Re}\sum_{i=1}^{n} a_i(t)e^{j\int \omega_i(t)dt} \tag{8.13}$$

Further define the marginal spectrum:

$$h(\omega) = \int_{-\infty}^{+\infty} H(\omega,t)dt \tag{8.14}$$

The marginal spectrum statistically characterizes the accumulated estimation distribution at each frequency point of the entire data set.

The wind power output in Fig. 8.2 is analyzed by HHT, as shown in Fig. 8.8. The obtained r term represents the trend of wind farm output power. Comparative analysis shows that the spectral energy obtained by wavelet analysis is widely distributed in the frequency range, which is a global analysis; while most of the energy of the HHT spectrum is concentrated in a certain time and frequency range, it can clearly reflect the distribution of signal energy with time and frequency, and it has strong locality, which is more suitable for analyzing time-varying, random, and intermittent wind power.

Compared with methods such as Fourier and wavelet transforms that depend on prior function bases, HHT does not have a fixed prior basis and is more suitable for processing non-stationary random signals. It is a more adaptive time-frequency local analysis method. The amplitude of the IMF is allowed to change, breaking through the limitation of defining the simple harmonic signal with constant amplitude as the basis, making the signal analysis more flexible and changeable. Each IMF can be regarded as an inherent vibration mode in the signal. The instantaneous frequency obtained by the Hilbert transform has a clear physical meaning, can express the local characteristics of the signal, and can accurately draw a time-frequency diagram. The instantaneous frequency is the derivative of the phase function, and the local frequency can be defined without the entire wave, so it can distinguish singular signals from low-frequency signals.

8.1.2 The Impact of Wind Power Fluctuations on System Frequency and Voltage

When the proportion of wind power integration reaches a high level, its output changes will affect the regional power grid and even the entire system, including voltage, frequency, and area control error, etc. This section is based on the Western United States WSCC The 3-machine 9-node power system and actual wind power data were used to simulate and analyze the impact of wind power fluctuations on the frequency and voltage of the power system at different time scales.

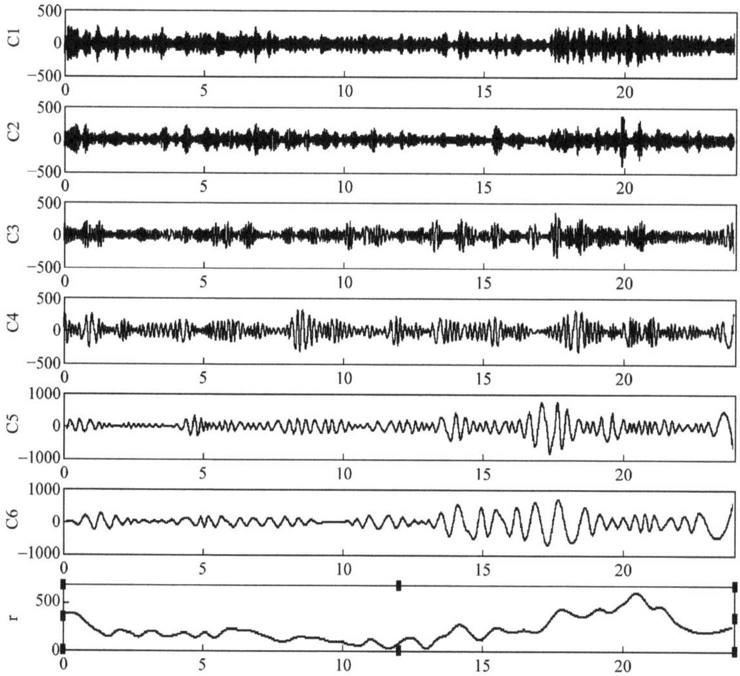

a) EMD results

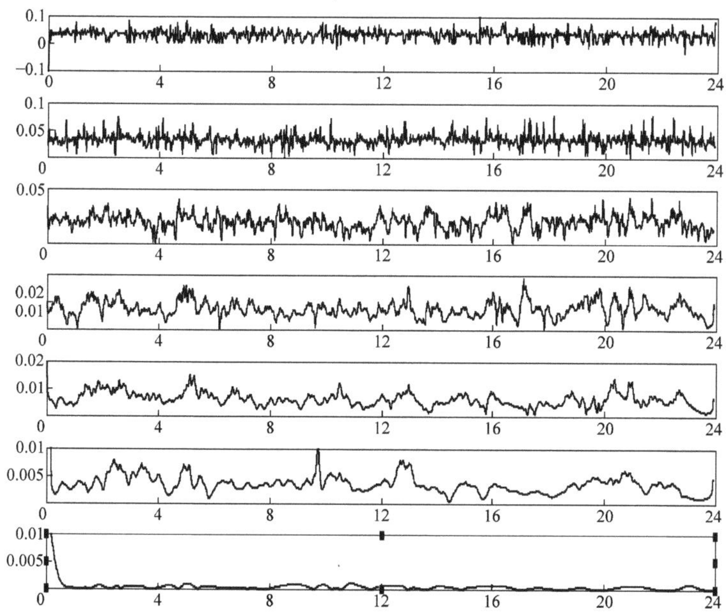

b) Instantaneous frequency of each IMF component

Fig. 8.8 HHT analysis of wind power. (**a**) EMD decomposition results. (**b**) Instantaneous frequency of each IMF component

In the WSCC 3-machine 9-node power system, a wind farm with an installed capacity of 100 MW is connected at Bus5, with a power penetration rate of 20%, as shown in Fig. 8.9.

Specific parameters of the WSCC 3-machine 9-node power system: The base value is $S_B = 100$ MVA, $U_B = 230$ kV, system frequency is 60 Hz. Generator G1 is set as the system's balance node Slack, with voltage and amplitude set to 1.04 pu, reference phase angle is 0°; G2 and G3 are set as PV nodes, with active power output set to 1.63 pu and 0.85 pu respectively, voltage and amplitude are both set to 1.025 pu.

8.1.2.1 Generator Parameters

G1: 247.5MVA, 16.5 kV, power factor is 1, Salient-Pole turbine, 180 rpm, $x_d = 0.146$, $x'_d = 0.0608$, $x_q = 0.0969$, $x'_q = 0.0969$, $x_l = 0.0336$, $T'_{d0} = 8.96$ s, $T'_{q0} = 0$ s, $H = 23.64$ s, $D = 0$

G2: 192MVA, 18 kV, power factor is 0.85, Round-Rotor turbine, 3600 rpm, $x_d = 0.8958$, $x'_d = 0.1198$, $x_q = 0.8645$, $x'_q = 0.1969$, $x_l = 0.0521$, $T'_{d0} = 6$ s, $T'_{q0} = 0.535$ s, $H = 6.4$ s, $D = 0$

G3: 128MVA, 13.8 kV, power factor is 0.85, Round-Rotor turbine, 3600 rpm, $x_d = 1.3125$, $x'_d = 0.1813$, $x_q = 1.2578$, $x'_q = 0.25$, $x_l = 0.0742$, $T'_{d0} = 5.89$ s, $T'_{q0} = 0.6$ s, $H = 3.01$ s, $D = 0$

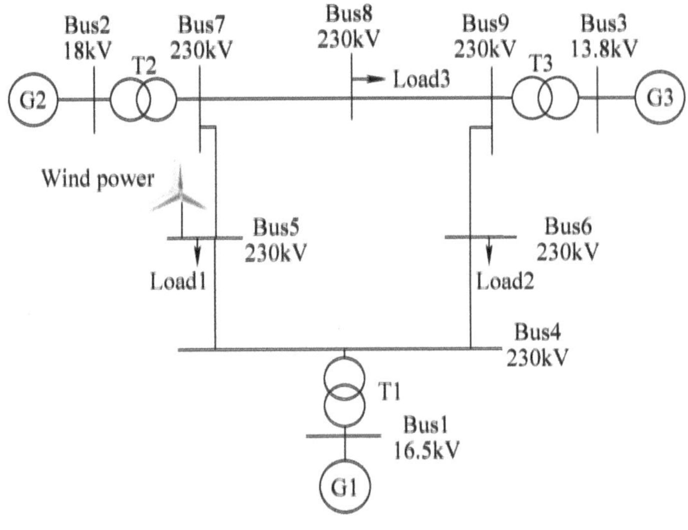

Fig. 8.9 Simulation model of power system with wind farm

8.1.2.2 Transformer Parameters

T1: 16.5/230 kV, $X_T = 0.0576$;
T2: 18/230 kV, $X_T = 0.0625$;
T3: 13.8/230 kV, $X_T = 0.0586$

8.1.2.3 Line Parameters

Line1: Z = 0.01 + j0.085, B/2 = j0.088;
Line2: Z = 0.032 + j0.161, B/2 = j0.153;
Line3: Z = 0.017 + j0.092, B/2 = j0.079;
Line4: Z = 0.039 + j0.17, B/2 = j0.179;
Line5: Z = 0.0085 + j0.072, B/2 = j0.0745;
Line6: Z = 0.0119 + j0.1008, B/2 = j0.1045

8.1.2.4 Load

LumpA (load1): 125 + j50MVA; LumpB (load2): 90 + j30MVA; LumpC (load3): 100 + j35MVA.

Figure 8.10 is a time-domain graph of the normalized active power output of the wind farm.

The signal is decomposed by EMD, the results are shown in Fig. 8.11, $c_1 - c_9$ shows the decomposed IMF components, r is the residue.

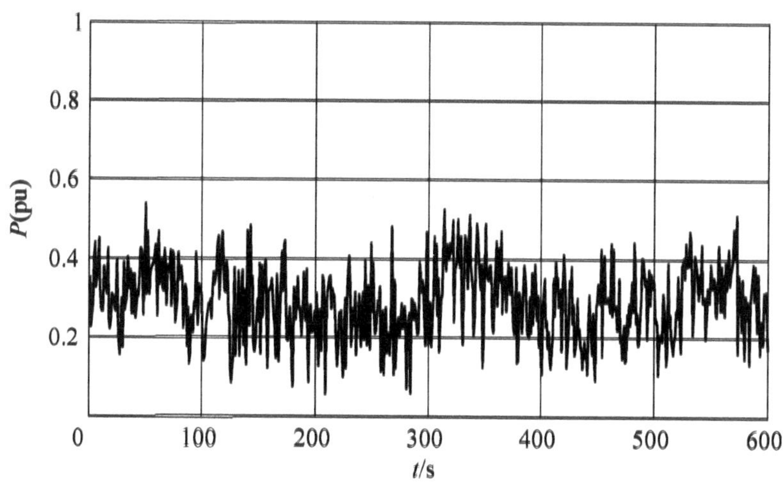

Fig. 8.10 Normalized active power output time-domain graph of the wind farm

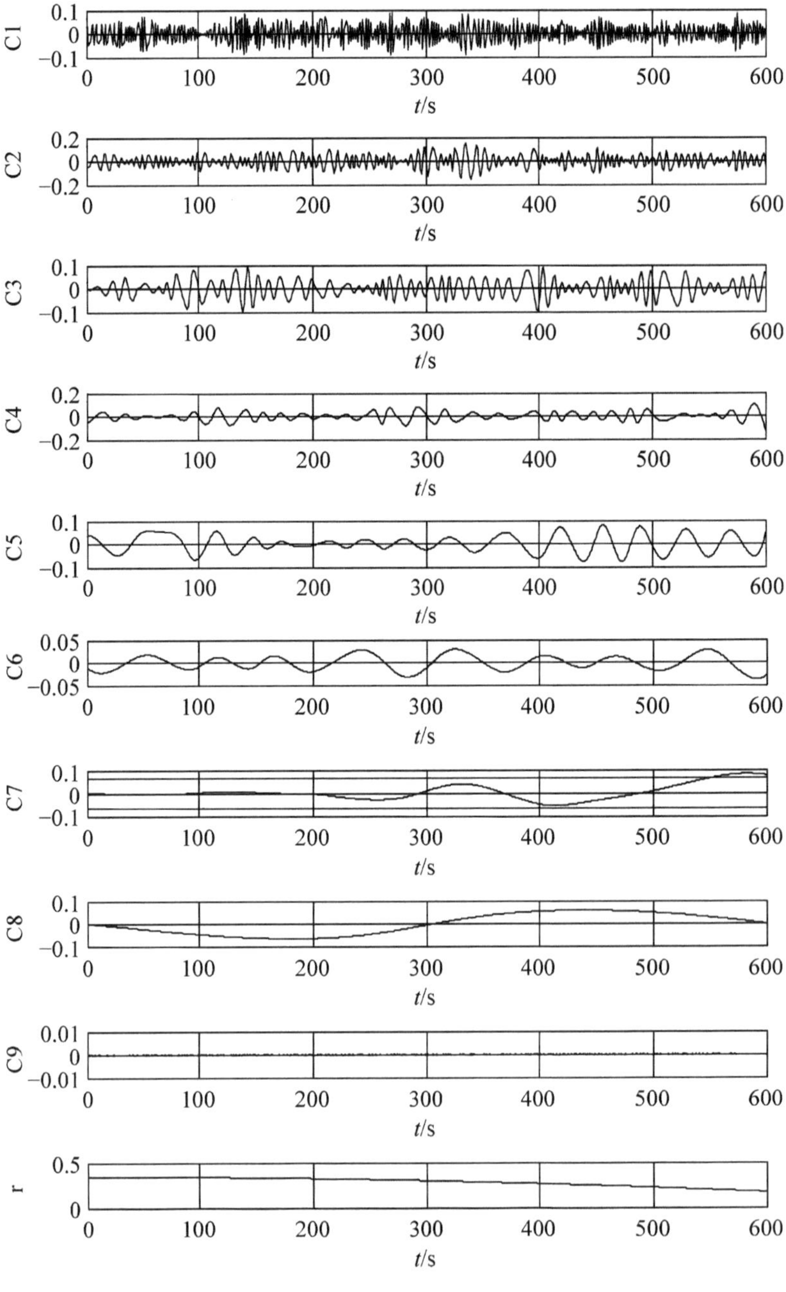

Fig. 8.11 Each IMF component after wind power EMD decomposition

The instantaneous frequencies of each IMF component are calculated, as shown in Fig. 8.12.

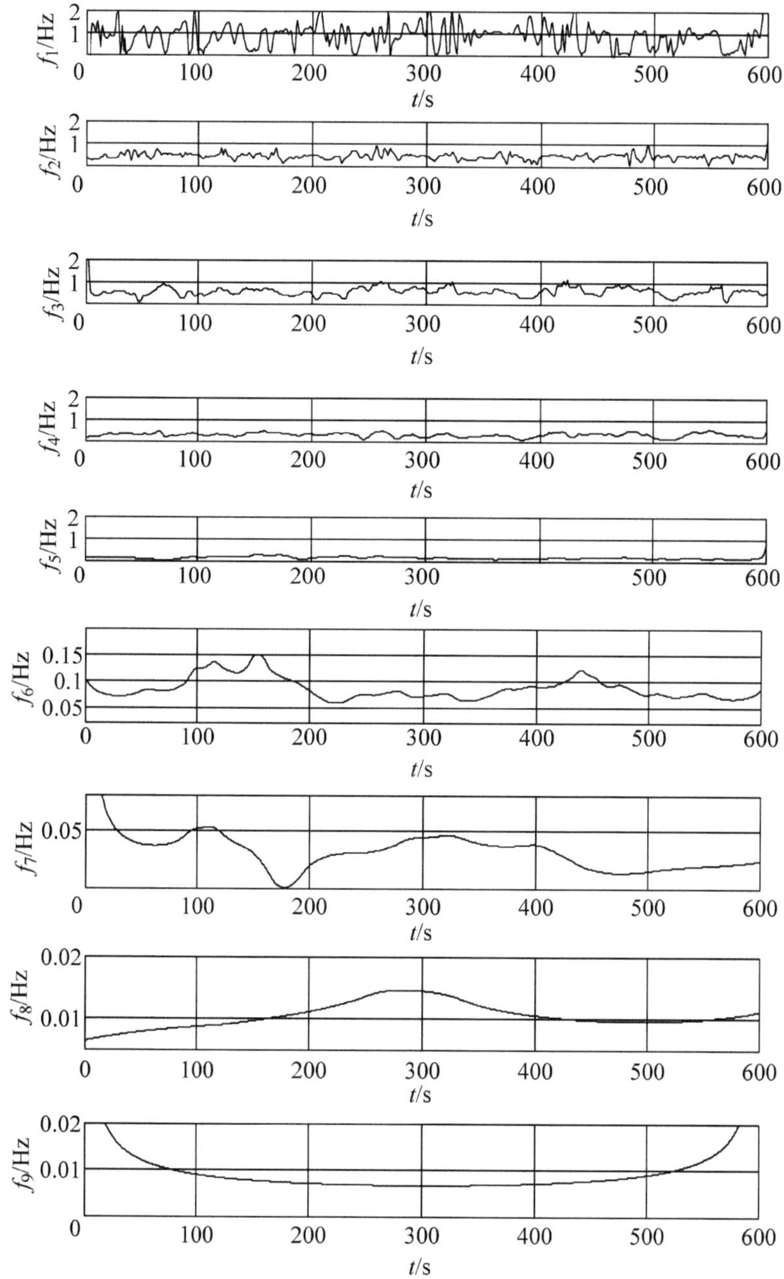

Fig. 8.12 Instantaneous frequencies of each IMF component

It can be seen that the IMF components obtained by EMD decomposition are strictly arranged in order of local frequency from high to low. Then, the concept of signal energy is introduced to analyze each IMF component. In practical applications, non-periodic signals are generally considered to be energy-limited signals, for non-periodic signals $X(t)$, its energy characteristics can be represented by the square of its modulus $|X(t)^2|$, known as energy density or instantaneous power, which represents the energy or intensity of the signal per unit time at moment t. Correspondingly, the total energy of the signal within Δt time is defined as:

$$E_X = \int \left|X(t)\right|^2 dt \tag{8.15}$$

For discrete signals, the total energy is:

$$E_X = \sum_{i=1}^{n} \left|X(n)\right|^2 \tag{8.16}$$

Considering the energy differences of each IMF component obtained from EMD decomposition, the energy of the original signal and each IMF component is taken as the feature quantity to analyze the relationship between different IMF components and the original signal. The energy of each IMF component is defined as $E_{IMF(i)}$, so the proportion of the energy of each IMF component in the total energy of the original signal is:

$$prop = \frac{E_{IMF(i)}}{E_X} \times 100\% \tag{8.17}$$

According to the above definition, the energy and ratio prop of each IMF component are shown in Table 8.1.

As can be seen, the energy values of each IMF component vary greatly, and their proportions in the total energy are also different. According to the instantaneous

Table 8.1 Energy and proportion of each IMF component

Item	Energy	*Prop*
Original data	56.4181	1.0000
IMF1	2.0172	0.0358
IMF2	0.8776	0.0156
IMF3	0.4535	0.0080
IMF4	0.3914	0.0069
IMF5	0.4547	0.0081
IMF6	0.1608	0.0029
IMF7	1.4443	0.0256
IMF8	1.1761	0.0208
IMF9	0.0051	0.0001
Res	50.8260	0.9009

frequency of each IMF component, they can be divided into high, medium, and low three frequency bands. f_1 is between 1 Hz–2 Hz, which is the high frequency band; f_2-f_7 is between 0.02–1 Hz, which is the medium frequency band; f_8, f_9 is below 0.02 Hz, which is the low frequency band.

The waveforms and energy ratios of each part are shown in Figs. 8.13 and 8.14 respectively. The fluctuation of wind power is mainly concentrated in the medium frequency band of 0.02-1 Hz, and the high frequency component (greater than 1 Hz) and low frequency component (less than 0.02 Hz) are very small. In addition, the proportion of res is the largest, accounting for about 90% of the total energy, representing the main part of wind power. Replacing different wind farm data for the above analysis, the results are basically similar, but the scale of the wind farm, wind resource conditions, etc. will have different impacts.

The power of high, medium, and low different frequency bands is input into the power system, and the system's frequency and voltage responses are shown in Figs. 8.15, 8.16, and 8.17.

As can be seen, the wind power fluctuations of different frequency bands have different effects on the system's frequency and voltage. For system frequency, the low-frequency wind power fluctuation has a small impact on the system frequency, and the system frequency fluctuation range is within 0.02 Hz; the high-frequency wind power fluctuation has a greater impact on the system frequency, with a frequency fluctuation range of about 0.05 Hz; the medium-frequency wind power fluctuation causes the system frequency fluctuation range to expand to 0.2 Hz, which has a significant impact on the system frequency. For system voltage, the wind power fluctuations of low and high frequencies have almost no effect on the system voltage, while the medium-frequency wind power fluctuation causes the system voltage to change by a maximum of about 20%.

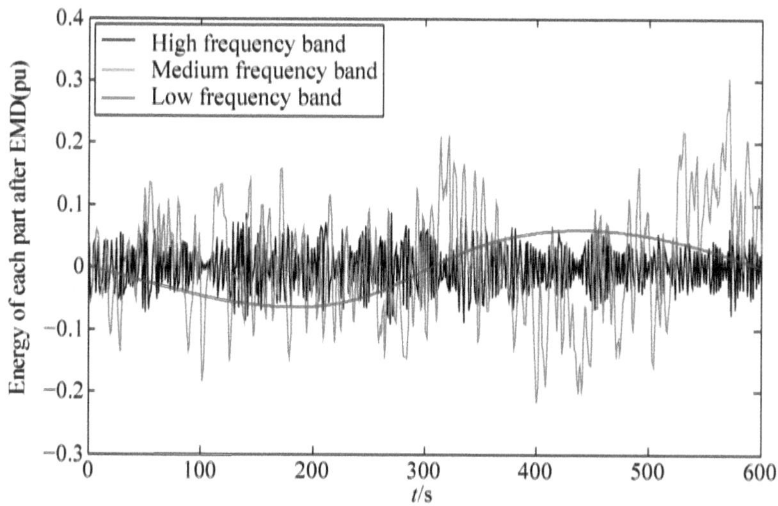

Fig. 8.13 Waveforms of each component in high, medium, and low frequency bands

Fig. 8.14 Pie chart of the energy proportion of each IMF component

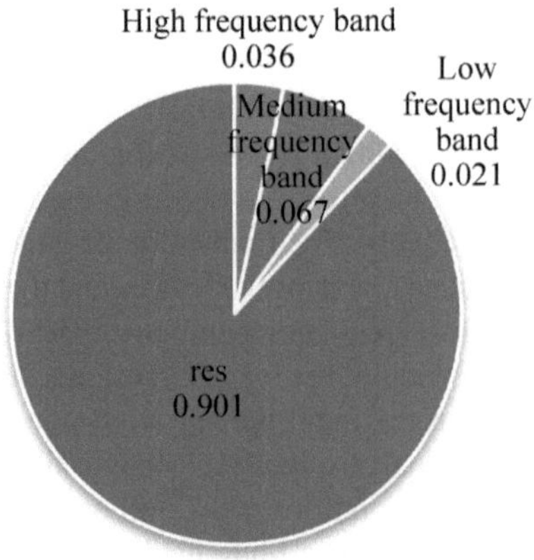

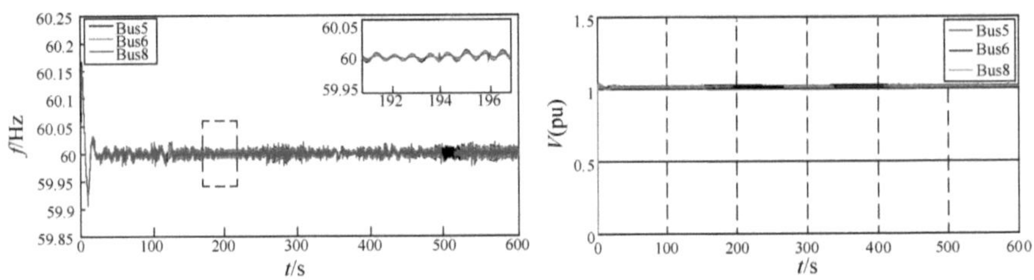

Fig. 8.15 System frequency and voltage response after the high-frequency fluctuation component is injected

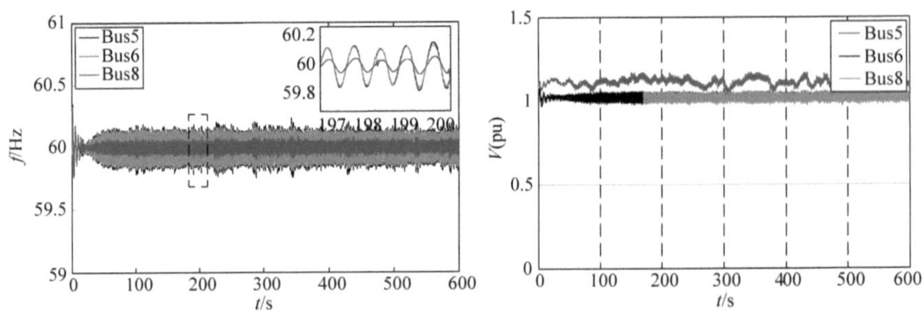

Fig. 8.16 System frequency and voltage response after the medium-frequency fluctuation component is injected

In summary, the wind power fluctuations in the medium frequency band have a more significant impact on the system frequency and voltage. Strengthening the suppression of power fluctuations in the medium frequency band of wind power can reduce the impact of wind power access on system frequency and voltage.

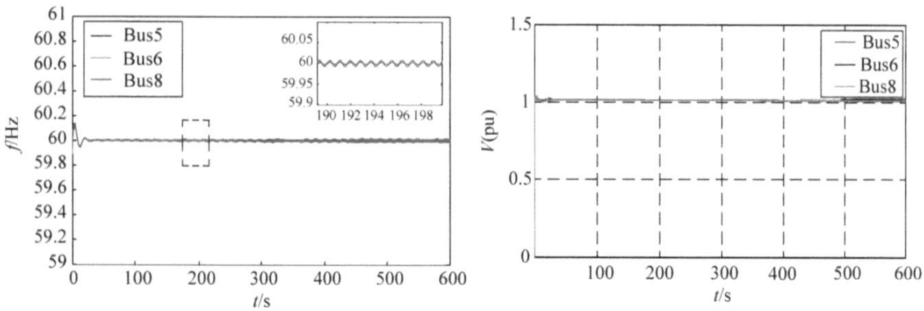

Fig. 8.17 System frequency and voltage response after the low-frequency fluctuation component is injected

8.1.3 The Impact of Wind Power Fluctuations on the System's Low-Frequency Oscillations

8.1.3.1 Low-Frequency Oscillation Analysis of the Power System Based on HHT

In the low-frequency oscillation of the power system, a certain oscillation modal component can be represented as:

$$x(t) = Ae^{-\lambda t}\cos(\omega t + \theta) \tag{8.18}$$

Where, A is the initial amplitude, λ is the attenuation coefficient, ω is the oscillation frequency, Θ is the initial phase.

The instantaneous amplitude after its Hilbert transform is:

$$a(t) = \sqrt{x(t)^2 + x_H(t)^2} = Ae^{\lambda t} \tag{8.19}$$

Furthermore,

$$\ln a(t) = \lambda t + \ln A \tag{8.20}$$

It can be seen that, lna(t) is a linear function of time. In order to accurately obtain the attenuation coefficient λ, a least squares fitting of the linear function is performed on lna(t). The fitted slope is λ, and the fitted intercept is lnλ.

In addition, the instantaneous phase is:

$$\varphi(t) = \omega t + \theta \tag{8.21}$$

In order to accurately obtain the angular frequency and initial phase, a least squares fitting of the phase angle is performed. The fitted slope is the angular frequency ω, and the fitted intercept is the initial phase Θ.

Moreover, according to the oscillation characteristics of the damped oscillation link:

$$x(t) = Ae^{-\xi\omega_0 t}\cos\left(\omega_0\sqrt{1-\xi^2}t+\theta\right) \tag{8.22}$$

Combined with formula (8.18), we can get,

$$\begin{cases} \xi\omega_0 = \lambda \\ \omega_0\sqrt{1-\xi^2} = \omega \end{cases} \tag{8.23}$$

The damping ratio can be obtained:

$$\xi = \frac{\lambda}{\sqrt{\omega^2+\lambda^2}} \tag{8.24}$$

Thus, the process of extracting low-frequency oscillation modes of the power system based on HHT can be obtained:

1. Perform EMD decomposition on the measured active power of wind power to obtain power fluctuation data in different frequency bands;
2. Connect the decomposed different wind power modes to the simulation system, and adjust the output of the system's power generation to balance the overall system;
3. Obtain the active power curve on the line near the wind power access point in the system;
4. Decompose the active power curve using EMD, obtain the IMF components of different frequency bands, perform HHT transform on them respectively, and obtain the instantaneous amplitude and instantaneous phase of each modal component;
5. Perform least squares fitting on the instantaneous amplitude and instantaneous frequency of each mode, obtain the amplitude, frequency and damping ratio of each mode, and then determine whether the system has undergone low-frequency oscillation.

8.1.3.2 Case Analysis

This section is based on the WSCC 3-machine 9-node power system, and builds a real-time simulation platform for the power system with wind power in the RT-LAB environment, analyzing the impact of wind power access on the system's low-frequency oscillation, as shown in Figs. 8.18 and 8.19.

The WSCC system includes 3 generator models, all of which use a three-stage model, covering excitation, speed regulation and other functions, and can achieve primary frequency regulation. The wind power model outputs equivalent output according to actual wind power data, and its access point is generator node 2.

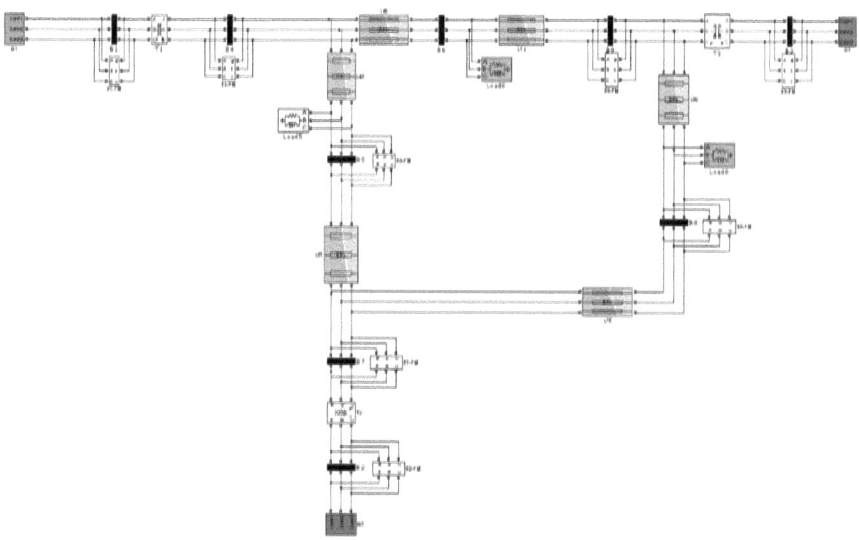

Fig. 8.18 WSCC 3-machine 9-node power system simulation model

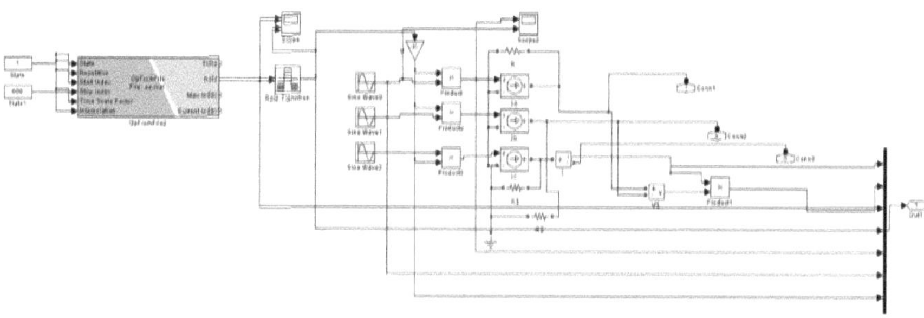

Fig. 8.19 Wind turbine simulation model

The actual wind power active power is decomposed into corresponding IMF components by EMD, which is used as the power output of the wind power model, and then the active power curve of line 5–7, which is closer to the access point and has a larger power flow, is measured. This power curve is further decomposed by EMD and HHT transformation to obtain the instantaneous amplitude and instantaneous frequency of each modal component. Through fitting calculation, the corresponding damping ratio is obtained, and then it is analyzed whether the system has undergone low-frequency oscillation. Based on the simulation platform, the system's response before and after wind power access is compared and analyzed, as well as the impact of high, medium, and low fluctuation frequency bands of wind power on the system's low-frequency oscillation.

Before the wind power is connected, the active power curve on system line 5–7 is measured, as shown in Fig. 8.20. This curve is approximately a straight line with a power value of 76.3 MW, and there is no power fluctuation and system oscillation phenomenon.

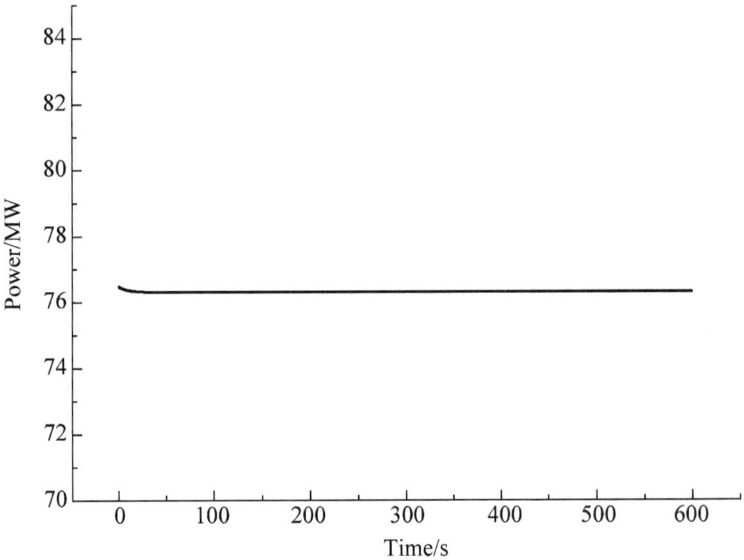

Fig. 8.20 Active power curve of line 5–7 when wind power is not connected

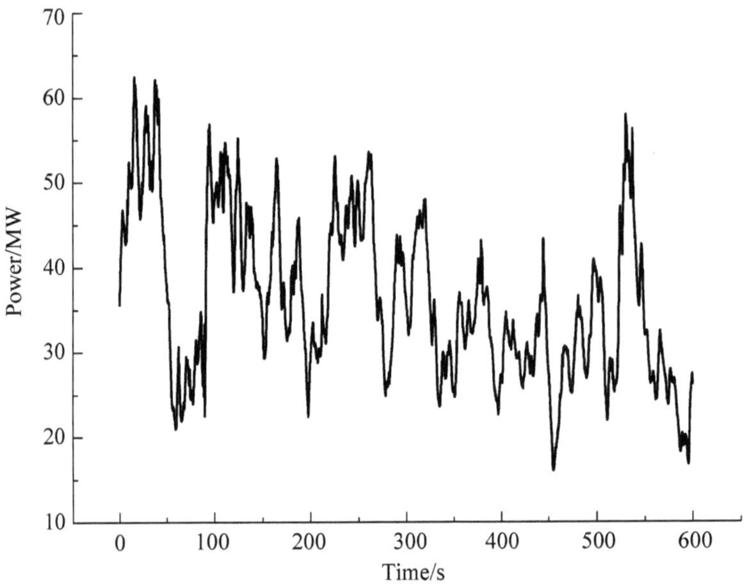

Fig. 8.21 Power curve of a wind farm

The actual operation data of a wind farm is connected to generator node 2, the capacity of the wind farm is 100 MW, the recording time span is 600 s, and the sampling time is 0.1 s, as shown in Fig. 8.21.

The EMD decomposes the wind power fluctuation into high, medium, and low frequency bands, the high frequency is greater than 1 Hz, the medium frequency is 0.03-1 Hz, and the low frequencyis less than 0.01 Hz. The decomposed wind power includes high frequency h, medium frequency m, and low frequency l IMF components and a residual component r, as shown in Fig. 8.22.

Perform HHT transformation on the high, medium, and low frequency bands of wind power to obtain the instantaneous amplitude and instantaneous frequency of the oscillation mode of each frequency band. The fitted amplitude and frequency are shown in Table 8.2. It can also be seen from this that the amplitude of the oscillation mode in the medium frequency band is greater than that in the high and low frequency bands.

The fluctuating components of wind power in the high, medium, and low frequency bands are superimposed with the residual components, resulting in three sets of wind power curves as shown in Fig. 8.23.

These three sets of wind power are connected to system node 2, and the active power curve of line 5–7 is obtained through simulation, as shown in Figs. 8.24, 8.25, and 8.26. The impact of fluctuations in different frequency bands on the system's low-frequency oscillations is analyzed.

After the obtained active power curves are decomposed by EMD and transformed by HHT, the corresponding damping ratios are obtained, as shown in Table 8. It can be seen that the damping ratio of the IMF component corresponding to the high frequency band is negative, the damping ratio corresponding to the medium frequency band is both positive and negative, and the damping ratio

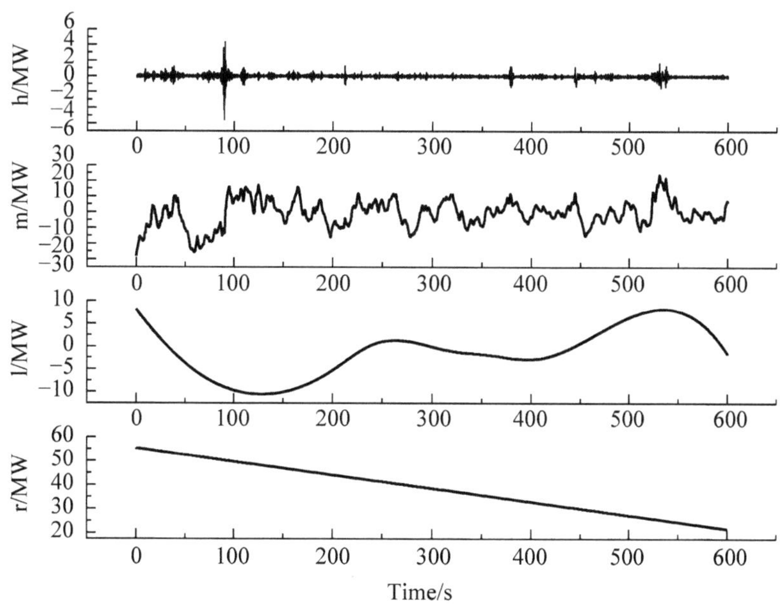

Fig. 8.22 Power components after EMD decomposition

Table 8.2 Amplitude and frequency of oscillation modes in high, medium, and low frequency bands

	High frequency band	Medium frequency band	Low frequency band
Amplitude A(MW)	0.1973	11.1441	7.9013
Frequency f(Hz)	1.0879	0.1324	0.0025

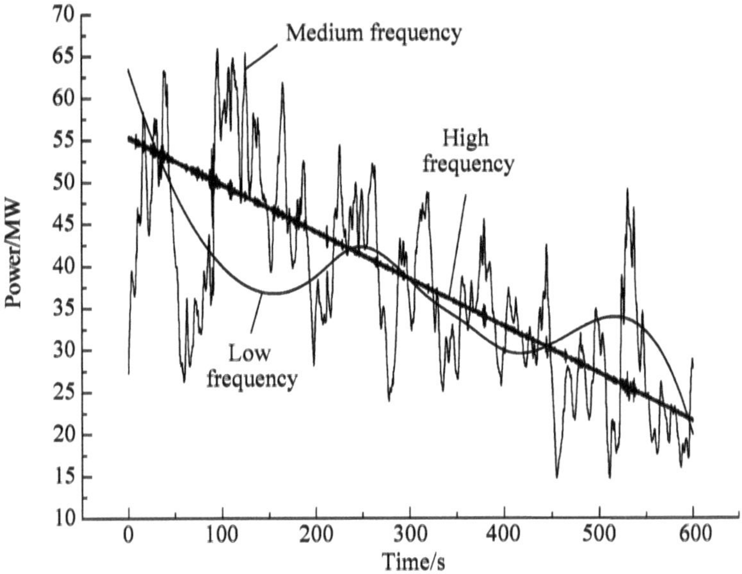

Fig. 8.23 Power curves of high, medium, and low frequency bands with residual components

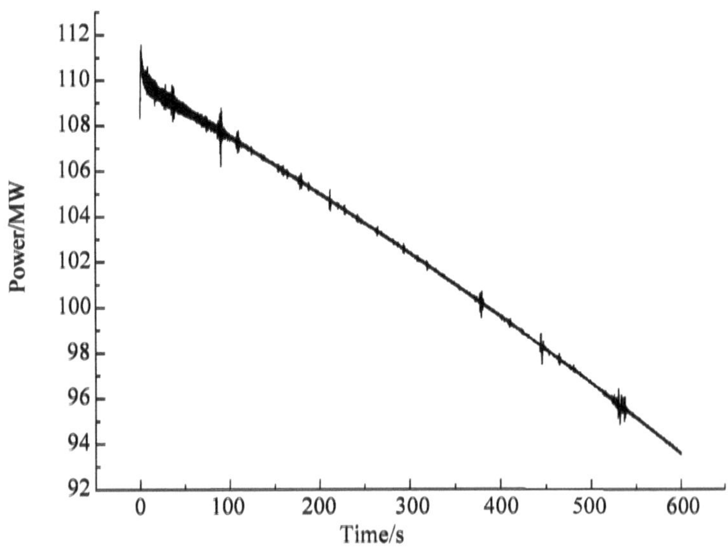

Fig. 8.24 Active power curve of line 5–7 when high frequency component is injected

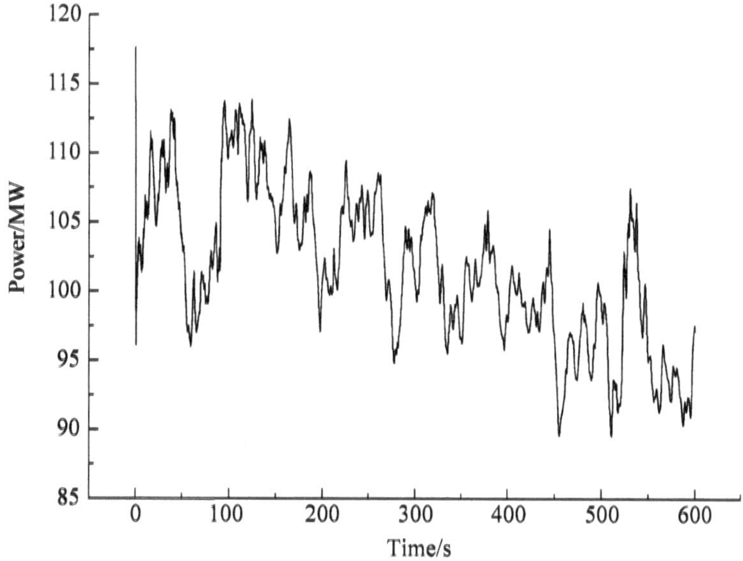

Fig. 8.25 Active power curve of line 5–7 when medium frequency component is injected

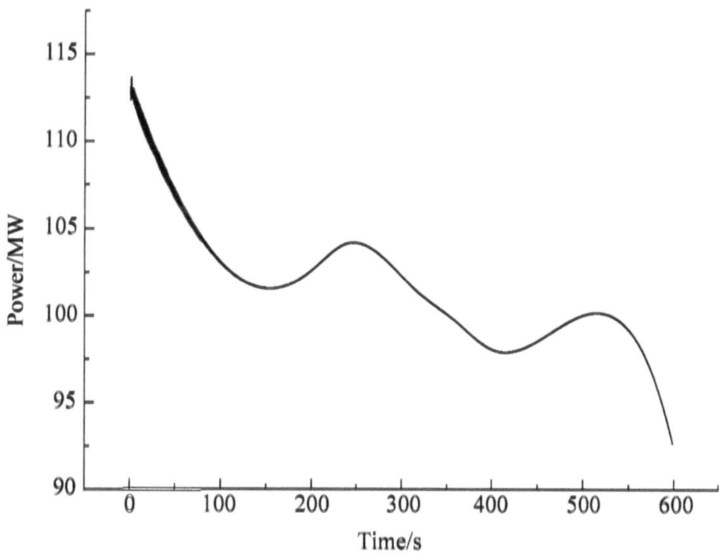

Fig. 8.26 Active power curve of line 5–7 when low frequency component is injected

Table 8.3 Damping ratios corresponding to high, medium, and low frequencies

IMF	High frequency	Medium frequency	Low frequency
1	−0.00031	0.000169	0.000351
2	−0.00019	0.000337	0.001296
3	−0.00026	0.000441	0.001899
4	−0.00031	0.000249	0.007537
5	−0.00052	0.003098	0.000051
6	−0.00056	−0.02357	
7	−0.00069	−0.00224	
8		−0.07098	
9		0.033107	

corresponding to the low frequency band is all positive. This indicates that the connection of wind power in the high and medium frequency bands will cause low-frequency oscillations in the system. In addition, the minimum damping ratio corresponding to the medium frequency band is −0.07098, which is much larger than the −0.00069 corresponding to the high frequency band, so the fluctuation of wind power in the medium frequency band has a more significant impact on the system's low-frequency oscillations.

In conclusion, the integration of large-scale wind power can potentially cause low-frequency oscillations in the power system, and the power fluctuations in the mid-frequency range have a significant impact on these oscillations. Therefore, it is crucial to suppress the mid-frequency range that has a significant impact on the system, which can reduce the harm caused by the grid connection of wind power. For specific power systems, due to differences in wind farm scale, wind resource conditions, grid structure at the connection point, and system operation conditions, the impact of wind power fluctuations will also vary. This section mainly focuses on the characteristics of wind power fluctuations and the analysis methods for causing low-frequency oscillations, providing a reference for the suppression of wind power fluctuations.

8.2 Energy Storage Smoothing Active Power Fluctuations of Wind Power

As analyzed in the previous section, wind power has fluctuation characteristics at different time scales, ranging from seconds to several minutes, or even several hours. In addition, different power systems have different requirements for wind power fluctuation limits due to their own operational control needs. Therefore, a single energy storage technology is difficult to meet the technical performance and economic performance requirements at the same time. To improve the economic efficiency of the entire life cycle of the energy storage system, a composite energy storage with complementary performance is an effective solution. This section focuses on the strategies and methods of applying multiple types of energy storage to smooth wind power fluctuations.

8.2.1 *Overall Architecture of Multi-Type Energy Storage Smoothing Wind Power Fluctuations*

As analyzed in Chap. 5, power-type energy storage represented by supercapacitors and flywheels has the advantages of high power density, fast response speed, and long cycle life, but low energy density; energy-type energy storage represented by lithium-ion batteries and lead-acid batteries has the advantage of high energy density, but limited cycle life. Therefore, two types of energy storage devices with complementary performance can be used to smooth the output power fluctuations of the wind farm, fully utilizing their complementary advantages. For short-term, small-scale high-frequency power fluctuations, power-type energy storage is used for smoothing; for long-term, large-scale low-frequency power fluctuations, energy-type energy storage is used for smoothing [136, 137].

The architecture of multi-type energy storage smoothing wind power output fluctuations is shown in Fig. 8.27.

The multi-type energy storage suppresses wind power fluctuations, and the key in control is the coordination of different types of energy storage, that is, how to determine the power that each energy storage device needs to compensate at each control moment. The common method is to use a first-order filtering algorithm to

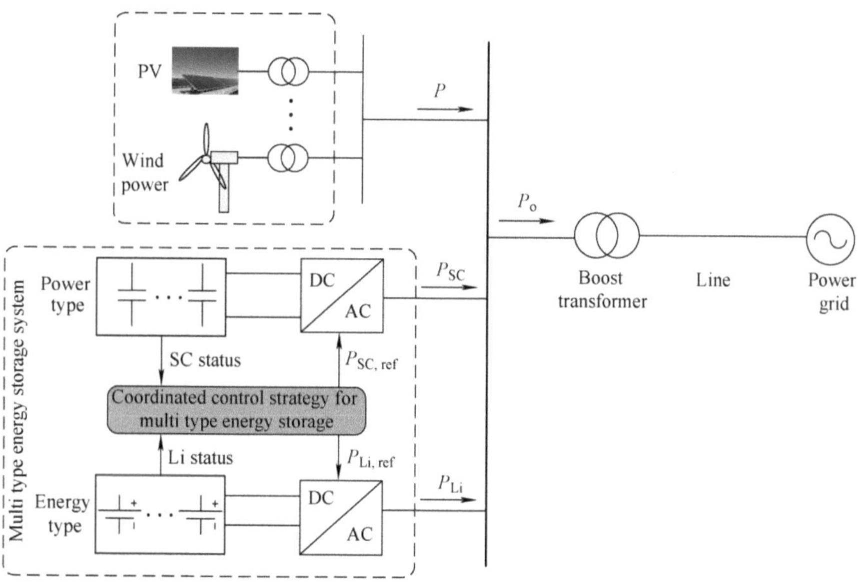

Fig. 8.27 Architecture diagram of multi-type energy storage smoothing wind power fluctuations

decompose the active power of wind power into high frequency and low frequency parts, and then allocate them to power-type energy storage and energy-type energy storage respectively. The first-order filtering algorithm is simple and easy to apply in engineering, but there is a problem of control lag, which can easily cause excessive smoothing of wind power and increase the investment cost of energy storage. This section first analyzes the application of the first-order low-pass filtering algorithm in wind power fluctuation smoothing. Considering the good multi-constraint handling ability and online optimization ability of Model Predictive Control (MPC), the focus is on the research of multi-type energy storage coordination control strategy based on MPC. On the basis of the control principle of MPC, the system state space equation and MPC control architecture for energy storage to smooth wind power fluctuations are established, and a comparative analysis is carried out.

8.2.2 Energy Storage Control Based on First-Order Low-Pass Filter

The basic idea of using a first-order low-pass filter algorithm to smooth wind power fluctuations is to filter the original wind power through a first-order low-pass filter and adjust the time constant of the filter so that the low-frequency part meets the grid-connected fluctuation limit of wind power. The high-frequency part serves as the energy storage load, which is undertaken by power-type and energy-type energy storage. Of course, the load allocation of power-type energy storage and energy-type energy storage can be achieved through another first-order low-pass filter. Depending on whether the time constant can be adjusted, it is divided into fixed time constant filtering algorithm and variable time constant filtering algorithm [138].

8.2.2.1 Principle of First-Order Low-Pass Filter Algorithm

The core of the first-order low-pass filter (Low-pass Filtering Algorithm, LFA) algorithm is the first-order low-pass filter model, and its equivalent circuit is shown in Fig. 8.28.

According to the input-output relationship of voltage, we can get:

$$RC\frac{du_o(t)}{dt}+u_o(t)=u_i(t) \tag{8.25}$$

Fig. 8.28 First-order low-pass filter circuit

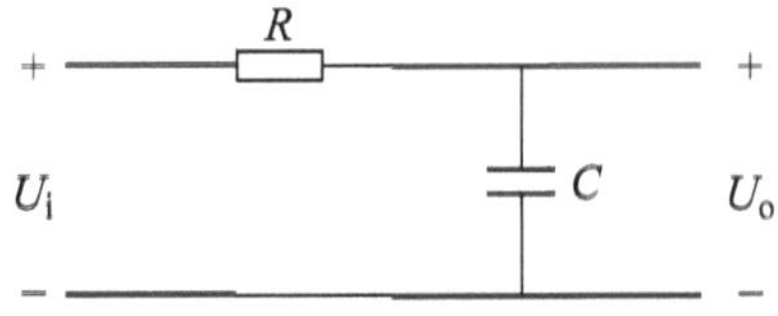

Further, the transfer function of LFA can be obtained:

$$H(s) = \frac{U_o}{U_i} = \frac{1}{RCs + 1} = \frac{1}{\tau s + 1} \quad (8.26)$$

Where, $\tau = RC$, is the time constant of LFA.

Using z-transform, the above formula can be discretized as follows:

$$H(z) = \frac{1}{1 + \frac{\tau}{T_c} - \frac{\tau}{T_c} z^{-1}} \quad (8.27)$$

Assume $X(k)$ represents the input quantity, $Y(k)$ represents the output quantity, the relationship between LFA output and input can be obtained:

$$Y(k) = \frac{\tau}{\tau + T_c} Y(k-1) + \frac{T_c}{\tau + T_c} X(k) \quad (8.28)$$

As can be seen, in the discretized LFA expression, k moment's output quantity $y(k)$ is not only related to k *is related to the state quantity at the moment, and also to the output quantity at the previous moment* $y(k-1)$. *The larger the time constant* τ, *the smaller the difference between* $y(k)$ *and* $y(k-1)$, *the better the filtering effect of the filter, but the more obvious the lag effect.*

Through the differential operator $s = j\omega$, it is transformed into the complex domain:

$$H(j\omega) = H(s)\big|_{s=j\omega} \frac{1}{\sqrt{\tau^2\omega^2 + 1}} e^{-j\mathrm{arctg}\omega\tau} \quad (8.29)$$

The Bode plot of LFA is shown in Fig. 8.29. The cut-off frequency f_c indicates that the signal amplitude begins to drop rapidly from this point. $e^{-j\theta}$ represents the phase lag $e^{-j\theta}$ ($\theta = \mathrm{arctg}\omega\tau$) at the cut-off frequency point f_c, which illustrates the lag characteristics of the low-pass filtering algorithm. It can be seen that the larger the time constant τ, the more obvious the lag effect.

8.2.2.2 Fixed Time Constant Filtering Algorithm

With the grid-connected fluctuation rate of the wind farm as the target, the control strategy of suppressing the power fluctuation of the wind farm through the fixed time constant LFA is shown in Fig. 8.30.

Selecting actual operation data of a wind farm for offline simulation, the rated capacity of the wind farm is 10 MW, the sampling time is 30 s, the data length is 2870, and the duration is 00:00–24:00. The suppression target is set so that the power fluctuation rate of wind power does not exceed 2% per minute. Figure 8.31

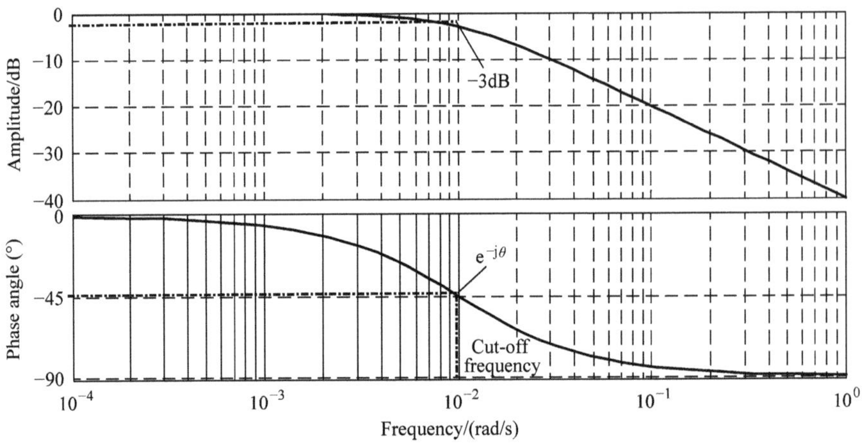

Fig. 8.29 Bode plot of first-order low-pass filtering algorithm LFA ($\tau = 100$)

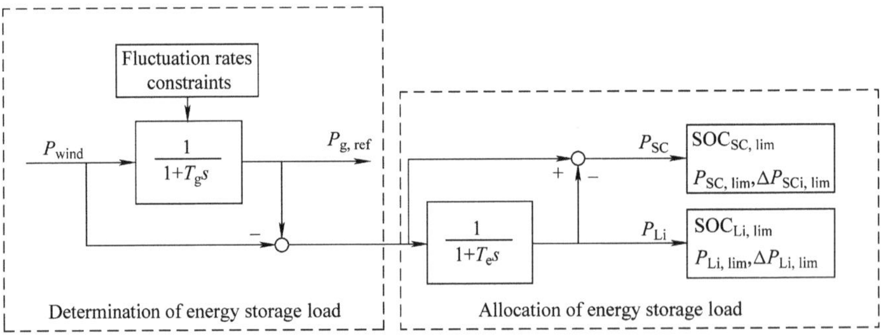

Fig. 8.30 Principle diagram of wind power suppression control implemented by fixed time constant LFA

shows the grid-connected power of wind power before and after suppression, and Figs. 8.32 and 8.33 compare the 1-min power fluctuation rate of wind power before and after suppression. It can be seen that after suppression, the power fluctuation of the wind farm is basically limited within 2%, but more than 90% of the fluctuation rate is concentrated below 1%, which is excessive suppression. In actual engineering, this will increase the configuration capacity and cost of energy storage, which is a shortcoming of the fixed time constant LFA.

8.2.2.3 Variable Time Constant Filtering Algorithm

To address the shortcomings of the fixed time constant algorithm, a variable time constant first-order low-pass filter algorithm is designed, as shown in Fig. 8.34. By considering factors such as the control target of wind power grid-connected

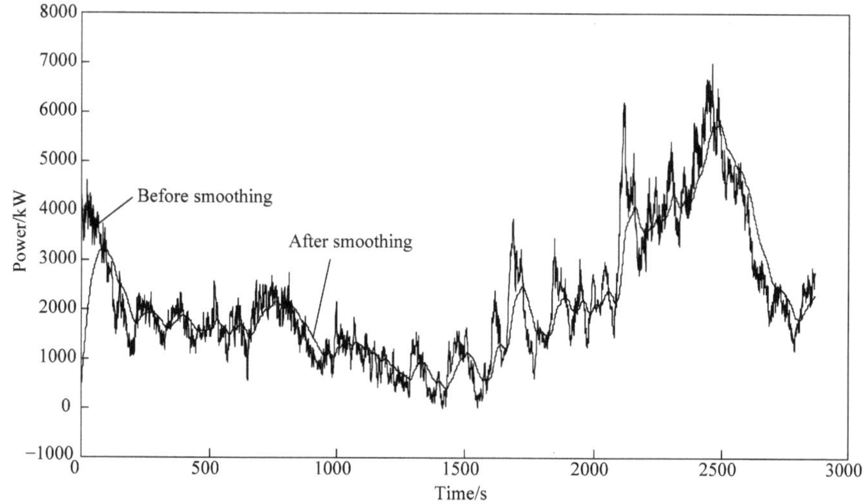

Fig. 8.31 Time domain diagram of wind farm power before and after suppression

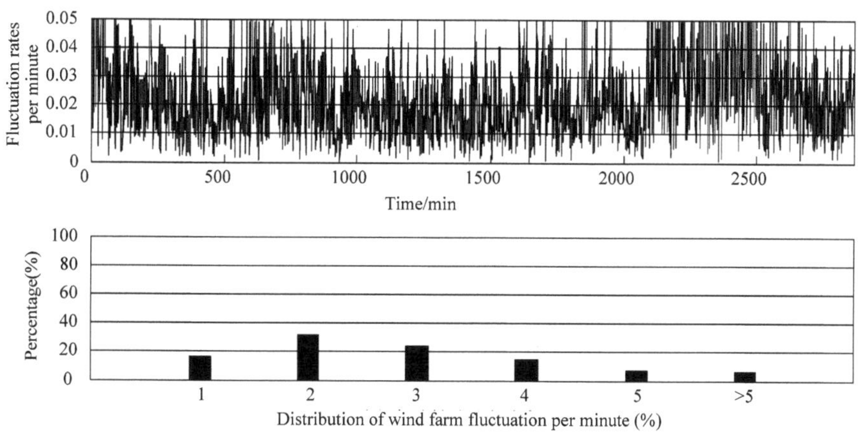

Fig. 8.32 Distribution of wind farm power fluctuation (1 min) before suppression

fluctuation rate and the state of energy storage charge, the filter time constant is dynamically adjusted to meet the requirements of grid-connected wave power, while improving the utilization rate of energy storage and reducing the cost of energy storage.

Still analyzing the actual operation data of the wind farm in the previous section, Fig. 8.35 is a comparison of the fluctuation rate of wind power before and after suppression for 1 min. It can be seen that the variable time constant filter algorithm can meet the control requirements of the wind power grid-connected fluctuation rate. As can be seen from Fig. 8.36, compared with the fixed time constant LFA, the variable time constant LFA can adjust the filter time constant according to the actual

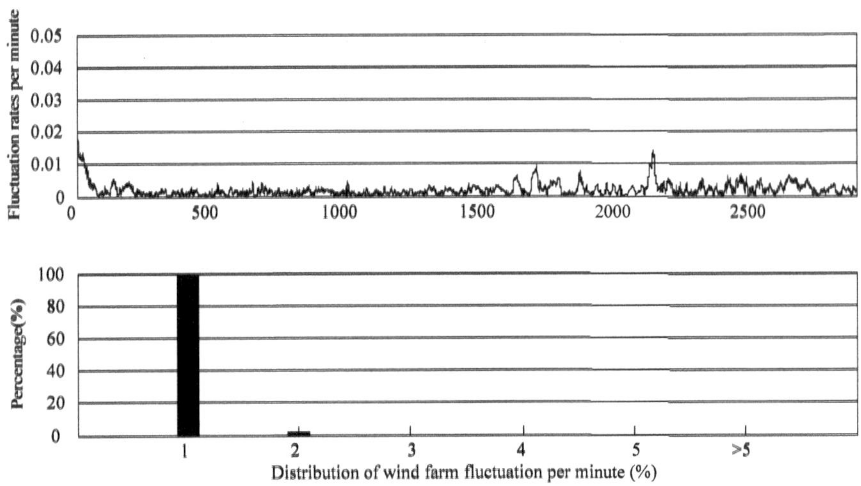

Fig. 8.33 Distribution of wind farm power fluctuation (1 min) after suppression

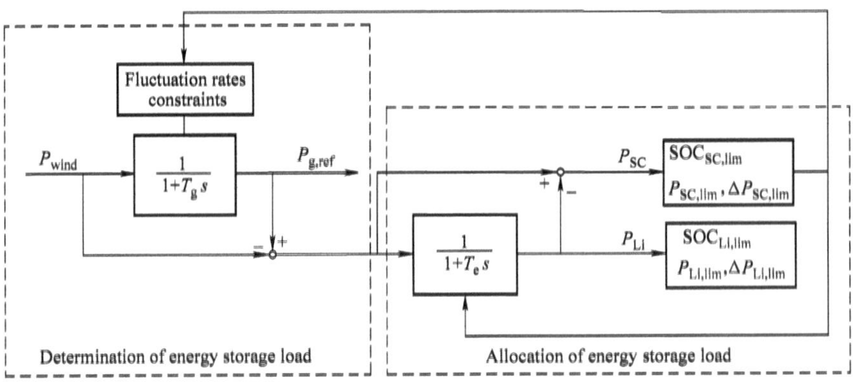

Fig. 8.34 Variable time constant LFA implementation of wind power suppression control principle diagram

situation, thereby appropriately relaxing the grid-connected fluctuation rate requirement when the wind power fluctuation rate is not large, thereby reducing the load of energy storage and reducing the overall configuration capacity and cost of energy storage.

Similarly, the total energy storage load is also adaptively adjusted in the distribution of power type energy storage and energy type energy storage according to the output situation and state of charge of each energy storage unit. Figure 8.37 shows the load allocation of energy storage, and Fig. 8.38 shows the state of charge of each energy storage unit. It can be seen that through the adjustment of the filter time constant, the energy storage load can be well distributed to the power type energy storage and energy type energy storage, so that the state of charge of each energy storage unit is maintained in a reasonable state.

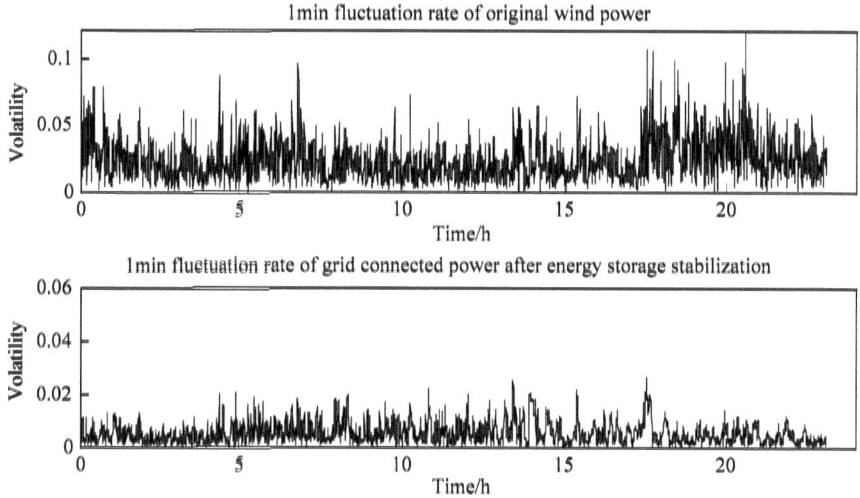

Fig. 8.35 Comparison of fluctuation rate before and after energy storage suppression for 1 min

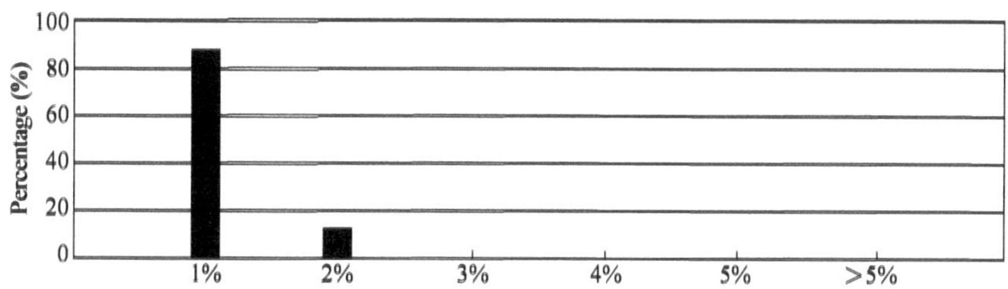

Fig. 8.36 Distribution of wind power fluctuation after suppression

8.2.3 MPC Based Energy Storage Control

Model Predictive Control (MPC) is a new type of computer optimization control algorithm that emerged in the late 1970s and is one of the effective methods for dealing with multivariable and multi-constraint systems. MPC uses the mathematical model of the controlled object to predict its future output, and obtains a series of control actions from the current time to the specified prediction time through optimization, but only implements the first control in the control sequence at each prediction cycle. MPC solves the online optimization control problem through the alternating process of continuous local optimization and continuous rolling implementation of control, and has a strong ability to deal with various disturbances and uncertainties. Compared with the first-order low-pass filter algorithm with delay effect, the MPC algorithm can obtain the control quantity one to several control cycles in advance, thereby effectively eliminating the impact of delay.

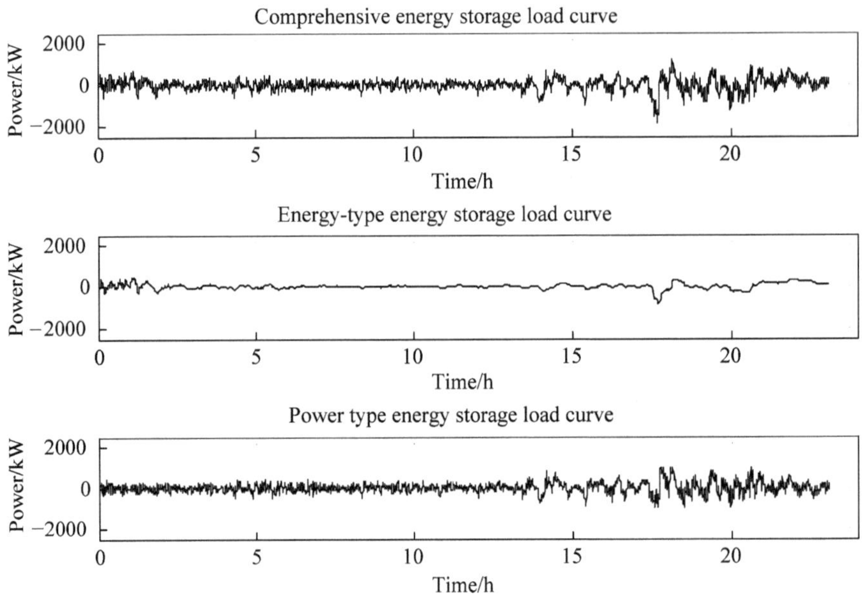

Fig. 8.37 Distribution of total energy storage load and its allocation between energy type storage and power type storage

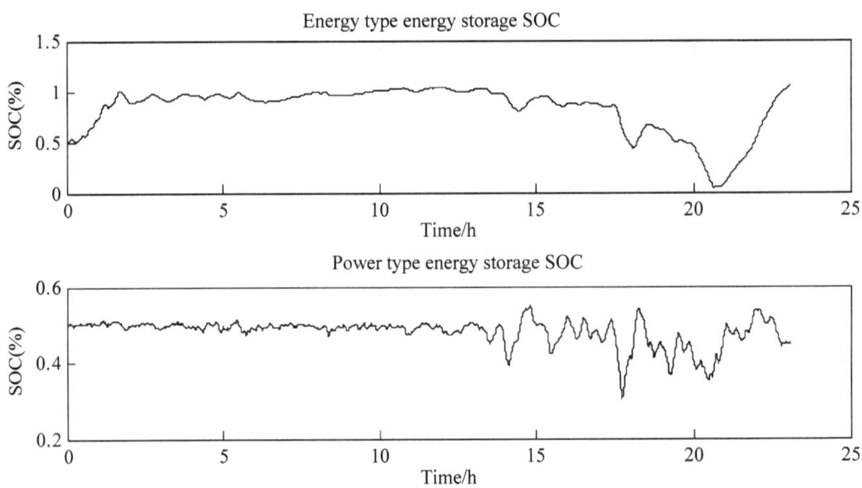

Fig. 8.38 State of charge of energy type storage and power type storage

8.2.3.1 MPC Algorithm Principle

The MPC algorithm includes several parts such as rolling time domain, state space model, and MPC transfer function.

Principle of Rolling Time Domain

The Receding Horizon is the core idea of MPC. Assuming the current state quantity is $x(k)$, the control input quantity is $u(k)$, and the output quantity is $y(k)$, then the rolling time domain mainly includes the following steps:

1. At the current time k and the current state $x(k)$, considering the current and future constraints, by solving the optimization problem, we obtain the control instruction sequence for future moments $k + 1, k + 2, \ldots k + M$;
2. Apply the first value of the control instruction sequence to the control system;
3. At moment $k + 1$, update the state to $x(k + 1)$, and repeat the above operation steps.

The concept of rolling time domain is shown in Fig. 8.39.

State Space Model (State Space Model)

According to state space theory, a state space equation can be established:

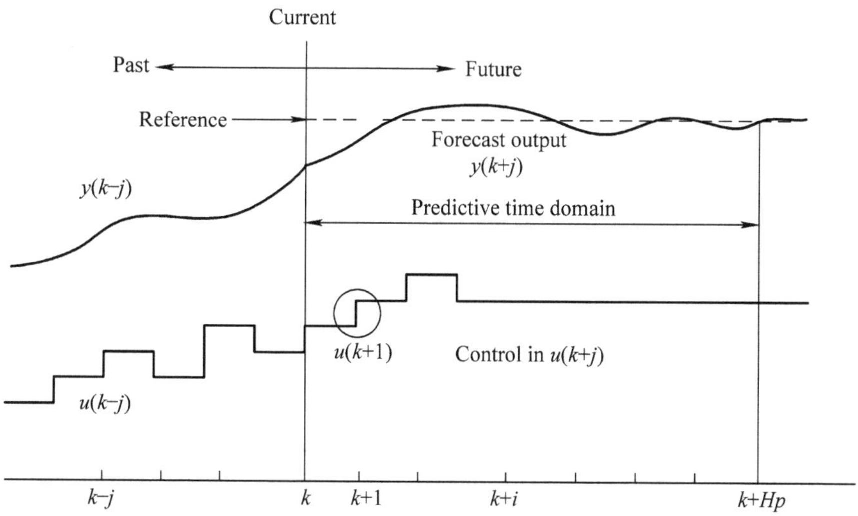

Fig. 8.39 Schematic diagram of the rolling time domain

$$x(k+1) = Ax(k) + B_1u(k) + B_2r(k) \tag{8.30}$$

$$y(k) = Cx(k) + D_1u(k) + D_2r(k) \tag{8.31}$$

Where, $r(k)$ is the disturbance input.

Combining the above equations, we get:

$$y(k+1) = C(Ax(k) + B_1u(k) + B_2r(k)) + D_1u(k+1) + D_2r(k+1) \tag{8.32}$$

Iterating like this until the prediction time domain length M steps, we can get the matrix equation:

$$\begin{bmatrix} y(k) \\ y(k+1) \\ y(k+2) \\ \cdots \\ y(k+M) \end{bmatrix} = \begin{bmatrix} C \\ CA \\ CA^2 \\ \cdots \\ CA^M \end{bmatrix} [x(k)] + \begin{bmatrix} D_1 & 0 & 0 & 0 & \cdots & 0 \\ CB_1 & D_1 & 0 & 0 & \cdots & 0 \\ CAB_1 & CB_1 & D_1 & 0 & \cdots & 0 \\ \cdots & \cdots & \cdots & \cdots & \cdots & \cdots \\ CA^{M-1}B_1 & CA^{M-2}B_1 & CA^{M-3}B_1 & \cdots & CB_1 & D_1 \end{bmatrix} \begin{bmatrix} u(k) \\ u(k+1) \\ u(k+2) \\ \cdots \\ u(k+M) \end{bmatrix}$$
$$+ \begin{bmatrix} D_2 & 0 & 0 & 0 & \cdots & 0 \\ CB_2 & D_2 & 0 & 0 & \cdots & 0 \\ CAB_2 & CB_2 & D_2 & 0 & \cdots & 0 \\ \cdots & \cdots & \cdots & \cdots & \cdots & \cdots \\ CA^{M-1}B_2 & CA^{M-2}B_2 & CA^{M-3}B_2 & \cdots & CB_2 & D_2 \end{bmatrix} \begin{bmatrix} r(k) \\ r(k+1) \\ r(k+2) \\ \cdots \\ r(k+M) \end{bmatrix} \tag{8.33}$$

This matrix is the State Space Model (SSM), which can reflect the relationship between the system's input, output, state variables, control variables, and disturbance variables, "predict" the future behavior of the system, and SSM is the basic model of the MPC algorithm.

MPC Transfer Function

According to the State Space Model (SSM), the structure of the control object is shown in Fig. 8.40.

It can be seen that the output at moment k $y(k)$ is only related to the state variable at moment k $x(k)$, the control input $u(k)$, and the disturbance input $r(k)$, and is not related to the output of the previous moment $y(k-1)$.

Based on the establishment of the state space model, the MPC algorithm can be implemented by controlling the increment of the control variable in the rolling time domain. At any moment k, as long as the initial predicted value of the output

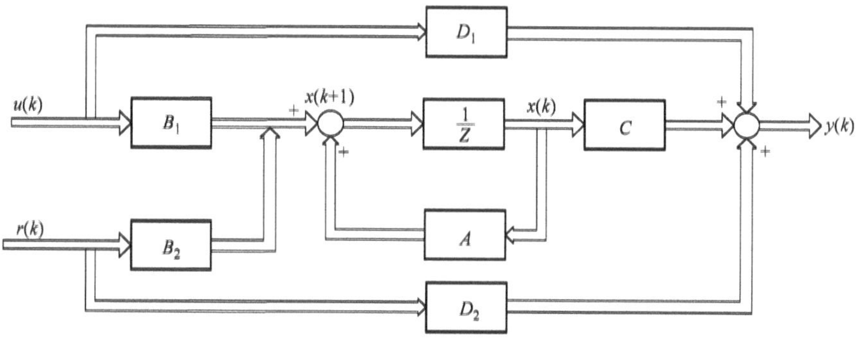

Fig. 8.40 Structure diagram of the State Space Model (SSM)

variable $y_0(k + i|k)$ is known, the future output can be calculated according to the future control variable increment Δu, that is:

$$y_M(k+i|k) = y_0(k+i|k) + \sum_{j=1}^{\min(M,i)} a_{i-j+1} \times \Delta u(k+j-1) \quad i = 1,\cdots,N \tag{8.34}$$

Assuming the expected output of the system at future time $k + j$ is R(k + j|k), the control goal is to minimize the error between the future output and the expected output, that is:

$$J = \min \sum_{j=1}^{M} \left[R(k+i|k) - y_0(k+i|k) - a_{i-j+1} \times \Delta u(k+j-1)\right]^2 \quad i = 1,\cdots,N \tag{8.35}$$

It can be seen that the essence of MPC is to compensate for the error between the expected output and the predicted output at future moments by predicting the control increment at future moments. Traditional PID control compensates for the error between the expected output and the actual output at the current moment through the control increment at the current moment. Obviously, MPC has better predictability than PID.

8.2.3.2 MPC Coordination Control Strategy

Figure 8.41 is a schematic diagram of the principle of smoothing wind power fluctuations using the MPC algorithm. First, the original wind power is taken as the disturbance, the predicted grid-connected power and energy storage SOC are taken as the state variables, and the energy storage power command is taken as the control variable. Through time-domain rolling and iteration, the M-step grid-connected power of wind power is obtained, so that its power fluctuation rate can be obtained. Then, an energy storage optimization model is constructed, with the minimum use of energy storage as the objective function. Through the constraints of wind power

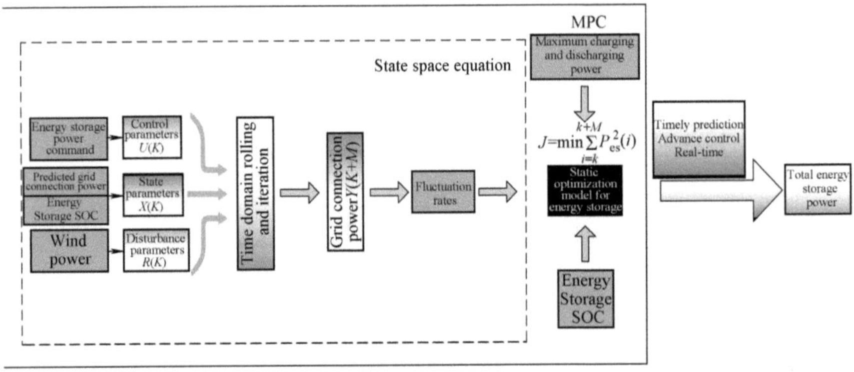

Fig. 8.41 Model Predictive Control Strategy for Smoothing Wind Power Fluctuations

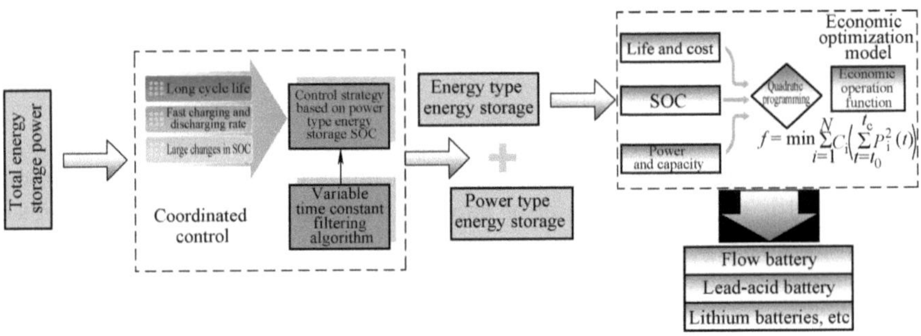

Fig. 8.42 Coordination Control Diagram of Multiple Types of Energy Storage

grid-connected fluctuation rate, energy storage SOC, and maximum charging and discharging power, the total power reference value of energy storage can be obtained.

Figure 8.42 is the coordination control among multiple types of energy storage. The variable time constant first-order filtering algorithm can be used for load distribution of energy-type storage and power-type storage, and the charge state of power-type storage is considered in actual control. For large-capacity energy-type storage, an economic optimization model can be constructed to achieve load distribution among energy-type storage units, and factors such as the charge state, health status, and operation and maintenance costs of each storage unit are considered to optimize the technical and economic performance of the energy-type storage system.

Overall Control Structure

The overall structure of the control system based on MPC is shown in Fig. 8.43. Where, P_w represents the output power of the wind farm, P_g represents the grid-connected power of the wind storage system, P_{es} represents the comprehensive load of energy storage, P_{psd} represents the load of power-type storage device (PSD), P_{esd} represents the load of energy-type storage device (ESD), SOC_{es} represents the comprehensive charge state of energy storage.

The control system contains two layers of energy storage power distribution units. The first layer determines the overall energy storage system's integrated load command based on the MPC algorithm, according to the known wind power data P_w and the corresponding fluctuation control requirements, the grid-connected power P_g and the energy storage integrated load P_{es} can be obtained; The second layer is the distribution of energy storage integrated load between PSD and ESD, which can be achieved through a first-order low-pass filter algorithm. The basic principle is to allocate the high-frequency part to PSD and the low-frequency part to ESD. In this system, a continuous prediction model is used for ultra-short-term prediction of wind power.

Control System Model Based on MPC

Combining the actual problem requirements, a control system model based on MPC for multi-type energy storage to smooth wind power fluctuations is established. The relationship between wind power, energy storage power, and grid-connected power is:

$$P_g(k+1) = P_{es}(k) + P_w(k) \tag{8.36}$$

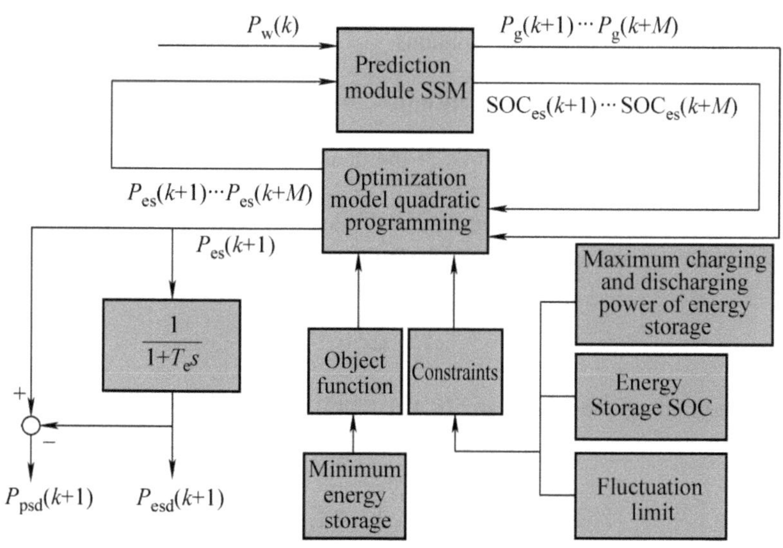

Fig. 8.43 Structure diagram of multi-type energy storage coordination control based on MPC-LFA

Set the energy storage control cycle to T_c, the total energy storage capacity to C_{es}, and the energy storage integrated load state to:

$$SoC(k+1) = SoC(k) - T_c \cdot P_{es}(k) / C_{es} \tag{8.37}$$

Take the grid-connected power P_g and the energy storage integrated load state SOC as state variables x_1, x_2; the energy storage power command P_{es} as the control variable u; the wind farm output power P_w as the disturbance input; P_g and SOC as output variables y_1, y_2, the state space equation can be obtained:

$$\begin{aligned} \begin{bmatrix} x_1(k+1) \\ x_2(k+1) \end{bmatrix} &= \begin{bmatrix} 0 & 0 \\ 0 & 1 \end{bmatrix} \begin{bmatrix} x_1(k) \\ x_2(k) \end{bmatrix} + \begin{bmatrix} 1 \\ -T_c \end{bmatrix} \left[u(k) \right] + \begin{bmatrix} 1 \\ 0 \end{bmatrix} \left[r(k) \right] \\ \begin{bmatrix} y_1(k) \\ y_2(k) \end{bmatrix} &= \begin{bmatrix} 1 & 0 \\ 0 & 1 \end{bmatrix} \begin{bmatrix} x_1(k) \\ x_2(k) \end{bmatrix} \end{aligned} \tag{8.38}$$

Iterate the above formula to get the matrix equation, use the quadratic programming method to optimize in each control domain to get the corresponding control command $u(k)$, $u(k + 1)$, $u(k + 2)$, ⋯, $u(k + M)$, and then combine the low-pass filter algorithm to achieve the secondary distribution of energy storage load.

To implement the MPC control method, it is necessary to standardize the optimization model, and the output of the control variables should follow the form below:

$$U^{OPT} = \left[u^{OPT}(k), u^{OPT}(k+1), \cdots, u^{OPT}(k+M) \right]^T \tag{8.39}$$

Control variables can be obtained through numerical calculation, that is, converted into solving static optimization problems:

$$U^{OPT} = \arg\min U^T W U + 2U^T V \tag{8.40}$$

Where, W is the coefficient matrix of the quadratic term U^2, V is the coefficient matrix of the linear term U.

In actual control, the objective function follows the principle of minimizing the amount of energy storage used in each optimization cycle, that is:

$$J = \min \sum_{i=k}^{k+M} P_{es}^{2}(i) \tag{8.41}$$

The corresponding constraints are: (8.42), (8.43) and (8.44)

$$0 \le P_g(k) \le P_{g\_\max}, \quad k = 1, 2, \cdots, M \tag{8.42}$$

$$0 \le SOC_{es} \le 1, \quad k = 1, 2, \cdots, M \tag{8.43}$$

$$P_{es\_\min} \le P_{es}(k) \le P_{es\_\max}, \quad k = 1, 2, \cdots, M \tag{8.44}$$

$$\frac{\max\limits_{i=1,2,\cdots n} P_g(k-i) - \min\limits_{i=1,2,\cdots n} P_g(k-i)}{P_{\text{wind\_rated}}} \leq \gamma,\quad k=1,2,\cdots,M \tag{8.45}$$

The above constraints are grid power constraints, energy storage charge state constraints, energy storage ramp rate limits, and wind storage grid power fluctuation rate limits, respectively, P_{wind_rated} is the rated installed capacity of the wind farm.

Distribution of Energy Storage Load Between PSD and ESD

The low-pass filter algorithm is used to allocate the energy storage load between PSD and ESD. The power output by the low-pass filter algorithm is allocated to the energy storage ESD, and the remaining power is allocated to the power storage PSD. The cutoff frequency of a low-pass filter is:

$$f_c = 1/2\,\pi\tau \tag{8.46}$$

The cutoff frequency f_c is inversely proportional to the time constant τ. If the time constant τ is larger, the cutoff frequency f_c will be lower, and PSD will compensate for more fluctuating power, while the power compensated by ESD will decrease; vice versa.

Control Algorithm Flowchart

The flowchart of the above-mentioned multi-type energy storage coordination control algorithm based on MPC is shown in Fig. 8.44.

Case Study Analysis

The actual operation data of a 100 MW wind farm is selected for simulation analysis. The system sampling interval is 1 s, the duration is 1 h from 7:00 to 8:00 in the morning, with a total of 3600 sampling points. The fluctuation limit of the wind power grid-connected power is not more than 2% per minute. The aforementioned MPC-LFA algorithm is used to smooth the output power fluctuations of the wind farm, while completing the load allocation of energy/power type storage.

As can be seen from Fig. 8.45, using the MPC-LFA and two-stage low-pass filtering algorithm storage control strategies, the fluctuation rate of the grid-connected power after smoothing is significantly reduced. Further comparing the control effects of the two methods, the MPC-LFA method has a smaller time delay between the grid-connected power and the original wind power, and the control is more immediate. This is because MPC considers the state of the next few steps at the current moment, achieving advanced optimization control, which can to some extent eliminate the delay effect of the traditional two-stage low-pass filtering algorithm, and realize the determination of the comprehensive load of energy storage.

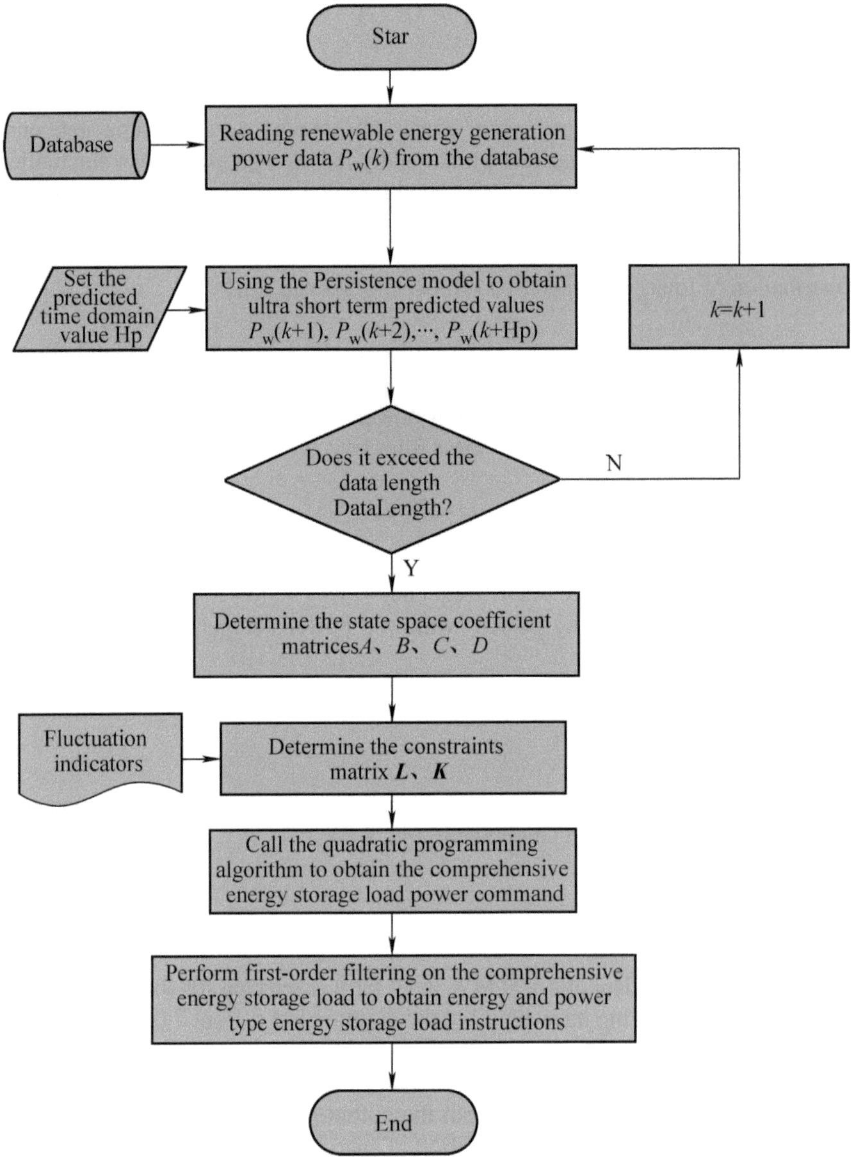

Fig. 8.44 Flowchart of multi-type energy storage coordination control based on MPC

Figure 8.46 shows the 1 min power fluctuation rate before and after energy storage smoothing. It can be seen that after smoothing by the MPC-LFA algorithm, the fluctuation rate of wind power grid-connected power can be well controlled within 2%, meeting the grid-connected requirements.

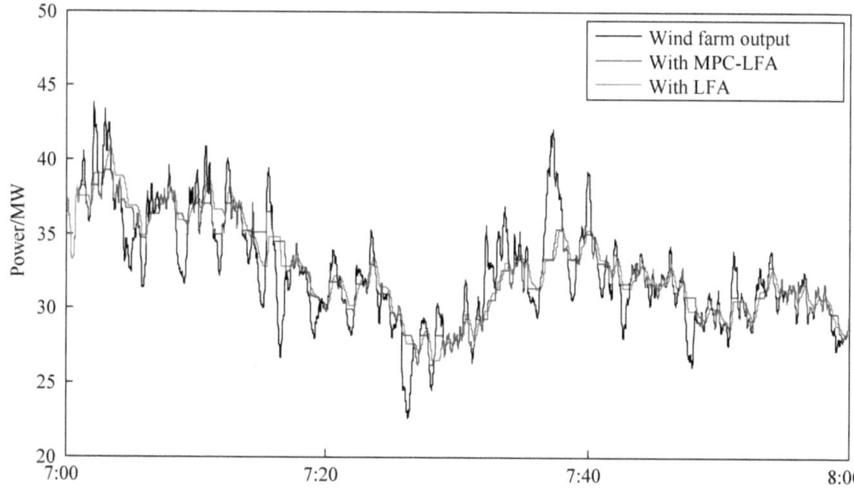

Fig. 8.45 Wind power grid-connected power under MPC-LFA and two-stage LFA algorithm

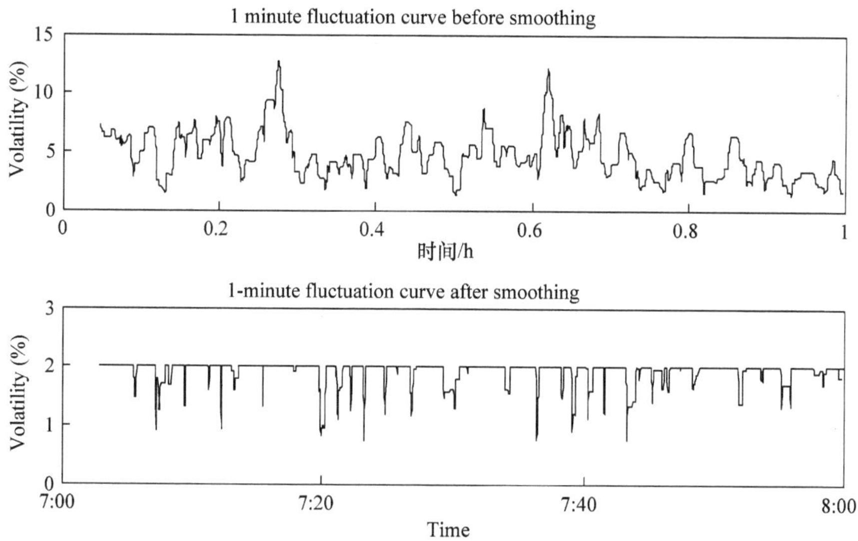

Fig. 8.46 Original wind power 1 min fluctuation rate and grid-connected power 1 min fluctuation rate

Under the MPC algorithm, the comprehensive load curve of energy storage is shown in Fig. 8.47. Through the low-pass filtering algorithm, the comprehensive load of energy storage is further allocated to obtain the energy storage load under high and low frequency bands, as shown in Fig. 8.48. It can be seen that the high-frequency part of the wind power fluctuation is compensated by the power type

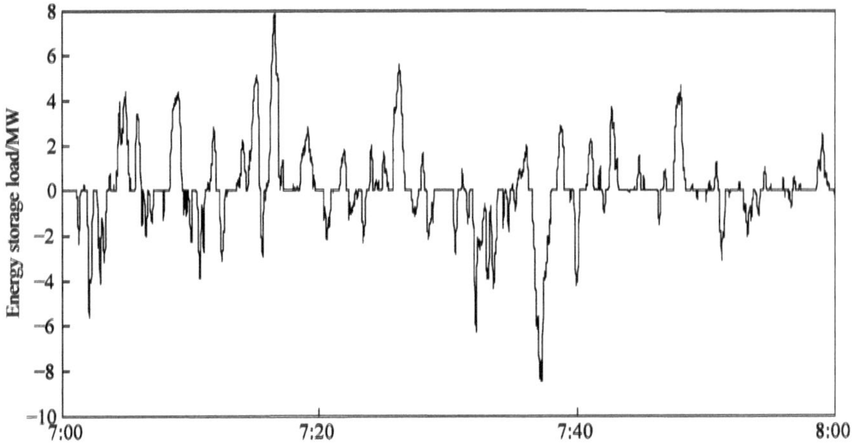

Fig. 8.47 Comprehensive load curve of energy storage

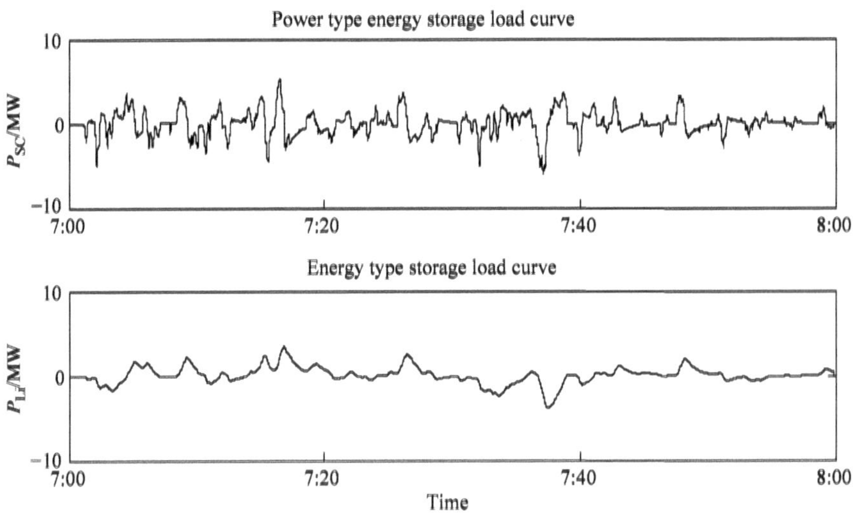

Fig. 8.48 Power type and energy type storage load curve

storage, and the low-frequency part is compensated by the energy type storage, which is conducive to fully exerting the complementary advantages of the technical economy of the two types of storage.

Table 8.4 shows the maximum energy storage power and energy storage capacity required under the two control strategies. It can be seen that the MPC-LFA control method saves 4.1% of the maximum energy storage power and 23.5% of the capacity compared to the LFA method.

Table 8.4 Comparison of energy storage requirements under different control strategies

Control strategy	LFA	MPC-LFA	Savings (%)
Power (MW)	8.8234	8.4635	4.1
Capacity (MWh)	0.3545	0.2713	23.5

Chapter 9
Wind and Storage Joint Participation in System Frequency and Voltage Regulation

With the large-scale development of renewable energy sources such as wind power and photovoltaics, it means that some conventional generators in the future power system will be phased out. The existing photovoltaic power generation systems and wind turbine units (the mainstream products are variable-speed constant-frequency doubly-fed wind turbines and direct-drive wind turbines) adjust the power output and synchronize with the grid through power electronic converters, and generally work in the maximum power tracking mode. This causes the operation of the power generation equipment to be decoupled from the system frequency and voltage, and it cannot provide auxiliary services such as frequency regulation and voltage regulation like traditional units when the system frequency and voltage change. In this case, if the missing part of the inertia response, frequency and voltage regulation capabilities are not supplemented, it will cause the overall inertia of the power system to decrease, and the frequency and voltage regulation capabilities to weaken, making the system's frequency and voltage change rate increase, the lowest point decrease, and the steady-state deviation increase under disturbances such as unit disconnection, line failure, and sudden load changes. Frequency and voltage stability issues occur more frequently, bringing new problems to system operation and scheduling. However, photovoltaics and wind power can adjust the operating mode of the power electronic converter to provide the system with certain active/reactive power support, thereby achieving regulation of the system frequency and voltage. Of course, since the output capability of its active/reactive power is directly related to its operating state (light intensity or wind speed), there is a problem of low credibility of frequency and voltage regulation capacity. Therefore, energy storage can be well combined with the self-regulation ability of photovoltaics or wind power to improve the frequency and voltage regulation capabilities of photovoltaics or wind power, making them similar to conventional generator units with auxiliary service functions.

X. Tang et al., *Electrical Energy Storage Technologies and Applications*,
https://doi.org/10.1007/978-981-96-5625-7_9

9.1 Wind-Storage Joint Participation in System Frequency Regulation

9.1.1 Power System Frequency Regulation

Frequency reflects the balance between power generation and electricity consumption in the AC power system and is one of the important indicators of the power system. On the one hand, frequency instability may cause power system frequency collapse, system disintegration and other accidents; on the other hand, when the system operates at low frequency conditions, due to the increase in excitation current of asynchronous motors and transformers and other equipment, the consumed reactive power increases, causing the system voltage level to drop, which may also cause voltage collapse and other accidents.

The frequency of the power system directly depends on the speed of the synchronous generator, and their relationship is:

$$f = \frac{np}{60} \tag{9.1}$$

Where, n is the speed of the synchronous generator, p is the pole pair number of the synchronous generator.

Therefore, to maintain the stability of the power system frequency, it is required that the speed of all generators in the system remain stable. The speed of the unit depends on the balance of the input power of the prime mover and the output power of the generator, and is constrained by the mechanical inertia of the rotor. When the impact of rotor mechanical damping is ignored, their relationship is:

$$\begin{cases} \dfrac{d\delta}{dt} = \omega - \omega_0 \\ \dfrac{d\omega}{dt} = \dfrac{\omega_0}{T_J}(P_t - P_e) \end{cases} \tag{9.2}$$

Where, ω is the electrical angular speed of the generator, ω_0 is the synchronous electrical angular speed, δ is the angle between the two electrical angular speeds, T_J is the inertia time constant of the generator set, P_t is the mechanical torque, P_e is the electromagnetic torque. It can be seen that the motion state of the generator rotor is determined by the difference between the mechanical power of the prime mover and the electromagnetic power of the generator. When the difference between the two is not zero, it will inevitably cause a change in the generator speed, which will then cause a change in frequency. The mechanical power of the prime mover depends on its own characteristics and the characteristics of the speed control system. Although it is not constant, it can be considered constant during the electromechanical transient process; the electromagnetic power output by the generator not only depends on its own electromagnetic characteristics, but also depends on the load characteristics of the power system, network structure, and other operating conditions of the generator, which are the main causes of frequency fluctuations in the power system.

The frequency stability of the power system refers to the ability of the system to maintain the frequency of the entire system or the frequency of the subsystem after decoupling within the allowable range or not to drop below the dangerous value when a large disturbance occurs, such as generator shutdown, load shedding, etc., under the action of the automatic regulation device.

After the power system is disturbed, frequency regulation is mainly divided into the following three stages:

First, in the initial stage after the disturbance, it is compensated by the rotor inertia kinetic energy of all operating units in the system. Considering that the rotor speed of the traditional generator set is coupled with the system frequency, the rotor of each unit will first actively respond to the change in system frequency, instantaneously release/store part of the kinetic energy to suppress the change in system frequency. This process does not require any regulation, it is naturally completed, it reflects the natural characteristics of the power system—inertia, also known as inertia response (Inertia Response), reflected as the system's equivalent inertia time constant:

$$H = \frac{J\omega_n^2}{2S} \tag{9.3}$$

When the speed is ω, the rotating energy possessed by the synchronous generator set is:

$$E_K = \frac{1}{2}J\omega^2 \tag{9.4}$$

When the system frequency changes, the speed of the synchronous generator changes with the frequency, and the kinetic energy released by the generator at this time is:

$$\Delta P = \frac{dE_k}{dt} = \frac{1}{2}J \times 2\omega\frac{d\omega}{dt} = J\omega\frac{d\omega}{dt} \tag{9.5}$$

Written in per unit form:

$$\Delta P^* = \frac{\Delta P}{S} = 2H\frac{d\omega^*}{dt} = 2H\frac{df}{dt} \tag{9.6}$$

Where, J is the unit inertia, ω is the rotor speed, S is the rated capacity of the unit. The inertia constant H of traditional units is generally 2–9 s, which varies with the type of unit, with thermal power units being larger and hydroelectric units being smaller.

When the disturbance lasts for a few seconds, if the frequency deviation exceeds a certain threshold f_d, the effect of adjusting the frequency solely based on inertia response is not good. At this time, the generator set will start the unit speed regulator to reduce the system frequency deviation. Under the regulation of the generator power frequency characteristics and the load's own regulation characteristics, the frequency will rise or fall. This process is called primary frequency regulation (Primary Frequency Regulation, PFR). However, primary frequency regulation is differential and cannot return the frequency to the rated value. The primary frequency regulation of the power system is independently completed by each generator set based on local frequency information. The degree of change in its active power depends on the system's frequency deviation value, which is reflected as the droop characteristic of the generator set R.

$$R = \frac{-\Delta f / f_r}{\Delta P / P_r} \tag{9.7}$$

Where: f_r is the system's rated frequency, P_r is the unit's rated power, Δf is the frequency deviation, is the power deviation. The R value of the units in the system is usually 4%–6%.

If the frequency change lasts for a long time, and the frequency continues to fluctuate for minutes, then the automatic generation control (Automatic Generation Control, AGC) or the generator's frequency regulator begins to act, further adjusting the frequency to restore it to the rated state. This adjustment process of the frequency is called secondary frequency regulation (Secondary Frequency Regulation, SFR).

The power system frequency also has a tertiary regulation, which mainly considers seasonal factors, power generation economic factors, etc., and redistributes the unit output according to the principle of economic dispatch.

With the continuous rapid development of wind power, the penetration of wind power in the power system is gradually increasing. The most commonly used wind turbine types are doubly-fed wind turbines and direct-drive wind turbines, which generally operate in the maximum power tracking mode, that is, at any wind speed, the output active power has reached the maximum value of the utilizable wind energy efficiency. Therefore, compared with conventional thermal or hydro units, wind turbine units cannot actively provide frequency regulation capability when the power system frequency changes.

Taking a power system with a total load of 1000 MW connected to wind power as an example, the connection of wind power will replace the installed power of thermal and hydro power in the same proportion. Set the load disturbance ΔP_L to be 60 MW (0.06p.u.), when the wind power penetration rate p is respectively 0, 10%, 20%, 30%, 40%, the system's frequency change process is shown in Fig. 9.1, and the comparison analysis of indicators such as the lowest frequency, steady-state frequency deviation, and frequency change rate is shown in Table 9.1.

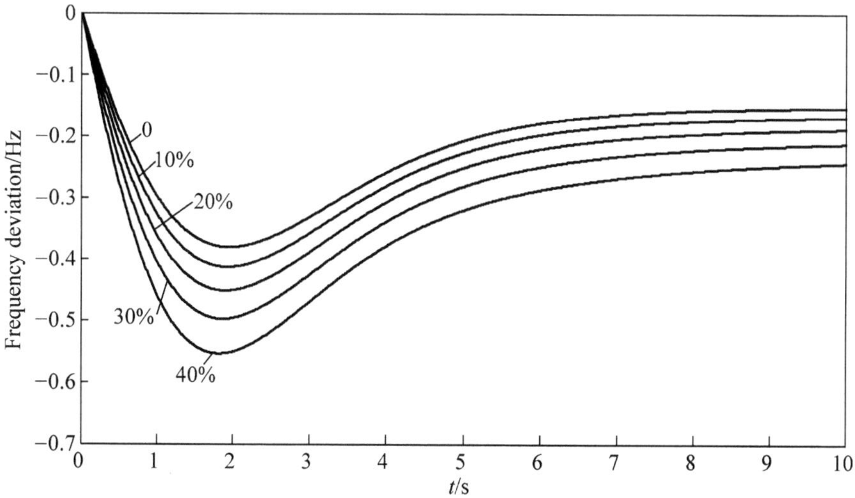

Fig. 9.1 System frequency characteristics at different penetration rates

Table 9.1 Comparison of frequency characteristic indicators at different penetration rates

Penetration rate	Lowest frequency (Hz)	Steady-state frequency deviation (Hz)	Rate of frequency change (Hz/s)
0	49.62	−0.153	−0.298
10%	49.59	−0.169	−0.326
20%	49.55	−0.188	−0.361
30%	49.50	−0.213	−0.404
40%	49.45	−0.244	−0.458

As can be seen, with the increase in wind power penetration, the lowest frequency decreases and the rate of frequency change increases after the system is disturbed, making the issue of frequency stability more prominent.

The above analysis process and results are also similar for photovoltaic power systems. It can be seen that in future power systems, as the penetration rate of renewable energy sources such as wind and photovoltaic power increases, the proportion of conventional thermal power units or hydropower units decreases relatively, which means a decrease in the equivalent inertia of the system and a reduction in frequency regulation reserve capacity, leading to worse frequency indicators during disturbances, and even after the disturbance disappears, the frequency is difficult to return to normal levels.

9.1.2 Frequency Regulation of Wind Turbine Units

Early wind turbines mostly used squirrel-cage asynchronous generators, which operated at a fixed speed. The rotor speed of the wind turbine unit was coupled with the system frequency, with a slip of about 1% ~ 2%. These turbines could actively provide inertia support for the system, but the capacity they could provide was small, and they basically had no contribution in the subsequent frequency regulation process.

The mainstream double-fed wind turbines and direct-drive wind turbines are variable-speed turbines. Due to the control of power electronic converters, the former's rotor electromagnetic speed can fluctuate within ±30% of the system synchronous speed, while the latter's fluctuation range is even larger. Therefore, the rotor of the variable-speed wind turbine is decoupled from the system frequency and cannot actively provide inertia support when the system frequency changes.

However, due to the control flexibility of variable-speed wind turbines, by adjusting the control objectives and strategies, the unit can actively respond to changes in system frequency, giving it inertia response and frequency regulation capabilities similar to traditional units. Currently, wind turbine units mainly participate in system inertia response and frequency regulation through rotor inertia control, overspeed deload control, and pitch angle control, etc. [139].

9.1.2.1 Inertia Control

Rotor inertia control is a process in the operation of wind turbine units, where the current set of the rotor-side converter is changed to control the temporary changes in rotor speed. This allows for the short-term release/absorption of some of the kinetic energy stored in the rotating mass of the wind turbine unit, in order to quickly respond to transient changes in system frequency, providing a rotational inertia similar to traditional units.

Taking a doubly-fed wind turbine as an example, by adding an auxiliary frequency control loop, the system's frequency change rate df/dt is detected in real time, and the kinetic energy stored in the wind turbine blades is controlled to provide short-term power support, as shown in Fig. 9.2 [140–143]. The wind turbine unit with the added auxiliary frequency control loop has a significant effect in supporting system frequency, increasing the equivalent inertia of the system and reducing the frequency deviation and frequency change rate after disturbance.

$$\Delta P = K_{\mathrm{df}} \frac{d\Delta f}{\mathrm{d}t} + K_{\mathrm{pf}} \Delta f \tag{9.8}$$

Where: K_{df} is the differential weight coefficient of frequency deviation; K_{pf} is the weight coefficient of frequency deviation; Δf is the system frequency deviation.

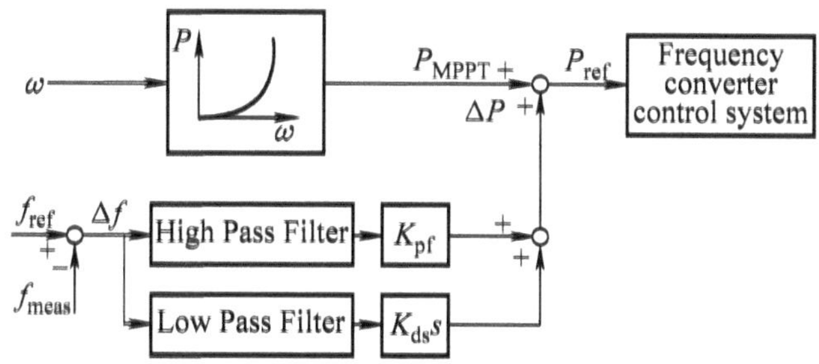

Fig. 9.2 Integrated Inertia Control Method

The function of the high-pass filter in the figure is to avoid the impact of persistent frequency deviation on the control strategy.

GE's WindINERTIA control system, by adding an inertia response control link in the control system of the wind turbine unit, temporarily increases the rated capacity output by 5% ≈ 10% when the system frequency drops, lasting a few seconds [144].

Although variable-speed wind turbine units can provide virtual inertia support similar to traditional units through control, there are several shortcomings in application:

1. Limited by the adjustable range of wind turbine speed, it is difficult to increase unit power output by reducing rotor speed when system frequency drops and wind speed is low; it is difficult to reduce unit output power by increasing rotor speed when system frequency rises and wind speed is high.
2. The control effect is related to its operating state. Due to the randomness and fluctuation of wind speed, it is difficult to guarantee a high credibility inertia response capacity. Even within the same frequency regulation process, the change in wind speed can increase uncertainty.
3. Since the rotor speed cannot be maintained at a reduced/increased state for a long time, during the wind turbine speed recovery process after inertia response, the absorption/release of some energy may cause a secondary decrease/increase in system frequency [141].

9.1.2.2 Overspeed Control

Overspeed control of wind turbines is to control the rotor of the wind turbine unit to run at overspeed, making it operate at a suboptimal point of power capture state, reserving a part of the active power for reserve, for inertia response and primary frequency regulation, as shown in Fig. 9.3.

As can be seen, at a certain wind speed, the output power of the wind turbine is determined by its speed, and the operating point can be changed by adjusting the speed of the wind turbine. When at point A, the output power of the wind turbine is

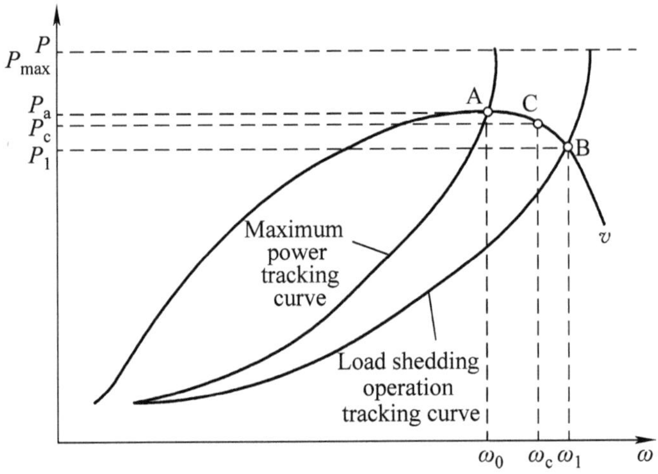

Fig. 9.3 Variable Speed Wind Turbine Overspeed Control Schematic

the maximum, corresponding to the speed at the maximum power point being ω_0, if the wind turbine speed is controlled to exceed point A to point B at this time, the output power of the turbine will decrease, achieving load reduction and retaining some spare capacity. If it is necessary to increase the output power of the turbine, the speed can be controlled to decrease to point C in the figure. In this process, on the one hand, the decrease in speed can directly release a certain amount of kinetic energy, and on the other hand, the unit operation point is also moving in the direction of increasing power generation. When the speed drops to point A, the power reaches its maximum, and further decrease will cause decrease in power. Therefore, to the left of the maximum power point A is the unstable area of active control, and it should be avoided to enter this area.

The overspeed control of the turbine is achieved by adding a frequency adjustment link in the control system. Its technical advantage is that the response speed of participating in the system's primary frequency adjustment is fast, and the impact on the mechanical stress of the turbine itself is not significant. The disadvantage of this technology is that when the wind speed reaches the rated value, the unit needs to achieve constant power operation through pitch angle control, so overspeed power generation is only applicable to operating conditions below the rated wind speed. Of course, overspeed control to some extent reduces the power generation benefits of the wind farm.

9.1.2.3 Pitch Control

The pitch control of the wind turbine is achieved by controlling the pitch angle, changing the angle of attack of the blade and the input mechanical energy, so that it is at a certain operating point below the maximum power point, thereby leaving a certain spare capacity. Under certain wind conditions, the larger the pitch angle, the greater the active reserve left by the unit. Its control characteristic curve is shown in Fig. 9.4.

As shown in the figure, the increase in the pitch angle of the turbine blades will cause the power-speed curve to move down as a whole, and the operating point will drop from point 1 to point 3, and the captured wind energy will decrease; on the contrary, if the pitch angle is reduced at this time, the energy captured by the turbine can be correspondingly increased [145]. In actual control, the pitch angle change rate limit value can be considered in the pitch angle control link of the direct-drive wind turbine [146], or the pitch angle reference value can be generated according to the power reduction demand and the frequency-active droop curve, and then the pitch angle actuator can be controlled [147].

The adjustment ability of pitch angle control is strong, the adjustment range is large, and power control under full wind speed can be achieved. However, because its actuator is a mechanical component, the response speed is relatively slow; and when the pitch angle changes too frequently, it is also easy to exacerbate the mechanical wear of the unit, shorten the service life, and increase maintenance costs. Generally, pitch control is mostly used for operating conditions above the rated wind speed, and it is more effective in supporting reserves when the system frequency drops. In this case, the turbine's participation in system frequency regulation lasts longer.

As can be seen from the above analysis, the inertia response control, overspeed control, and pitch control of the wind turbine each have a certain range of application and operating condition constraints, as shown in Table 9.2. In order to meet the system's requirements for the speed and continuity of wind power frequency regulation, the above frequency regulation methods of wind turbine can be combined to complement each other and improve the frequency regulation capability and economic benefits of wind power. However, the combined control of wind turbine is inevitably subject to the influence of wind speed changes and operating conditions, and the credibility of the capacity to participate in system primary frequency regulation and inertia response under all wind conditions is not guaranteed.

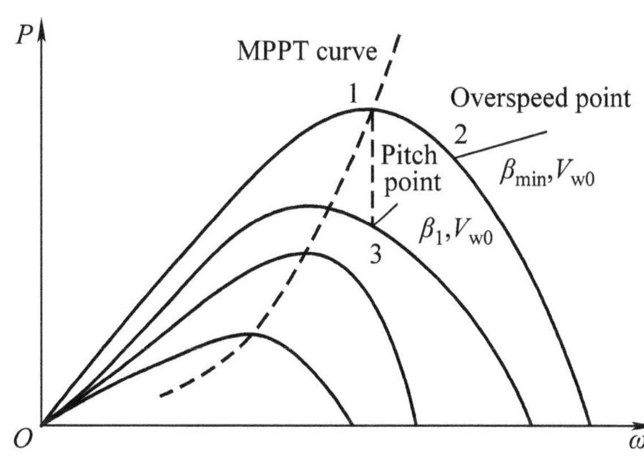

Fig. 9.4 Pitch control characteristic curve of the turbine

Table 9.2 Comparison of wind turbines frequency regulation methods

Frequency control method	Advantages	Disadvantages
Inertia control	Can operate under all wind conditions Significant power increase Fast response speed	Short duration Rotor speed recovery process can affect frequency recovery Power change rate is limited by mechanical stress and converter capacity Control blind spots exist due to speed limit restrictions
Overspeed control	When wind speed is low Overspeed control is used Fast response speed	Unable to overspeed at high wind speeds, control blind spots exist Wind speed fluctuations affect the reliability of reserve capacity Derated power generation, wind farm efficiency decreases
Pitch control	When wind speed is high Pitch control is used Large power regulation range	Slow response speed Frequent actions can cause mechanical fatigue Wind speed fluctuations affect the reliability of reserve capacity Derated power generation, wind farm efficiency decreases

9.1.3 Energy Storage Participation in Wind Power Frequency Control

By configuring a certain capacity of energy storage system in the wind farm, using its fast response, precise control, bidirectional regulation, flexible and controllable, and not constrained by the operation status of the wind turbine, it can be used as a means for wind power to participate in system frequency regulation.

Energy storage can be directly combined with wind power turbine to optimize the operation process. For example, flywheel energy storage combined with doubly-fed wind power units [148, 149], battery energy storage directly connected to the STATCOM DC bus [150], supercapacitor energy storage connected to the common DC bus of multiple direct-drive wind turbines [151], etc., to smooth the active output of the wind turbine and frequency regulation and other auxiliary functions, but it will also make the structure and control of the wind power unit complex.

However, whether energy storage is combined with wind turbines or independently configured in wind farms, if it solely relies on energy storage to undertake all the active control and frequency regulation needs of the wind farm, it will inevitably lead to problems of large energy storage capacity configuration, high cost, and poor economic benefits. If the wind power's own frequency regulation means are combined with energy storage, using the technical advantages of energy storage to compensate for the former's deficiencies in response speed and capacity reliability, it can enable wind power to have inertia response and frequency regulation capabilities under all wind conditions, improving the overall technical and economic performance of the system [152].

Therefore, this section proposes a wind-storage coordinated control strategy that combines rotor control (including inertia control and overspeed control), energy storage, and pitch control at the wind farm level, as shown in Fig. 9.5 [153]. Taking advantage of the fast and flexible response of wind turbine rotor speed control, it first responds to system frequency changes; pitch control participates in system frequency regulation for a longer time after a certain period; energy storage timely compensates for the blind spots of wind turbine frequency regulation and the lack of reserve capacity caused by wind speed changes, avoiding problems such as limited energy supply from rotor speed control, high energy storage cost, slow response time of pitch control, frequent actions reducing its lifespan, etc. While meeting the system's frequency regulation needs, it reduces the cost of energy storage and unit wear.

9.1.4 Wind-Storage Joint Frequency Regulation Control

9.1.4.1 Wind Turbine Power Reserve Strategy

Considering the technical and economic aspects of frequency regulation, four wind speed values are set, and different derating reserve strategies are given for different wind speed set values. The frequency regulation reserve capacity is set as a certain percentage of the maximum power generation of the wind turbine at a specific wind speed, and the wind speed at which the maximum output power of the wind turbine is 40% of the rated power is set as the threshold wind speed. Considering the rotor speed limit, from the threshold wind speed to the cut-out wind speed set value is divided into three stages: low wind speed, medium wind speed, and high wind speed. Among them, the upper limit of low wind speed is the wind speed at which overspeed control can fully provide reserve capacity; the upper limit of medium wind speed is the wind speed at which the speed reaches the maximum speed when

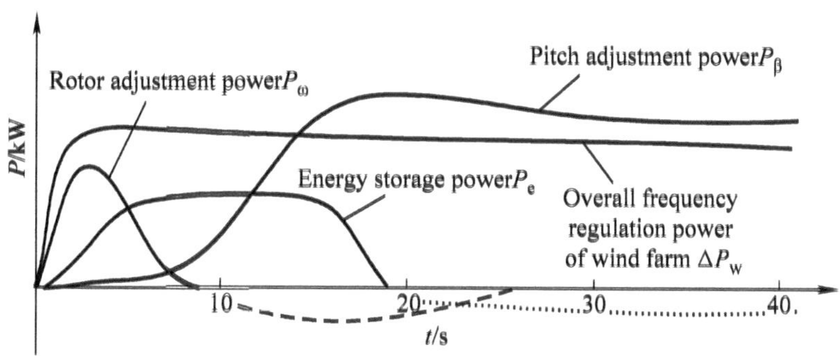

Fig. 9.5 The effect of energy storage participating in wind power frequency regulation

following the maximum power point. Under different wind speed set values, the corresponding overspeed control strategies are different, and the overspeed control power reserve curve is shown in Fig. 9.6.

When the wind speed is lower than the threshold wind speed, the power that the wind turbine can generated is small, at this time, the overspeed reserve power is small, and the frequency regulation capacity provided is too small. In this stage, the wind turbine can operate according to maximum power tracking. When the wind speed is in the low wind speed segment, overspeed control can provide a certain frequency regulation capacity, at this time, the wind turbine should operate in the derating reserve mode. When the wind speed is in the medium wind speed segment, due to speed limitations, overspeed cannot be performed after reaching the maximum speed, and the insufficient reserve capacity is provided by pitch control. When in the high wind speed segment, the frequency regulation reserve capacity is mainly provided by wind turbine pitch control.

In the medium and high wind speed segments, overspeed control needs to cooperate with pitch control, and the pitch angle is related to the wind energy utilization coefficient. In the medium and high wind speed segments, the wind turbine speed is the maximum speed, at a certain wind speed, the tip speed ratio is a fixed value, the pitch angle is a function of the wind energy utilization coefficient, and its control strategy is shown in Fig. 9.7.

In the figure, R_W for the radius of the wind turbine, λ for the tip speed ratio, C_P for the wind energy utilization coefficient, $P_{\mathrm{OSP.UL}}$ for the power output of the wind turbine after overspeed control, $C_{P.\mathrm{Pich.UL}}$ for the wind energy utilization coefficient that needs to be achieved by adjusting the blade pitch.

In different wind speed sections, according to the wind farm conditions and control requirements, the initial blade pitch angle and rotor speed of the system can be obtained, as shown in Fig. 9.8.

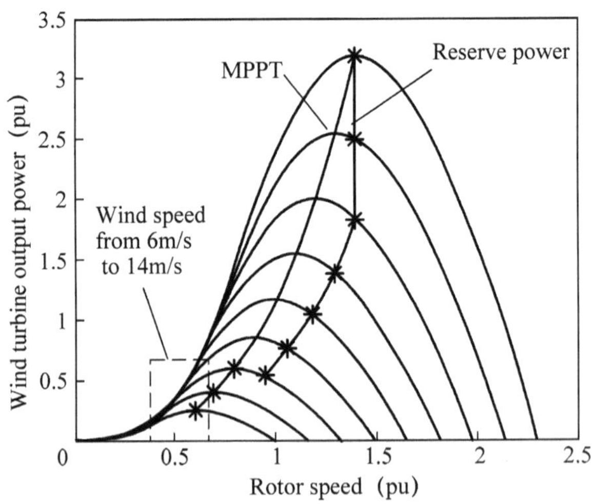

Fig. 9.6 Overspeed control power curve

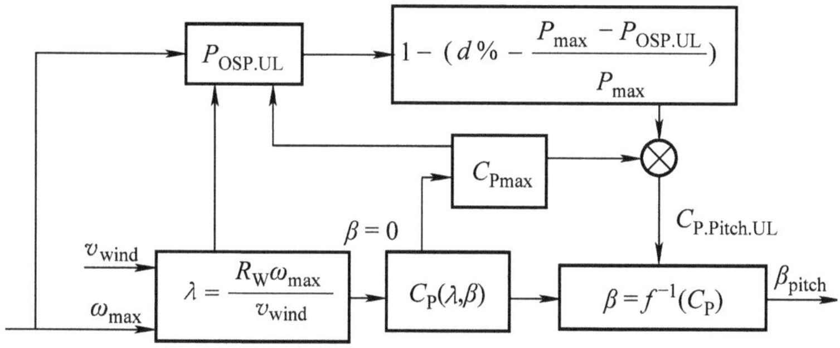

Fig. 9.7 Pitch control frequency regulation reserve strategy

In the figure, $d'\%$ represents the percentage of wind turbine reserve capacity that still needs to be provided after overspeed reserve. Through the system deloaded reserve scheme of Figs. 9.6, 9.7, and 9.8, the rotor speed, blade pitch angle, and initial active and reactive power of the wind turbine can be obtained during normal operation of the system.

9.1.4.2 Frequency Regulation Power Demand Determination Strategy

The frequency regulation power demand determination module mainly includes two control loops. The first loop is to achieve frequency difference response through proportional control, which is a differential regulation, similar to the primary frequency regulation of conventional power sources, as shown in the following formula:

$$P_1^* = K_{pf}\left(f_{grid} - f_{ref}\right) = K_{pf}\Delta f \tag{9.9}$$

Fig. 9.8 Initial blade pitch angle and speed requirements

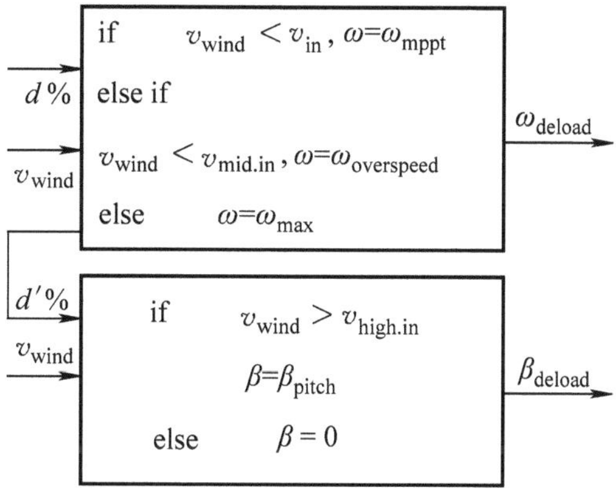

The other loop is to achieve frequency rate of change response through $d\Delta f/dt$ proportional control, similar to the inertia control of conventional power sources, as shown in the following formula:

$$P_2^* = K_{df} d\Delta f / dt \tag{9.10}$$

Therefore, through the above two control loops, the system frequency changes are detected, the wind turbine rotor speed is changed, the kinetic energy in the rotor is released or increased to change the active output, and the frequency regulation power demand determination strategy is shown in Fig. 9.9. In the figure, f_{grid} is the measured frequency of the power grid, f_{ref} is the rated or reference frequency.

9.1.4.3 Frequency Regulation Power Allocation Strategy

When the system experiences frequency changes, the frequency regulation power allocation module will determine the specific allocation of the frequency regulation power shortage according to the system frequency demand and the real-time conditions of the wind turbine. The frequency regulation power allocation strategy is shown in Fig. 9.10.

When the system frequency is detected to be lower than the rated frequency, the wind storage system needs to provide a certain active power to participate in frequency upward adjustment. At this time, the energy of the system deloaded reserve is released first. When the adjustment ability of the wind turbine reaches the limit, the remaining part is provided by the energy storage. When the system frequency is detected to be higher than the rated frequency, the wind storage needs to absorb a certain active power to participate in frequency downward adjustment. At this time, the wind turbine output power can be reduced by adjusting the wind turbine speed and blade pitch angle, and the energy storage can charge to absorb a certain active power, thus playing a role in frequency regulation.

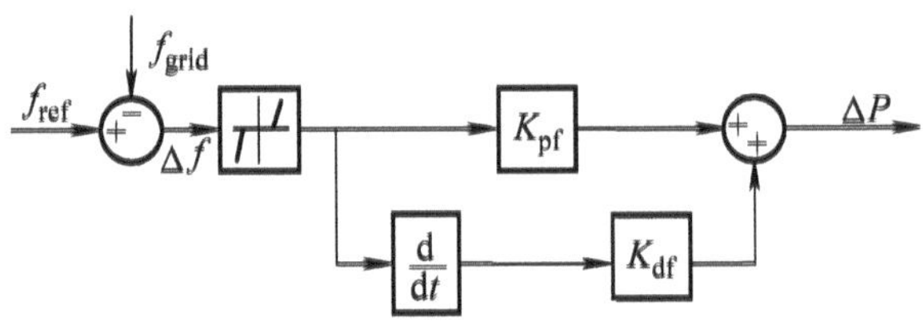

Fig. 9.9 Frequency regulation power determination strategy

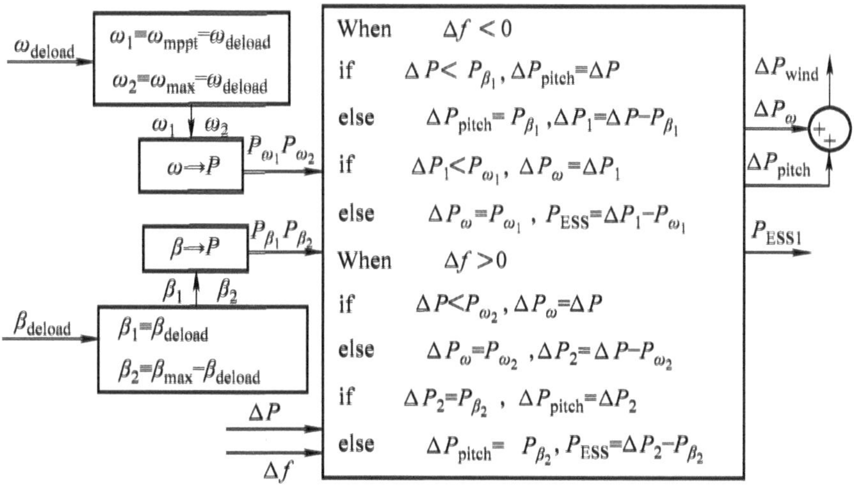

Fig. 9.10 Frequency regulation power allocation strategy

9.2 Wind-Storage Combined Participation in System Voltage Regulation

9.2.1 Voltage Regulation in Power Systems

Voltage stability in a power system refers to the ability of all buses in the system to continuously maintain an acceptable voltage after the system is disturbed. Conversely, if there is a gradual, uncontrollable voltage drop, the system enters a state of voltage instability. There are many reasons for the decline or instability of the voltage quality in the power system, such as:

1. A large increase in load, causing the system's transmission capacity to exceed the maximum power it can withstand.
2. Weakening of the power grid structure, due to line faults causing some parts of the transmission line to be cut off, leading to an increase in reactive power loss in the line.
3. A decrease in the source voltage, causing the system to be unable to maintain a normal voltage.

Voltage is related to the balance of reactive power. By controlling the generation, consumption, and transmission of reactive power in the power system, the bus voltage in the system can be controlled. Common voltage regulation methods include the following:

1. Changing the excitation current of the generator. The Automatic Voltage Regulator (AVR) of the generator maintains a constant terminal voltage by adjusting the size of the excitation current. That is, increasing the excitation cur-

rent of the generator will increase the voltage of the generator; reducing the excitation current of the generator will reduce the voltage of the generator.
Synchronous generators are equipped with automatic excitation adjustment devices AVR, which automatically adjust the terminal voltage of the generator, distribute reactive power, and improve the stability of the synchronous operation of the generator. The generator can maintain its rated power operation within 95%–105% of its rated voltage. Since the reactive power regulation capability of the generator itself can be fully utilized, no additional equipment and investment are needed, so it is the first choice among various voltage regulation methods.

2. Using reactive voltage regulation devices. These include, 1) parallel capacitors, parallel reactors, synchronous condenser, and Static Var Compensators (SVC), etc. Among them, synchronous condenser and SVCs can automatically adjust the size of reactive power to ensure that the bus voltage of the connected node remains unchanged, and can also work with the generator to maintain a constant voltage. 2) Series capacitor compensators.
   The configuration principle of reactive compensation equipment follows the principle of "layered and regional, local balance" of reactive power. In systems with insufficient reactive power, the primary issue is to increase reactive power compensation equipment, with a large number of parallel capacitors used as reactive compensation equipment. In special cases, static compensators and synchronous condenser are used. For 500 kV, 330 kV, and some 220 kV lines, sufficient inductive reactive compensation equipment should be installed to prevent overvoltage in the grid caused by excessive charging power during light load.
3. Changing the distribution of reactive power in the system by adjusting the tap changer of the transformer. This includes, (1) on-load voltage regulation, switching the transformer tap changer under load, the adjustment range can reach more than 30% of the rated voltage. (2) Off-load voltage regulation, adjusting the tap changer position to change the transformation ratio when the transformer is powered off, suitable for substations with seasonal power outages.
   In addition, voltage regulation can also be achieved by increasing the cross-sectional area of the grid conductors, changing the wiring method of the grid, changing the number of transformers running in parallel, and series capacitor compensation of transmission lines. For example, increasing the cross-sectional area of the conductor can reduce voltage loss, or by cutting off or putting into one of the double circuits, cutting off or putting into some of the transformers running in parallel in the substation, etc., to change the wiring method of the grid.
   Many wind farms in China are connected to relatively weak power grids. The startup and operation of wind power require a large amount of reactive power, but the self-regulation ability of wind turbine units for reactive voltage is insufficient. This leads to voltage fluctuations at the wind farm connection points, which can easily cause problems with grid voltage stability. The mechanical voltage adjustment methods in wind power collection areas, such as switching capacitors and reactors, are limited by adjustment speed and frequency, leading to unqualified

voltage at the connection points, or even exceeding limits, posing a risk of triggering high and low voltage disconnection of wind turbines. On the other hand, voltage stability and quality issues in wind farms can also affect the normal operation of wind turbines. The stator of a doubly-fed wind turbine is directly connected to the grid, and the rotor winding is connected to the grid through a converter. The converter capacity of the doubly-fed wind turbine is approximately equal to the generator slip power, generally 20%–30% of the unit's rated capacity. Under the impact of a grid short-circuit fault, the sudden drop in grid voltage can induce a large transient voltage in the rotor winding, which can generate a large surge current and possibly damage the rotor-side converter. To solve the impact of grid voltage drop on wind turbines, rotor-side control strategies such as crowbar control can be adopted. However, while crowbar control achieves low voltage ride-through of the wind turbine to some extent, when crowbar control is in effect, the rotor-side converter is blocked. If the wind turbine is to operate normally, it needs to absorb a certain amount of reactive power from the grid, which adversely affects the recovery of grid voltage. Exploring the reactive regulation capabilities of doubly-fed turbines and direct-drive wind turbines, forming voltage regulation schemes suitable for power systems with high penetration of wind power, and leveraging the reactive support capabilities of wind power for the grid, are of great significance.

9.2.2 Wind Turbine Voltage Regulation

The voltage regulation capability of a wind turbine is the ability of the unit to generate reactive power. From the perspective of motor operation, wind turbines are divided into two types: constant speed constant frequency wind turbines and variable speed constant frequency wind turbines. Constant speed constant frequency wind turbines use squirrel cage induction motors. When the rotor speed of the generator is less than the synchronous speed, the motor absorbs active power from the grid. When the rotor speed of the generator is greater than the synchronous speed, it converts the mechanical energy of the wind turbine into electrical energy and outputs active power to the grid. This type of wind turbine needs to absorb reactive power from the grid to establish a magnetic field during operation, cannot output reactive power to the system, and needs to be equipped with additional reactive compensation devices such as capacitors. The mainstream variable speed constant frequency wind turbines are doubly-fed wind turbine units. Under normal operating conditions, doubly-fed wind turbines (Double Fed Inductive Generator, DFIG) have flexible power regulation characteristics, including maximum power tracking, output power decoupling control, bidirectional flow of active power in the rotor-side converter, etc. By adjusting the rotor current, active and reactive power decoupling control is achieved, the reactive regulation range is expanded, and it plays a role in supporting the voltage level of the wind farm grid connection point to a certain extent, but its reactive regulation capability is not stable and reliable. The limit range of its reactive regulation changes with its active output, and the wind power has a strong randomness and fluctuation, which inevitably causes its reactive regulation capability to fluctuate. Doubly-fed wind turbines can generate reactive power by controlling their rotor-side and grid-side converters. The output of reactive power on the rotor side mainly relies on the generation of excitation current, while the grid-side converter generates it by changing the power factor. The active power and reactive power output from the stator side of the doubly-fed wind turbine are:

$$\begin{aligned} P_s &= -\frac{x_m}{X_{ss}}\left|\dot{U}_s\right| I_{qr} \\ Q_s &= -\frac{\left|\dot{U}_s\right|^2}{X_{ss}} + \frac{x_m}{X_{ss}}\left|\dot{U}_s\right| I_{dr} \end{aligned} \tag{9.11}$$

In the formula, P_s represents the active power output from the stator of the wind turbine, Q_s represents the reactive power output from the stator of the wind turbine, $\left|\dot{U}_s\right|$ represents the terminal voltage; x_m represents the excitation reactance; I_{qr}, I_{dr} represent the torque component and excitation component of the rotor current; x_s represents the stator side reactance; $X_{ss} = x_s + x_m$ represents the equivalent reactance of the stator circuit.

The active power output from the stator is related to the mechanical power of the doubly-fed unit P_{mec}. If the losses of the stator and rotor windings of the motor are ignored, the relationship between the mechanical power of the doubly-fed motor and the active power output of the fixed rotor satisfies:

$$P_{mec} = P_s - P_r \tag{9.12}$$

Where, P_r represents the active power on the rotor side, which is equal to the slip power of the active power on the stator side, and satisfies:

$$P_r = sP_s = sP_{mec} / (1 - s) \tag{9.13}$$

In the formula, s represents the slip rate.

When the terminal voltage is fixed, the adjustment of the reactive power on the stator side is mainly affected by the excitation component of the rotor current I_{dr}, and the rotor current is affected by the thermal limit current of the rotor winding and the maximum current limit of the converter. Therefore, the main factor affecting the reactive power output of the doubly-fed wind machine is the maximum current value of the rotor.

From this, the power limit circle equation of the doubly-fed wind machine can be obtained:

$$P_s^2 + \left(Q_s + \frac{\left|\dot{U}_s\right|^2}{X_{ss}} \right)^2 = \frac{\left|\dot{U}_s\right|^2 x_m^2}{X_{ss}^2} \left(I_{qr}^2 + I_{dr}^2 \right) \le \frac{\left|\dot{U}_s\right|^2 x_m^2}{X_{ss}^2} I_{r\max}^2 \tag{9.14}$$

In the formula, $I_{r\max}$ represents the maximum current value of the rotor.

The upper and lower limits of the reactive power output on the stator side of the doubly-fed wind machine are:

$$\begin{aligned} Q_{s\min} &= -\frac{\left|\dot{U}_s\right|^2}{X_{ss}} - \sqrt{\left(\frac{x_m}{X_{ss}} \left|\dot{U}_s\right| I_{r\max} \right)^2 - P_s^2} \\ Q_{s\max} &= -\frac{\left|\dot{U}_s\right|^2}{X_{ss}} + \sqrt{\left(\frac{x_m}{X_{ss}} \left|\dot{U}_s\right| I_{r\max} \right)^2 - P_s^2} \end{aligned} \tag{9.15}$$

Obviously, at a given wind speed (at this time, the active power output from the stator side of the wind turbine is fixed), the ability of the stator to output and absorb reactive power is not symmetrical.

The grid-side converter is essentially a voltage-type PWM rectifier, with the AC side having controlled current source characteristics, capable of four-quadrant operation. The power design of the grid-side converter is generally based on the maximum power difference of the wind power system, taking into account line losses, switch losses, etc. The maximum power of the grid-side converter is set to $P_{c\max}$, then the reactive power capability of the grid-side converter should meet:

$$P_c^2 + Q_c^2 \le P_{c\max}^2 \tag{9.16}$$

From this, the reactive power limit of the grid-side converter can be calculated as:

$$\begin{aligned} Q_{c\min} &= -\sqrt{P_{c\max}^2 - P_c^2} \\ Q_{c\max} &= \sqrt{P_{c\max}^2 - P_c^2} \end{aligned} \tag{9.17}$$

Where, $P_{c\max}$ is the capacity of the grid-side converter; P_c is the active power output value of the grid-side converter, this output power is related to the slip rate of the wind turbine unit and the mechanical power of the wind turbine P_{mec}:

$$P_c = sP_{mec} / (1 - s) \tag{9.18}$$

If system losses are ignored, the active power P_g input from the wind turbine unit to the grid is equal to the mechanical power P_{max} generated by the wind turbine.

The reactive output limit of the doubly-fed wind turbine is:

$$\begin{aligned} Q_{g\min} &= Q_{c\min} + Q_{s\min} \\ Q_{g\max} &= Q_{c\max} + Q_{s\max} \end{aligned} \tag{9.19}$$

Without considering the reactive power regulation output of the grid-side converter, the P-Q characteristic curve of the stator side of the doubly-fed wind turbine is shown as the dashed circle in Fig. 9.11. Since the grid-side converter of the doubly-fed wind turbine is generally designed according to the maximum power difference, when the unit's active output is small, the adjustable reactive output range of the grid-side converter is larger; when the unit's active output is close to or at the rated state, the reactive output of the grid-side converter can be ignored. Therefore, the P-Q characteristic curve considering the reactive output of the grid-side converter is shown as the solid circle in Fig. 9.11.

In the figure, the dashed circle is the reactive power regulation range using only the stator of the wind turbine unit for reactive regulation, with the power circle centered at $(-\frac{|\dot{U}_s|^2}{X_{ss}}, 0)$ and radius $\frac{|\dot{U}_s|x_m}{X_{ss}}I_{r\max}$. The solid circle is the overall reactive power output range of the wind turbine unit, which is obtained by adding the stator output reactive power and the grid-side converter output reactive power. In Fig. 9.11, P_0 and P_n are both theoretical values obtained through calculation, and the acquisition of this active power value only considers the rotor excitation factor of the wind turbine unit. When the active power is P_0 When the active power is P_n, the wind turbine will no longer have the ability to emit reactive power; when the active power is P_n, the rotor excitation current of the wind turbine reaches its maximum.

It can be seen that the reactive power regulation range of the wind turbine changes with the change in active output. When the wind speed is low, the active output of the wind turbine is small, and the reactive output and adjustable range of the wind turbine are large; on the contrary, when the wind speed is high, the active output of the wind turbine increases, and the reactive output and adjustable range of the wind turbine decrease. When the active output of the wind turbine is in the range of [0,

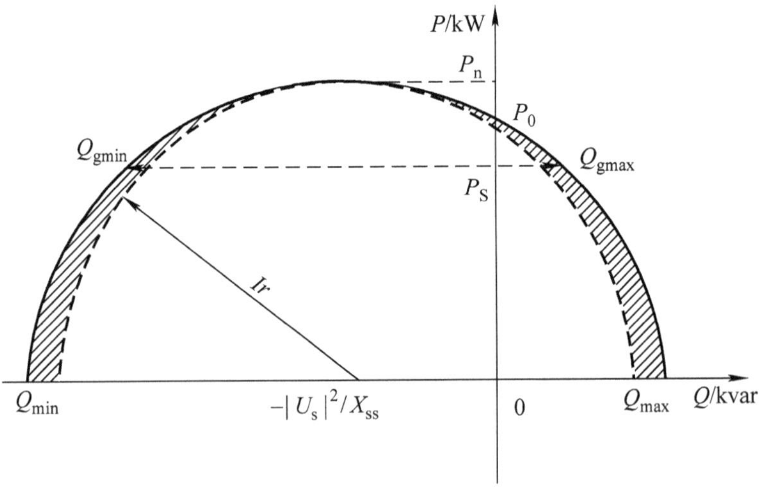

Fig. 9.11 PQ characteristic curve of doubly-fed wind turbine

P_0], the wind turbine can provide a certain amount of reactive power; when the active output is in the range of [P_0, P_n], the wind turbine no longer has the ability to output reactive power, and the reactive power it absorbs gradually approaches $-\frac{|U_s|^2}{X_s}$.

When the active output of the wind turbine is in the range of [P_0, P_n], the doubly-fed wind turbine must absorb a certain amount of reactive power. The reason is that when the active output of the wind turbine is close to the rated value, the rotor current torque component I_{qr} is large, limited by the maximum rotor current, the rotor current excitation component I_{dr} decreases, and its reactive power regulation range also decreases. At the same time, the doubly-fed wind turbine must absorb a certain amount of reactive power from the system to maintain a sufficient rotor current excitation component to establish a rotating magnetic field. It should be noted that according to the design standard, the rated power of the wind turbine is less than P_0, that is, during normal operation, the active output of the wind turbine is in the [0, P_0] segment, the doubly-fed wind turbine has a certain reactive power regulation capability, which is related to the size of the active power.

9.2.3 Wind-Storage Combined Voltage Regulation Control

9.2.3.1 Voltage Regulation Power Demand Determination Strategy

The control of wind-storage on system voltage regulation aims to ensure that the grid-connected point or the bus voltage at the end of the outgoing line is within a reasonable range. Fig. 9.12 shows a schematic diagram of voltage drop.

The vertical component of voltage drop:

$$\Delta U = \frac{PR + QX}{U} \tag{9.20}$$

In the high-voltage power grid, $R << X$, so, $\Delta U \approx \frac{QX}{U}$.

Assuming the reference value of the line end voltage is U_{2ref}, then we have,

$$U_1 - U_{2ref} = \frac{Q_{ref} X}{U} = kQ_{ref} \tag{9.21}$$

If the detected line end voltage is U_2, then we have,

$$U_1 - U_2 = \frac{QX}{U} = kQ \tag{9.22}$$

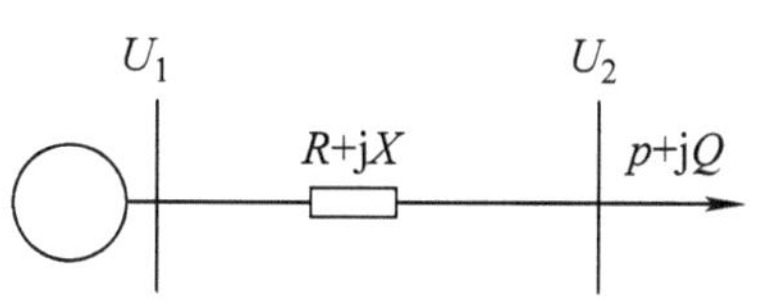

Fig. 9.12 Schematic diagram of voltage drop

In order to adjust U_2 to U_{2ref}, it is necessary to control Q to Q_{ref}. At this time,

$$U_2 - U_{2ref} = k\left(Q_{ref} - Q\right) = kQ_C \Rightarrow Q_C = {U_2 - U_{2ref}}/{k} \tag{9.23}$$

The capacity that needs to be compensated is Q_c. The control diagram is shown in Fig. 9.13. Where, U_{grid} is the measured voltage of the grid, U_{ref} is the reference voltage of the system.

9.2.3.2 Voltage Regulation Power Distribution Strategy

When the system voltage changes, the voltage regulation power distribution module will determine the specific distribution of voltage regulation power demand according to the system voltage regulation demand and the real-time working condition of the wind turbine. Considering that the reactive power of the doubly-fed wind turbine includes the reactive power on the stator side and the reactive power of the grid-side converter, in order to minimize the converter power consumption, the reactive power on the stator side is used first when distributing reactive power. For the wind-storage system, it tries to use the voltage regulation capability of the wind turbine itself as much as possible during voltage regulation, and uses energy storage for voltage regulation when it cannot complete the voltage regulation task. The voltage regulation power distribution strategy is shown in Fig. 9.14.

When a change in system voltage is detected, the maximum reactive power regulation range of the wind turbine is first determined. While ensuring that the capacity of the wind turbine converter and the rotor current do not exceed the limit, the reactive power regulation capability of the wind turbine is used first for reactive power regulation, and the insufficient part is supplemented by energy storage.

9.3 Wind-Storage Joint Participation in System Frequency and Voltage Regulation

From the above analysis, it can be seen that configuring an appropriate amount of energy storage in the wind farm, utilizing the ability of energy storage to quickly charge and discharge, flexibly control active/reactive power, and combining it with the frequency and voltage regulation capabilities of wind power itself, can effectively improve the frequency and voltage regulation performance of the wind farm as a main power source.

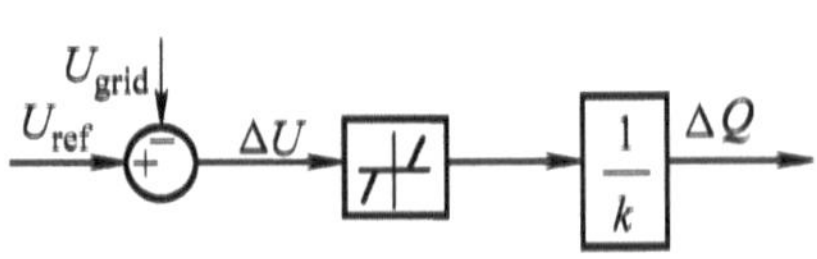

Fig. 9.13 Voltage regulation demand power determination module diagram

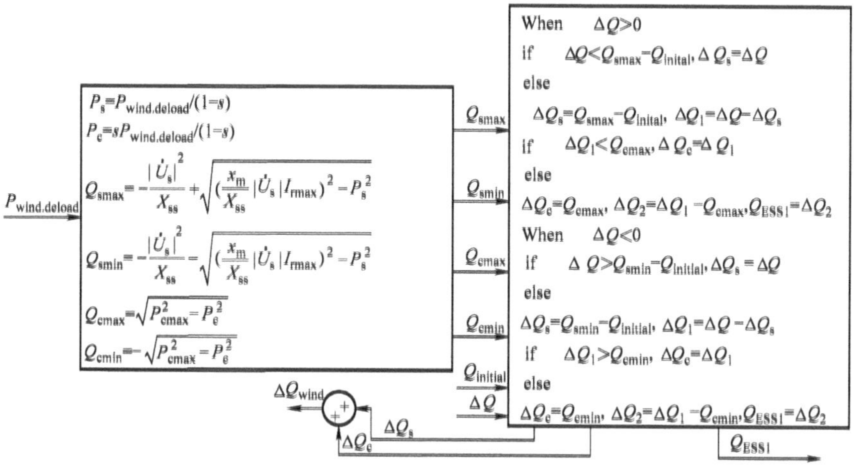

Fig. 9.14 Voltage regulation power distribution strategy

9.3.1 Wind-Storage Joint Frequency and Voltage Regulation Scheme

The combination of wind power and energy storage can fully exploit the advantages of wind turbines, such as rapid overspeed control response, strong blade control adjustment ability, wide adjustment range, and flexible and controllable energy storage. It avoids the impact of factors such as limited energy of wind turbine overspeed control, long response time of blade control, frequent participation in adjustment reducing service life, and high cost of energy storage. While meeting the system's frequency regulation needs, it improves the overall economic benefits of the wind-storage system, with simple control and easy application in engineering practice.

The participation of wind-storage combination in system frequency and voltage regulation can be divided into three modules in control, including frequency regulation module, voltage regulation module, and collaborative control module, as shown in Fig. 9.15. Among them, the frequency regulation module consists of the wind turbine power reserve module, frequency regulation demand power determination unit, and frequency regulation power distribution module. According to the system frequency status or dispatching demand, it determines the active power output of the wind-storage combined system; the voltage regulation module consists of the voltage regulation demand power determination unit and voltage regulation power distribution module. According to the system voltage status or dispatching demand, it determines the reactive power output of the wind-storage combined system; the collaborative control module consists of the wind turbine active/reactive power output reference value determination module and energy storage active/reactive power output reference value determination module. According to the operation status of

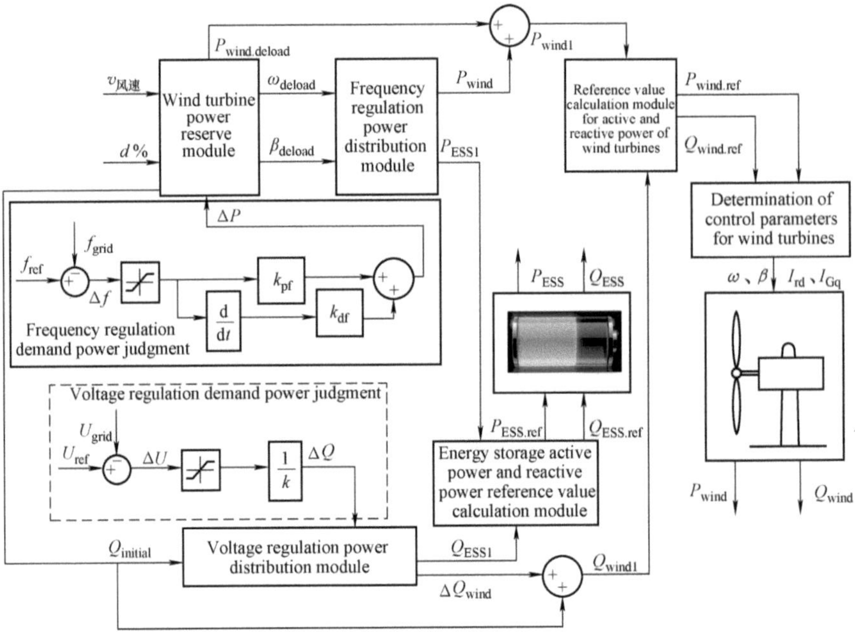

Fig. 9.15 Wind-storage combined participation in frequency and voltage regulation control scheme

the wind turbine and energy storage, it determines the corresponding active power and reactive power output.

In the figure, $v_{风速}$ represents the wind speed of the wind farm; $d\%$ represents the set percentage of wind turbine reserve capacity; Δf represents the detected system frequency difference; ΔU represents the detected system voltage difference; ΔP and ΔQ represent the demand for active and reactive power when the system frequency and voltage change; $P_{wind.\,deload}$, $\omega_{wind.\,deload}$ and $\beta_{wind.\,deload}$ represent the active power, corresponding rotor speed, and blade pitch angle emitted by the wind turbine after the deloaded reserve module; $Q_{initial}$ represents the initial value of the wind turbine's reactive power at the corresponding wind speed.

When a system frequency change is detected, the overspeed reserve capacity stored in the wind turbine rotor is first released through inertia control for inertia response.

Subsequently, the wind turbine deloaded control begins to operate for frequency regulation, including overspeed control and blade control. The overspeed control method can reduce the mechanical wear caused by frequent blade changes, control speed quickly, and its regulation efficiency is better than blade control. However, when the wind speed is high, due to the speed limit of the wind turbine rotor, overspeed operation cannot be performed, and blade control needs to be adopted; when the wind speed is too low (such as close to the cut-in wind speed), the frequency regulation ability provided by the wind turbine itself through overspeed or blade change is limited, and energy storage can supplement a certain amount of frequency regulation capacity.

When system voltage changes are detected, the wind turbine can output or absorb reactive power by adjusting the excitation current of the turbine and the power factor of the grid-side converter, participating in system voltage regulation. Similarly, due to the limitations of the wind turbine converter capacity and rotor current, when the system's reactive power demand is large, energy storage can supplement a certain voltage regulation capacity.

9.3.2 Wind-Storage Combined Frequency and Voltage Regulation Control Strategy

Considering that wind power is limited by the capacity of the wind turbine converter and rotor current, energy storage systems are also subject to capacity constraints. When the system experiences voltage or frequency changes, the frequency and voltage regulation capacity of wind turbines and energy storage systems should be reasonably allocated according to the operation status of wind-storage, to achieve the optimal regulation effect.

The frequency of the power system depends on the balance of active power and is maintained by the balance between the total active power output by all generator units in the system and the total active power consumed by all loads (including grid losses). Since the active load in the system is constantly changing, leading to frequency changes, it is necessary to timely adjust the active power output of all operating units in the system to ensure the frequency is within the allowable range. The national standard for frequency fluctuation range in China is shown in Table 9.3.

The voltage quality of the power system is one of the important indicators of power quality, and its fluctuations often occur due to some disturbances in the grid, such as the switching action of a large capacity load, or changes in the system state, etc. The national standard for voltage fluctuation range in China is shown in Table 9.4.

According to the frequency and voltage fluctuation ranges given in Tables 9.3 and 9.4, the operating state zoning of the power system can be obtained as shown in Fig. 9.16.

Figure 9.16 divides into five areas according to the frequency of the x-axis and the voltage of the y-axis, namely the dead zone, Area A, Area B, Area C, and outside the area. When the system status is detected in the dead zone, there is no need to

Table 9.3 National Standard for Frequency Fluctuation Grading

Frequency grade	Fluctuation range
Grade A	Between ±0.05 Hz
Grade B	Between ±0.5 Hz
Grade C	Between ±1 Hz

Table 9.4 National Standard for Voltage Fluctuation Grading

Voltage grade	Fluctuation range
Grade A	Between ±5%
Grade B	Between −10% and + 7%
Grade C	Between ±10%

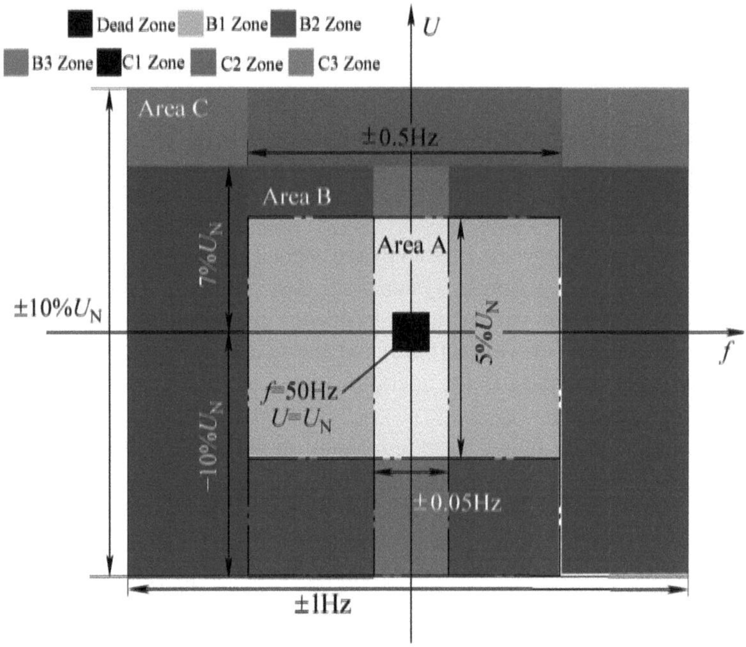

Fig. 9.16 Power System Operating State Zoning Diagram

enable wind power and energy storage to participate in frequency and voltage regulation; when the system status is in Area A, the quality of the system's electrical energy is relatively good, and in situations where the quality of electrical energy is highly demanded, wind storage can be used for minor adjustments; when the system status is in Area B or Area C, the quality of frequency and voltage needs to be improved, and wind storage needs to participate in frequency regulation and voltage regulation; when the system status is outside Area C, the quality of frequency and voltage is very poor, and wind storage needs to provide sufficient frequency regulation and voltage regulation capacity, but due to equipment and operational status limitations, active/reactive power, wind turbine/energy storage coordination is required.

Obviously, when the state of the power system varies, the requirements for frequency and voltage regulation are different. For convenience of analysis, the joint frequency and voltage regulation of wind storage is divided into three control modes, namely frequency priority mode, voltage priority mode, and no priority mode.

According to the power system state partition shown in Fig. 9.16, when the system status is detected in Areas B1 and C1, the system's frequency regulation demand is more apparent; at this time, the frequency regulation demand should be prioritized, and wind storage mainly meets the active power demand, that is, it is in the frequency priority mode, represented as Case1. When the system status is detected in Areas A, B3, and C3, the system's voltage regulation demand is more apparent; at this time, the voltage regulation demand should be prioritized, and wind storage mainly meets the reactive power demand, that is, it is in the voltage priority mode, represented as Case3. When the system status is detected in Areas B2, C2, and outside Area C, the system has demands for both frequency and voltage regulation; at this time, the priority levels of active power and reactive power of wind storage are the same, and the active and reactive power can be allocated according to the set ratio based on the actual situation, that is, it is in the no priority mode, represented as Case2.

The coordinated control strategy of wind storage participating in system frequency and voltage regulation is shown in Figs. 9.17 and 9.18. Among them, Fig. 9.17 is the wind turbine active/reactive power reference value determination module, and Fig. 9.18 is the energy storage active/reactive power reference value determination module.

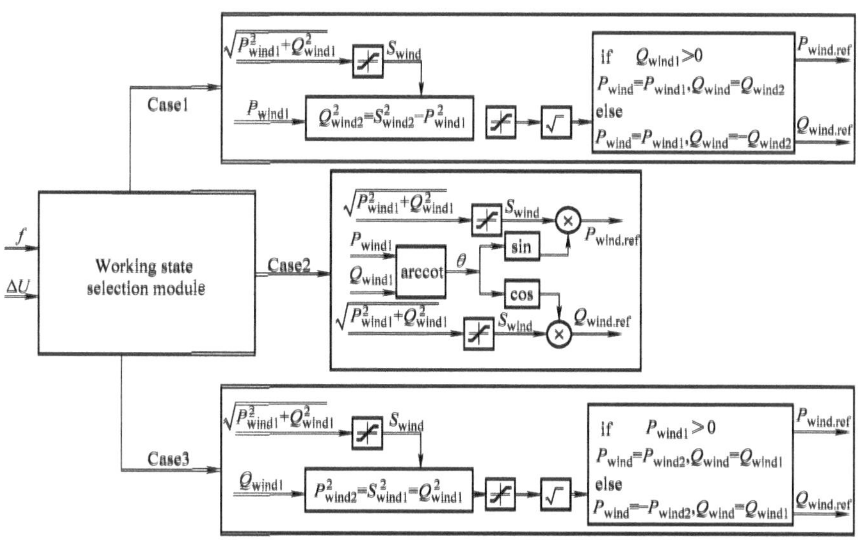

Fig. 9.17 Wind turbine active/reactive power reference value determination module

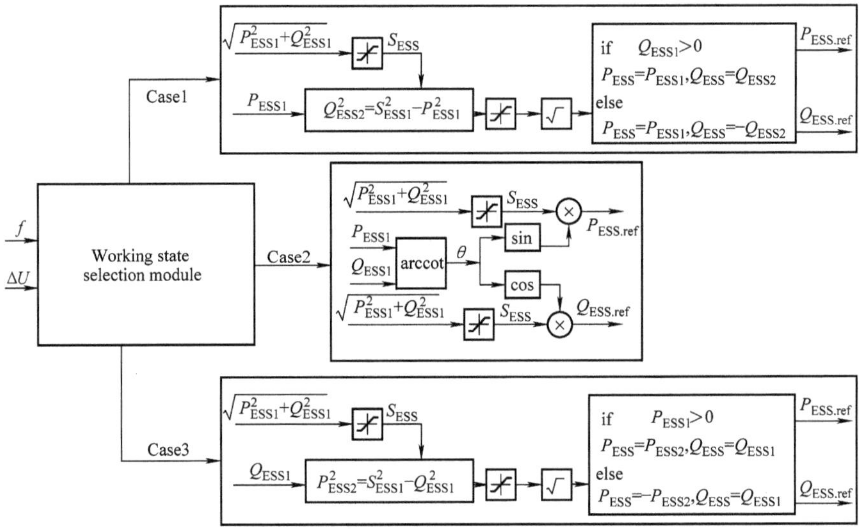

Fig. 9.18 Energy storage active/reactive power reference value determination module

9.3.3 *Case Study of Wind Storage Joint Frequency and Voltage Regulation*

The RTDS real-time simulation platform is set up to verify the effectiveness of the wind storage joint frequency and voltage regulation scheme. The example is based on the WSCC 3-machine 9-node power system in the western United States, with a wind storage joint system, the main parameters of the system are:

The base values are $S_B = 100$MVA, $U_B = 230$ kV, the system frequency is 50 Hz.

Generator Parameters

G1: 247.5MVA, 16.5 kV, power factor is 1, Salient-Pole turbine set, 180 rpm, $x_d = 0.146$, $x'_d = 0.0608$, $x_q = 0.0969$, $x'_q = 0.0969$, $x_l = 0.0336$, $T'_{d0} = 8.96$ s, $T'_{q0} = 0$ s, $H = 23.64$ s, $D = 0$

G2: 192MVA, 18 kV, power factor of 0.85, steam turbine unit (Round-Rotor), 3600 rpm, $x_d = 0.8958$, $x'_d = 0.1198$, $x_q = 0.8645$, $x'_q = 0.1969$, $x_l = 0.0521$, $T'_{d0} = 6$ s, $T'_{q0} = 0.535$ s, $H = 6.4$ s, $D = 0$

G3: 128MVA, 13.8 kV, power factor of 0.85, steam turbine unit (Round-Rotor), 3600 rpm, $x_d = 1.3125$, $x'_d = 0.1813$, $x_q = 1.2578$, $x'_q = 0.25$, $x_l = 0.0742$, $T'_{d0} = 5.89$ s, $T'_{q0} = 0.6$ s, $H = 3.01$ s, $D = 0$

Transformer Parameters

T1: 16.5/230 kV, $X_T = 0.0576$; T2: 18/230 kV, $X_T = 0.0625$; T3: 13.8/230 kV, $X_T = 0.0586$.

Line Parameters

Line1: $Z = 0.01 + j0.085$, $B/2 = j0.088$; Line2: $Z = 0.032 + j0.161$, $B/2 = j0.153$; Line3: $Z = 0.017 + j0.092$, $B/2 = j0.079$; Line4: $Z = 0.039 + j0.17$, $B/2 = j0.179$; Line5: $Z = 0.0085 + j0.072$, $B/2 = j0.0745$; Line6: $Z = 0.0119 + j0.1008$, $B/2 = j0.1045$.

Load LumpA: 125 + j50MVA, LumpB: 90 + j30MVA, LumpC: 100 + j35MVA.

Generator G1 is set as the system's balance node Slack, with a voltage amplitude of 1.04pu and a reference phase angle of 0°; G2 and G3 are set as PV nodes, with active power outputs of 1.63pu and 0.85pu respectively, and voltage amplitudes are both set to 1.025pu.

1. The corrected generator reactance $x = x_B \dfrac{S_r}{S_B}$, S_r is the rated capacity of the generator;
2. The corrected generator inertia $H = H_B \dfrac{S_B}{S_r}$;
3. The corrected transformer reactance $x = x_B \dfrac{S_r}{S_B}$, S_r is the rated capacity of the transformer.

The wind turbine and energy storage are installed at bus 8, with a single wind turbine capacity of 2.2 MW, a total of 70 units, a single energy storage capacity of 2.2 MW, a total of 30 units, and the simulation system is shown in Fig. 9.19.

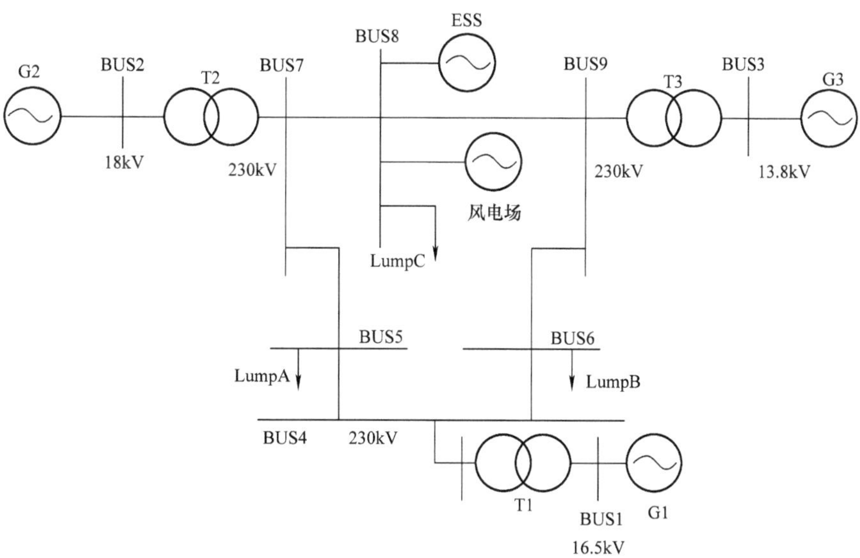

Fig. 9.19 Wind-storage joint frequency and voltage regulation simulation system diagram

9.3.3.1 Load Sudden Change

Choose the medium wind speed for simulation (wind speed 10 m/s), at bus 8, a sudden increase of 30 MW active load and 30Mvar reactive load occurs simultaneously, observe the effect of wind-storage participation in system frequency and voltage regulation, the results are shown in Fig. 9.20. Three operating schemes are set for comparison, scheme 1 is the case where both the wind turbine and energy storage do not participate in frequency-voltage regulation (black curve), scheme 2 is the

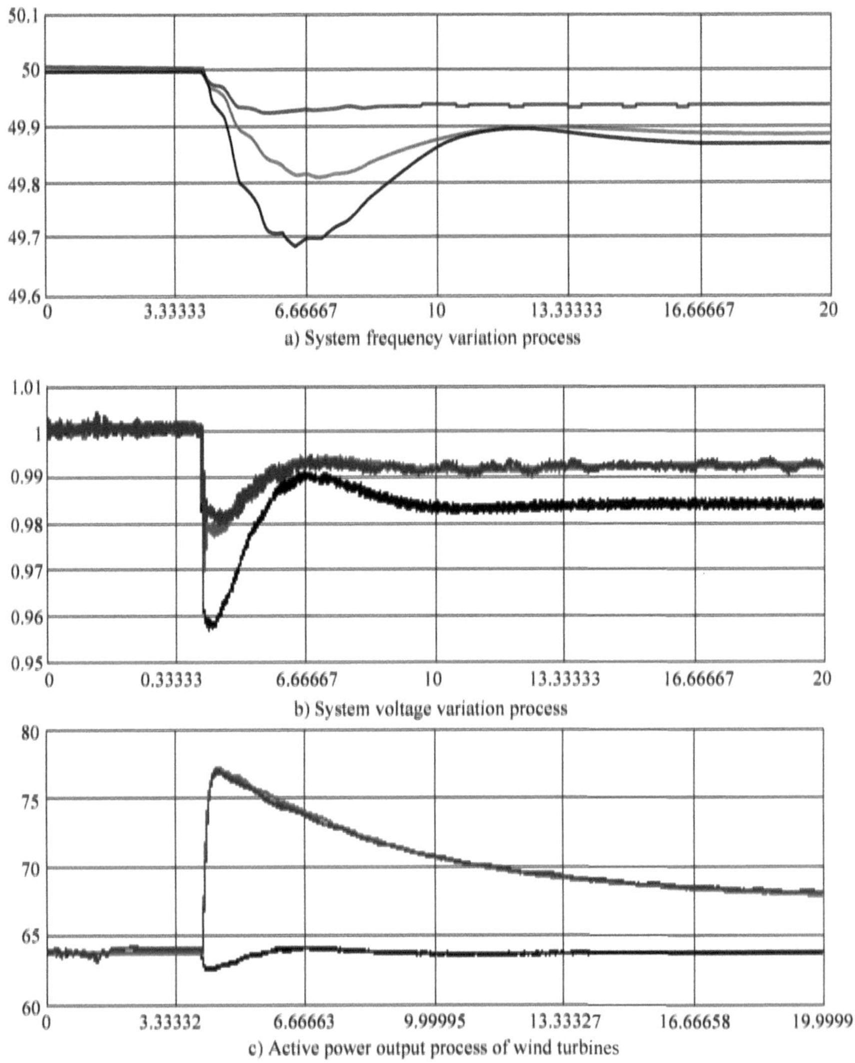

Fig. 9.20 Wind-storage joint frequency and voltage regulation process during load sudden change

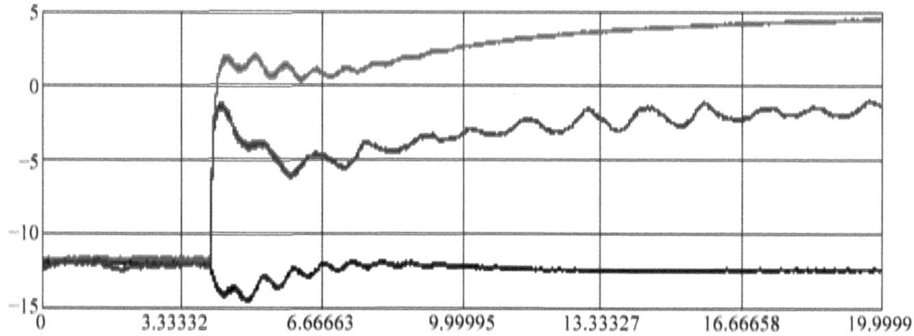

d) Reactive power output process of wind turbines

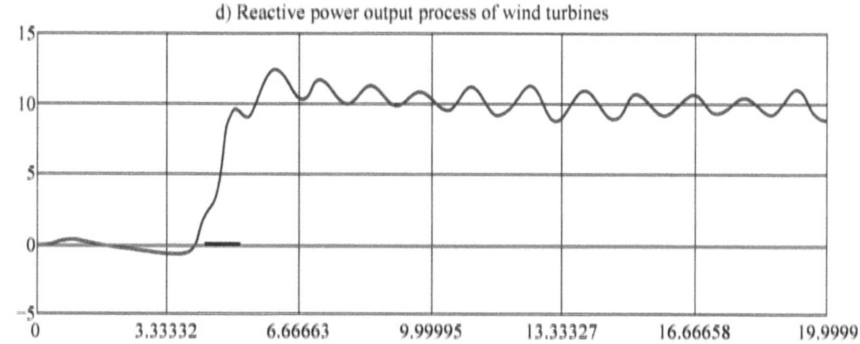

e) Active output process of energy storage

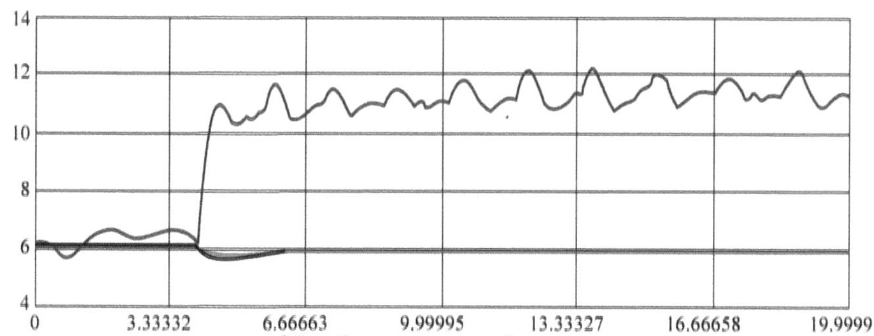

f) Reactive output process of energy storage

Fig. 9.20 (continued)

case where only the wind turbine participates in frequency-voltage regulation (red curve), scheme 3 is the case where wind-storage jointly participates in frequency-voltage regulation (blue curve).

(a) System frequency change process
(b) System voltage change process
(c) Active power output process of the wind turbine
(d) Reactive power output process of the wind turbine
(e) Active power output process of energy storage

(f) Reactive power output process of energy storage

It can be seen that when the active load and reactive load suddenly increase, the system's frequency regulation demand is large, at this time, frequency regulation is prioritized, as can be seen from Fig. 9.20, the frequency regulation effect is quite obvious. At the same time, since the active power output of wind-storage has not yet reached the upper limit, the wind-storage system also performed well in voltage regulation operation while regulating the frequency.

9.3.3.2 Short Circuit Fault

Similarly, choose the medium wind speed (10 m/s) scenario for simulation, set a typical single-phase metallic ground fault at bus 9, the results are shown in Fig. 9.21.

Scheme 1 is where both the wind turbine and energy storage do not participate in frequency and voltage regulation (black curve), scheme 2 is where only the wind turbine participates in frequency and voltage regulation (red curve), scheme 3 is where both the wind turbine and energy storage participate in frequency and voltage regulation (blue curve).

(a) System frequency change process
(b) System voltage change process
(c) Active power output process of the wind turbine
(d) Reactive power output process of the wind turbine
(e) Active power output process of energy storage
(f) Reactive power output process of energy storage

As can be seen from Fig. 9.21, when the system experiences a single-phase metallic grounding fault, the demand for voltage regulation is high. At this time, the coordinated control scheme prioritizes the voltage regulation demand. Wind-storage supports the system's fault transition process well through rapid reactive power output, achieving quick fault recovery.

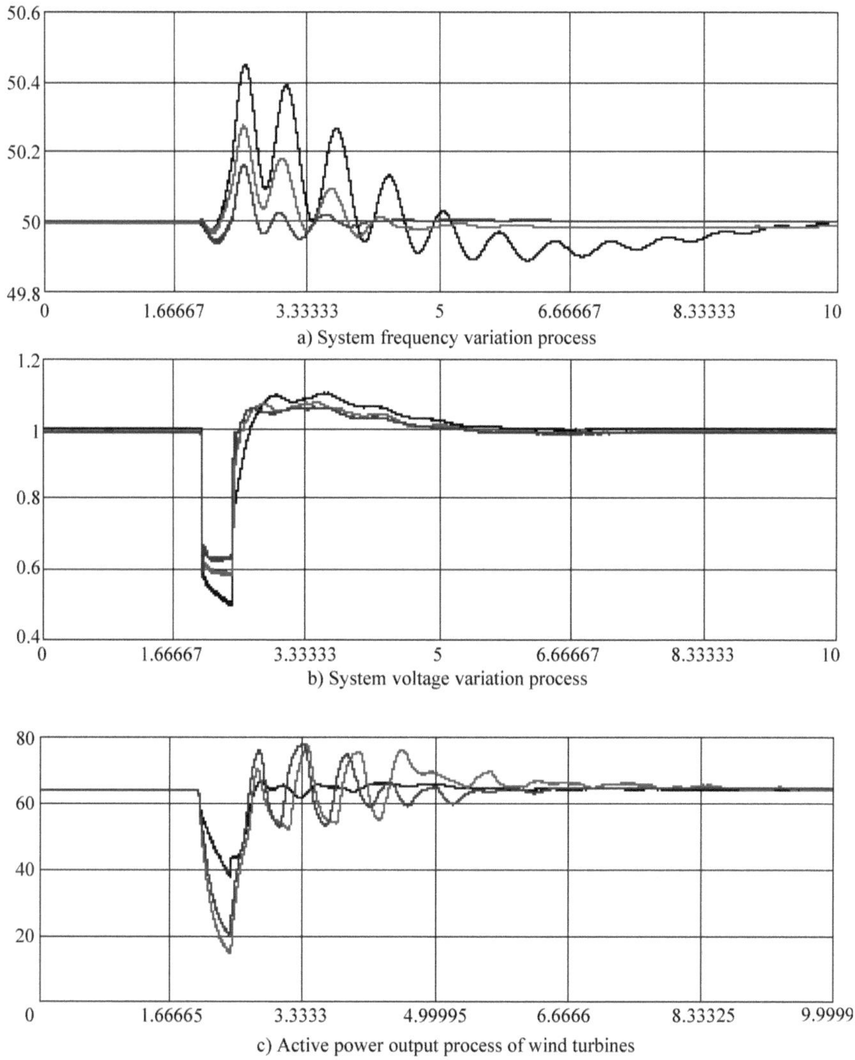

Fig. 9.21 Process of wind-storage combined frequency and voltage regulation when a single-phase metallic grounding fault occurs at bus 9

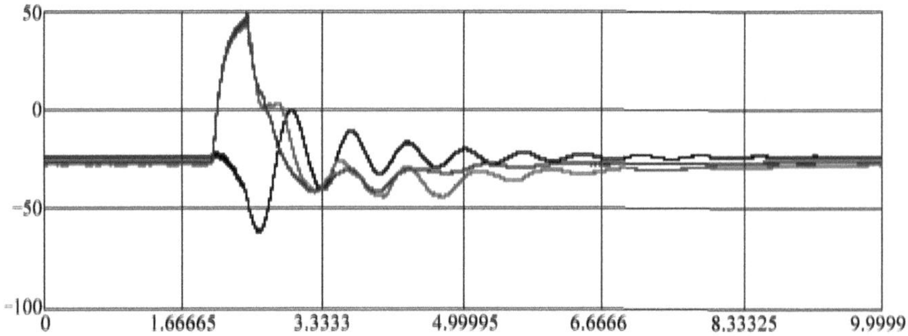

d) Reactive power output process of wind turbines

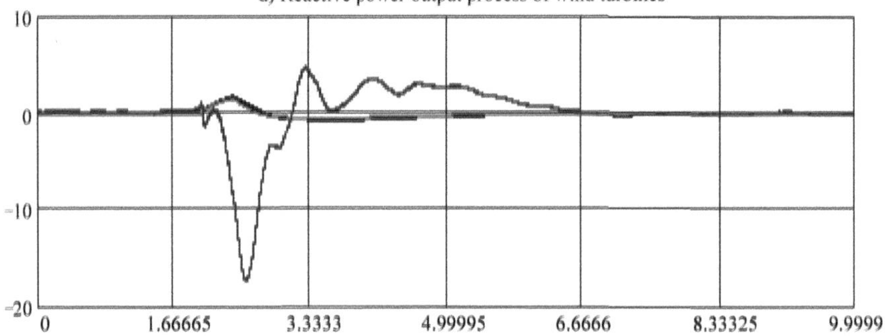

e) Active output process of energy storage

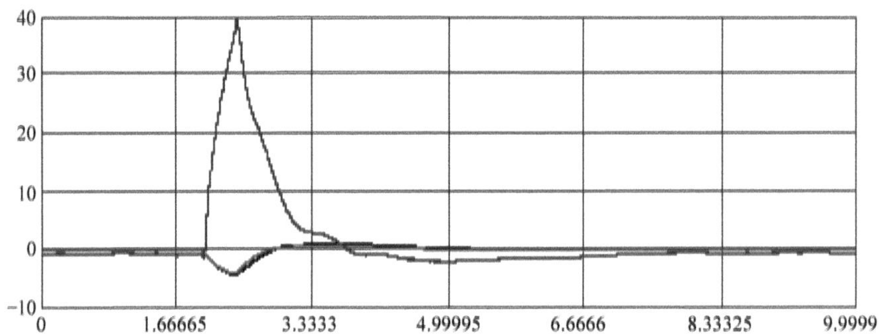

f) Reactive output process of energy storage

Fig. 9.21 (continued)

Chapter 10
Energy Storage-Based Virtual Power Plant

With the increasing deployment of energy storage in various scenarios of the power system, new participants and control methods are provided for virtual power plants, enhancing the performance of virtual power plants in response speed, regulation accuracy, and capacity reliability, making them more competitive in the market. This chapter analyzes the composition, modelling, and optimization scheduling method of virtual power plants considering energy storage and distributed renewable energy generation.

10.1 Overview of Virtual Power Plant

In the face of the large-scale integration of distributed renewable energy sources such as wind power and photovoltaics, and their output randomness and volatility, the traditional power system will inevitably undergo changes in structure, form, and operation mode to adapt to new requirements. With today's advanced network communication, real-time monitoring, big data processing, and other technical means, the "Virtual Power Plant" (VPP) has emerged and developed rapidly [154].

Simply, a virtual power plant is an organic combination of power sources, energy storage, and controllable loads. Power sources can include traditional thermal power, hydropower, and various renewable energies such as photovoltaics and wind power; energy storage includes various battery storage and pumped storage [155]. Users can join the virtual power plant to participate in demand response programs, including price-based and incentive-based demand response. Power sources, energy storage, and loads are connected to the control center through advanced communication technology, realizing two-way communication between the control center and each unit. Figure 10.1 is the typical architecture of a virtual power plant.

The rapid development of energy storage also brings new resources to the ancillary market. The energy storage-based VPP has the ability and demand to

X. Tang et al., *Electrical Energy Storage Technologies and Applications*,
https://doi.org/10.1007/978-981-96-5625-7_10

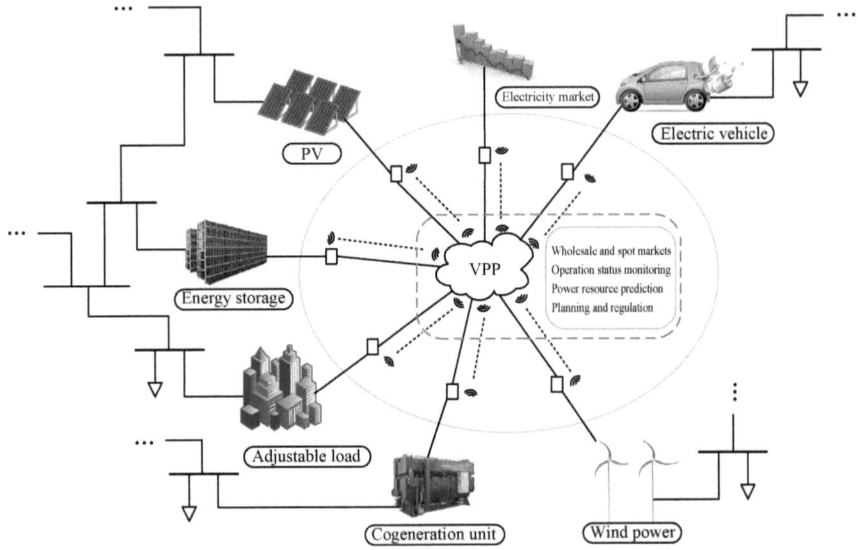

Fig. 10.1 Typical architecture of a virtual power plant

participate in the energy market and ancillary market at the same time. Controllable loads, through VPP participation in demand response, receive corresponding rewards, which is the main reason why virtual power plants can participate in various power markets. The advancement of communication and control technology has broken through the scale limitation of early virtual power plants participating in demand response programs, and has made significant progress in VPP capacity, response speed, and regulation accuracy.

As a technical means that can effectively manage dispersed power resources, VPP has some similarities with microgrids in terms of aggregated resource types, but there are significant differences between the two, as shown in Table 10.1.

In terms of operation mode, microgrids focus on "autonomy", and their top-down structure mostly adopts the method of local energy use, achieving grid-connected operation in normal times and island operation in case of faults. VPP, on

Table 10.1 Comparison between Microgrid and VPP

	Microgrid	VPP
Operating mode	Emphasizes "autonomy", energy is used locally	Emphasizes "participation", aggregated resources participate more in scheduling and trading
Physical structure	Relies on the integration of power components and power lines, needs to expand the original grid structure	Based on communication and software, no need to expand the original grid structure
Operational characteristics	"Grid-connected mode" and "islanded mode"	Only operates in "grid-connected mode", performance indicators are stricter

the other hand, emphasizes "participation" in its design concept, attracting and aggregating various distributed resources to form a whole to participate in grid dispatch or power market transactions.

In terms of physical structure, a microgrid relies on the integration of various power components and power lines, requiring structural expansion of the grid at the actual physical level. Therefore, the coverage of the microgrid is limited by geographical location. On the other hand, a Virtual Power Plant (VPP) is based on communication and software technologies, and its aggregation range and market interaction range depend only on the coverage and reliability of communication. It remotely monitors distributed resources within its range through an intelligent metering system, and the collected information is interacted through the communication network. Therefore, a VPP does not need to expand the original grid structure, and its coverage is much larger than that of a microgrid. In terms of operational characteristics, a microgrid is a controllable power component relative to the external grid, controlled by the switch at the public connection point. It can operate in "grid-connected mode" under normal conditions, and in "islanded mode" in case of failure. A VPP, however, can generally only operate in "grid-connected mode", so its performance requirements for interaction with the grid system are stricter.

With the continuous construction of distributed energy worldwide in recent years, research and application of virtual power plants have received increasing attention. Europe pays particular attention to the use of renewable energy, especially wind energy, so research on virtual power plants started earlier in European countries [156]. In the fifth Framework Programme of the European Union from 2001 to 2005, 11 companies from 5 countries including Germany, Netherlands, and Spain jointly proposed the VFCPP project. This project comprehensively coordinated and managed 31 dispersed and independent residential fuel cell cogeneration systems, forming a virtual power plant. After aggregating all fuel cells and determining the specific system load operation curve, the VPP control center communicates in real time with the on-site energy manager to formulate corresponding optimized scheduling plans, coordinate the heating and power supply of each unit, and achieve no-delay tracking of each predetermined load curve. At the same time, it can optimize the operation plan of each unit within the VPP, reduce production costs and load peak-valley difference, and improve the overall operation efficiency of the system. In the subsequent European Union's sixth Framework Programme, 20 research institutions and organizations from 8 countries including the UK, Spain, and France proposed the FENIX project based on the VFCPP. Conceptually, this project provides a set of power system technology systems and business frameworks that can aggregate a large number of distributed resources and manage them in a coordinated manner, thereby helping the EU power supply system achieve comprehensive goals of high cost-effectiveness, high security, and sustainable development.

In September 2014, the German company Younicos was commissioned by the German Grid Commission to provide a 5 MW/5MWh lithium-ion battery storage system, establishing Germany's first industrial battery storage project, and participating in the "primary reserve" power market, providing frequency regulation, voltage support, and black start services for the grid. At the same time, Younicos has a large number of distributed storage facilities through the secondary use of EV batteries for storage. Relying on its large storage resources, in 2018, it proposed a new business model of "storage as a service" for external leasing of storage facilities. This business uses virtual power plant technology, with the storage system as the basis, integrates many users' distributed resources, manages them in a coordinated manner, and can further participate in the primary frequency regulation and other auxiliary service markets [157].

In 2016, the UK National Grid released a 200 MW fast frequency response service procurement plan, stipulating that frequency regulation resources should respond to grid frequency fluctuations within 0.5 s, maintain the grid frequency around 50 Hz through precise control, and also stipulated that the minimum capacity of the storage system participating in the bidding should be 1 MW, the storage capacity should guarantee 100% service provision, and the charging and discharging time should not be less than 15 min. The project sizes of the final 8 winning bidders ranged from 10 MW to 49 MW, and while using virtual power plant technology to participate in frequency regulation services, they also participated in the reserve market to provide reserve capacity services for the grid.

At the end of 2018, the UK renewable energy developer Anesco and the battery storage integrated operator Limejump collaborated to build a 185 MW storage series project in the UK [158]. Limejump's virtual power plant platform, developed based on big data technology, is responsible for managing this project, fully utilizing the flexibility of distributed generation and battery storage, enabling it to achieve rapid power response, and participating in the balancing mechanism market of the UK National Grid.

In recent years, a number of virtual power plants have also been put into use in China. In May 2017, in response to the large-scale consumption of a large amount of clean energy such as wind power and photovoltaics, Jiangsu Power Company established the first "source-grid-load smart grid" system, which is the world's largest "virtual power plant" [159]. Through this VPP system, a large number of scattered uncontrollable power devices can be transformed into controllable resources from milliseconds to seconds, and 1370 enterprise users can stop power at any time, and the power capacity that does not affect the normal production and life of the enterprise has reached 3.76 million kilowatts. In June 2018, this "large-scale source-grid-load friendly interaction system" included storage in the management and control scope, upgraded to a "large-scale source-grid-load-storage friendly interaction system", effectively utilizing the storage power station resources on the grid side and the user side, enhancing the power regulation ability and safety of the power system. At the same time, China has also taken the lead in carrying out the standard-setting work related to virtual power plants. The "Virtual Power Plant" "Use Cases" and "Architecture and Functional Requirements" were officially launched in the International Electrotechnical Commission in March 2018.

As can be seen, the large-scale development of distributed power sources such as wind power, photovoltaics, and storage, as well as the advancement of power market reforms, have become a strong driving force for the continuous promotion and application of virtual power plants, and the corresponding demonstration and application systems are gradually increasing. The virtual power plant technology framework and optimized control strategy for the above needs have become important supporting technologies, which is also the focus of this chapter's analysis.

10.2 VPP Resource Model

VPP relies on advanced communication control hardware and intelligent computing software to effectively aggregate geographically dispersed distributed resources. The types of resources that can participate in the aggregation are diverse, including various distributed power sources, energy storage systems, dispatchable loads, etc.

Among various distributed power sources, the typical ones are wind power and photovoltaics, which are uncontrollable power sources, and their power generation mainly depends on wind speed and light conditions. At the same time, there are also some controllable distributed power sources, including various combined cooling, heating and power units (CCHP), fuel cells, etc., deployed on the user side. Their adjustable and controllable features have improved the regulation performance of VPP.

Energy storage is a very important regulation method and resource for VPP, ensuring that VPP can integrate various distributed resources in the control area and provide rapid and accurate power response capabilities. With the continuous development and rapid deployment of energy storage technology, the ability of VPP to absorb, integrate, and manage power resources will become stronger and stronger. For this reason, many VPP projects at domestic and abroad are carried out based on large-scale energy storage systems. Energy storage is mostly based on batteries, and with the emergence of a large number of electric vehicles, orderly charging and discharging management of electric vehicles can also serve as distributed energy storage resources for VPP regulation, adding flexible regulation methods to the virtual power plant.

This section models the main types of power resources involved in the virtual power plant according to the regulation needs of VPP.

10.2.1 Controllable Source Model

The controllable distributed generators (Distributed Generator, DG) in VPP are mainly gas turbines and small diesel engines. Their output is controllable, they have quick start-up and climbing capabilities, and can track and respond to load changes,

The cost of DG includes operating costs and start-stop costs [14]:

$$C_t^{dg} = \sum_{i=1}^{n_{dg}} \left[\left(K_i P_{i,t}^{dg} + K_i^f \mu_{i,t}^o \right) + \left(\lambda_i^{su} \mu_{i,t}^{su} + \lambda_i^{sd} \mu_{i,t}^{sd} \right) \right] \tag{10.1}$$

Where, n_i is the number of DG units; $P_{i,t}^{dg}$ is the power of the t period's The output power of the i th DG unit, as a decision variable; K_i, K_i^f are the fuel cost and fixed cost of the i th DG unit; λ_i^{su}, λ_i^{sd} are the start-up and shut-down costs respectively; $\alpha_{i\ t}^{o}$, $\alpha_{i\ t}^{su}$, $\alpha_{i\ t}^{sd}$ are three Boolean variables, respectively indicating whether the t period's i th DG unit is working, starting, or shutting down (yes is 1, no is 0). The operating cost of DG and the relationship with output power can be expressed as a characteristic parameter expression:

$$C\left(P_{dg}\right)=\alpha_{dg}P_{dg}^{2}+\beta_{dg}P_{dg} \tag{10.2}$$

Where, α_{dg}, $\beta_{dg}$are the characteristic parameters of DG cost, which can be obtained by fitting the characteristic curve of power cost [160, 161].

10.2.2 Uncontrollable Power Source Model

10.2.2.1 Wind Power Model

The output power of a wind turbine mainly depends on the wind speed, and its power uncertainty also comes from the uncertainty of the wind speed. The model of natural wind can usually be described by the Weibull distribution function:

$$f\left(v\right)=\frac{k}{c}\left(\frac{v}{c}\right)^{k-1}\exp\left[-\left(\frac{v}{c}\right)^{k}\right] \tag{10.3}$$

Where, v is the wind speed; >0, is the shape parameter; c >0, is the scale parameter [17].

The relationship between wind power output and wind speed can be expressed as:

$$P_{w}=\begin{cases}0\ 0\le v\le v_{ci}\ or\ v\ge v_{co}\\ av^{3}+b\quad v_{ci}<v<v_{r}\\ P_{w}^{n}\quad v_{r}\le v<v_{co}\end{cases} \tag{10.4}$$

Where in the formula, v_{ci} is the cut-in wind speed, v_{co} is the cut-out wind speed, v_r is the rated wind speed, P_w^n is the rated power of the wind turbine unit. a and b are the characteristic parameters of wind power, obtained by fitting the power-wind speed characteristic curve.

10.2.2.2 Photovoltaic Power Generation Model

Similar to wind power, the output power and its uncertainty of photovoltaic power generation depend on the intensity of solar radiation. The intensity of solar radiation is generally expressed by the Beta distribution function:

$$f(I)=\frac{1}{I_{\max}}\frac{\Gamma(\alpha+\beta)}{\Gamma(\alpha)\Gamma(\beta)}\left(\frac{I}{I_{\max}}\right)^{\alpha-1}\left(1-\frac{I}{I_{\max}}\right)^{\beta-1} \tag{10.5}$$

Where in the formula, I represents the solar radiation intensity; $\alpha > 0$ and $\beta > 0$ are the two parameters of the Beta distribution; Γ represents the gamma function [162].

The power of photovoltaic power generation is approximately proportional to the solar radiation intensity, so it can also be written in a similar Beta distribution function:

$$f\left(P_{pv}\right)=\frac{1}{P_{pv}^{n}}\frac{\Gamma(\alpha+\beta)}{\Gamma(\alpha)\Gamma(\beta)}\left(\frac{P_{pv}}{P_{pv}^{n}}\right)^{\alpha-1}\left(1-\frac{P_{pv}}{P_{pv}^{n}}\right)^{\beta-1} \tag{10.6}$$

Where in the formula, P_{pv}^{n} is the rated power of the photovoltaic power generation system.

10.2.3 Interruptible Load Model

Interruptible load is achieved through contracts between users and dispatchers, following the Demand Response (DR) mechanism, providing load interruption services within a specified time or under certain conditions to receive interruption compensation, while the dispatcher can maintain system stability by cutting off a specified load. The cost of demand response for a virtual power plant is essentially the compensation for interruptible load that the virtual power plant pays to users. Considering the different impacts of different interruptible loads on users, levels are generally set for load interruption. The higher the importance of the load interruption, the higher the compensation received, so the interruption compensation price is linked to the interruption level:

$$C_t^{DR}=\sum_{m=1}^{n_m}\lambda_m^{curt}L_{m\ t}^{curt}, \tag{10.7}$$

Where: n_m is the number of interruption levels; λ_m^{curt} is the compensation price for the mth level interruptible load; $L_{m\ t}^{curt}$, is the interruptible load of the mth level at time t, which is a decision variable.

When the interruptible level is sufficiently detailed, the curve representing the compensation price will change from a discrete multi-segment function to a continuously increasing quadratic function, as shown in Fig. 10.2 [162]:

Therefore, the cost of DR can be expressed as:

$$C\left(P_{dr}\right) = \alpha_{dr} P_{dr}^{2} + \beta_{dr} P_{dr} \tag{10.8}$$

Where, α_{dr} and β_{dr} is the controllable load compensation parameter.

10.2.4 Energy Storage System Model

The cost of energy storage systems, mainly composed of electrochemical batteries, is primarily due to the wear and tear on the battery life during the charging and discharging process. When not considering the impact of battery aging on its efficiency and other performance, its cost is approximately linearly related to the charging and discharging power:

$$C_{t}^{str} = \sum_{v}^{n_{str}} \left[a_{v}^{es} \left(P_{v,t}^{esc} + P_{v,t}^{esd} \right) + b_{v}^{es} \right] \tag{10.9}$$

Where, n_{str} is the number of energy storage units, $P_{v,t}^{esc}$, $P_{v,t}^{esd}$ are the charging and discharging power of the v th energy storage unit in the t time period, a_{v}^{es}, b_{v}^{es} are cost coefficients, which can be calculated from the cost curve or given by experience.

Considering that the cost of charging and discharging energy storage batteries is approximately the same, it can also be expressed as:

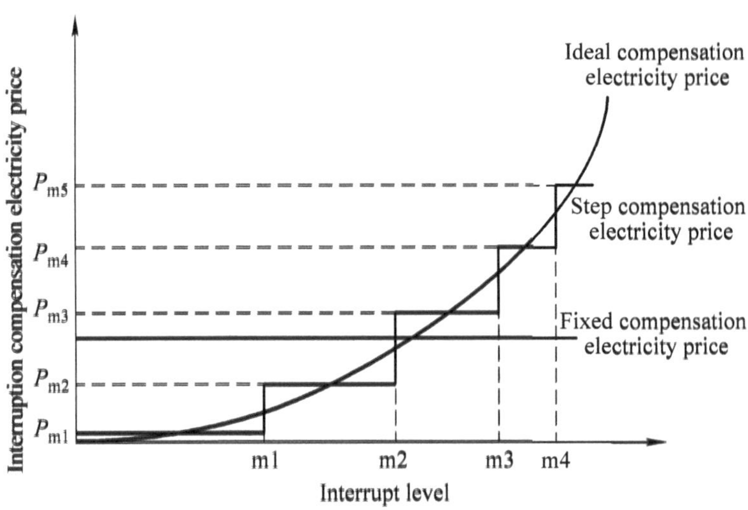

Fig. 10.2 Compensation price curve for interruptible load

$$C\left(P_{str}\right) = \alpha_{str}\left|P_{str}\right| + \beta_{str} \tag{10.10}$$

Where in the formula, α_{str} and β_{str} are the cost parameters of energy storage [160].

10.2.5 Other Resources

Relying on advanced communication and control technologies, as well as effective market mechanisms, the controllable resources of future VPPs can continue to expand. Various new types of distributed power generation or energy supply technologies, such as fuel cells, new energy storage, ground/air source heat pumps, orderly charging of electric vehicles and V2G(Vehicle to Grid), and combined heat and power integrated energy systems, provide rich control resources for VPPs. A unified, open, and scalable VPP communication and control architecture, as well as efficient energy models, will provide an important foundation for the integration and control of these resources.

10.3 VPP Optimization Scheduling Considering Uncertainty

The electricity market generally includes the Energy Market (EM) and auxiliary service markets. The Energy Market includes contract markets, day-ahead markets, and real-time balancing markets, while auxiliary service markets include the Frequency Regulation Market (FRM) and Spinning Reserve Market (SRM). VPP has the technical conditions to participate in the above-mentioned electricity markets, and the optimization control problem has significant differences in strategy for different market functions.

The literature [163] proposed a virtual power plant optimization dispatch model that takes into account demand response, integrating power output with demand response to increase the revenue of VPP participating in the day-ahead electricity market. Based on this, literature [164] integrated electric vehicles into the VPP bidding model, analyzing the coordinated optimization strategy of VPP participating in the contract market, day-ahead market, and balancing market at the same time. Furthermore, some studies have extended the types of markets that VPP participates in to auxiliary markets. Literature [165] established an optimization dispatch model for VPP participating in the energy market and peak shaving market at the same time based on time-of-use electricity prices, and analyzed the economy of VPP after participating in the peak shaving market. Literature [166] established an optimization dispatch model for VPP participating in the reserve market, which includes fuel-fired power generation, energy storage systems, and interruptible loads.

The uncertainty factors in the operation process of VPP are the main issues to be dealt with in VPP scheduling, such as the uncertainty of renewable energy output such as wind power and photovoltaics, and changes in electricity prices. Typical algorithms include point estimation method, stochastic programming method, robust optimization method, etc. Among them, the robust optimization method can quickly calculate the optimization results without knowing the probability distribution of uncertain parameters, and is widely used in power optimization scheduling, demand response planning, power market bidding strategy and other issues. Literature [167] established a corresponding robust optimization model for VPP containing cogeneration units, solving the scheduling problem when VPP participates in the energy market and reserve market at the same time. However, the results obtained by the robust optimization method are generally the worst-case scenario, and the results are too conservative.

In recent years, some studies have combined robust optimization methods with stochastic optimization methods. By obtaining the distribution probability of the worst-case scenario and then obtaining the optimal result, a two-stage distribution robust optimization method with higher economy is formed. Literature [168] proposed a static security correction control method for power systems considering wind power output uncertainty based on distribution robust linear optimization theory, and literature [169] built an optimization model considering wind power uncertainty using a two-stage distribution robust optimization method. This section applies the two-stage distribution robust optimization algorithm to the optimization scheduling of VPP, and establishes a max-min-max two-stage distribution robust optimization scheduling model.

10.3.1 Two-Stage Distribution Robust Optimization Model

Based on the established VPP resource model, the uncertainty of electricity load and energy market EM, spinning reserve market SRM, and frequency regulation market FRM electricity prices are considered in the model. With the maximum net profit of VPP as the optimization goal, combined with the simulation scenarios generated by the Monte Carlo method, K discrete scenarios and the initial probability of each discrete scenario are selected to establish the objective function:

$$K:\max P=\max P=\left(\begin{array}{l}\sum_{t=1}^{T} C_t^{m}-C_t^{dg}-C_t^{dr}-C_t^{str}+\min\limits_{\{p_k\}\in©}\\ \left(\sum_{k=1}^{K} p_k \max\left(\sum_{i=1}^{n_{dg}} c_{i,t}^{dg}\Delta P_{i,t,k}^{dg}+\sum_{i=1}^{n_{dr}} c_{i,t}^{dr}\Delta P_{i,t,k}^{dr}+\sum_{i=1}^{n_{str}} c_{i,t}^{str}\Delta P_{i,t,k}^{str}\right)\right)\end{array}\right) \tag{10.11}$$

Where in the formula, Ω is the feasible domain for the operating scenario, $\{p_k\}$ represents the probability value of each discrete scenario, $c_{i,t}^{dg}$, $c_{i,t}^{dr}$, $c_{i,t}^{str}$ are respectively the adjustment penalty coefficients for distributed power sources, demand-side response resources, and energy storage in the k scenario t time period, $\varnothing P_{i,t,k}^{dg}$,

$\varnothing P_{i,t,k}^{dr}$, $\varnothing P_{i,t,k}^{str}$ are the output adjustment amounts of the corresponding resources in the k scenario during the t time period.

As can be seen, Eq. (10.11) is a max-min-max three-layer two-stage optimization problem. The deterministic model in the first part characterizes the VPP's day-ahead economic dispatch situation, while the latter part represents the total profit change resulting from the output adjustments made by each VPP unit under the consideration of parameter uncertainty. The selected data can be generated using the Monte Carlo method, or by extracting relevant historical data, then filtering out a set number of discrete scenarios, and thus obtaining the initial probability values of each discrete scenario.

Generally speaking, traditional robust optimization methods will optimize uncertain parameters to the extreme points of parameter fluctuation range to ensure that the optimization objective can satisfy the constraint feasibility for all elements within the uncertain set, but such results are overly conservative. By adopting a two-stage distribution robust optimization method, through the inner layer's min-max decision variable $\{p_k\}$, it seeks the worst probability distribution situation of all discrete scenarios and obtains the minimum profit in this situation. In this way, it can comprehensively utilize the ability of robust optimization algorithms to handle uncertain parameters, as well as the reference role provided by historical data.

Constraint conditions:

1. Distributed Generation (DG) constraint conditions:

$$\mu_{i,t}^{o} - \mu_{i,t-1}^{o} \le \mu_{i,t}^{su} \tag{10.12}$$

$$\mu_{i,t-1}^{o} - \mu_{i,t}^{o} \le \mu_{i,t}^{sd} \tag{10.13}$$

$$P_{i,t}^{dg} \ge P_{i}^{\min} \mu_{i,t}^{o} \tag{10.14}$$

$$P_{i,t}^{dg} + R_{i,t}^{dg} + F_{i,t}^{dg} \le P_{i}^{\max} \mu_{i,t}^{o} \tag{10.15}$$

$$R_{i,t}^{dg} + F_{i,t}^{dg} \le r_{i}^{u} t^{r} \tag{10.16}$$

$$-r_{i}^{d} \le P_{i,t}^{dg} - P_{i,t-1}^{dg} \le r_{i}^{u} \tag{10.17}$$

$$t_{i}^{u} \mu_{t}^{su} \le \sum_{h=t}^{t+t_{i}^{u}-1} \mu_{i,t}^{o} \left(\forall t \le T - t_{i}^{u} + 1\right) \tag{10.18}$$

$$t_{i}^{d} \mu_{t}^{sd} \le \sum_{h=t}^{t+t_{i}^{d}-1} \left(1 - \mu_{i,t}^{o}\right)\left(\forall t \le T - t_{i}^{d} + 1\right) \tag{10.19}$$

$$\sum_{t=1}^{t_{i}^{u}-t_{i}^{ui}} \left(1 - \mu_{i,t}^{o}\right) = 0 \tag{10.20}$$

$$\sum_{t=1}^{t_{i}^{d}-t_{i}^{di}} \mu_{i,t}^{o} = 0 \tag{10.21}$$

$P_{i}^{\min}$ *and* $P_{i}^{\max}$ are the minimum and maximum output power of the ith DG unit respectively; $R_{i,t}^{dg}$ and $F_{i,t}^{dg}$ are the capacities of the ith DG unit participating in EM and FRM at time, which are decision variables; r_i^u and r_i^d are the upward and downward ramp rates of the ith DG unit respectively; t_i^u and t_i^d respectively represent the minimum startup operation and shutdown downtime; t^r represents the reserve service time.

2. Demand Response DR constraint conditions:
   Interruptible load can achieve the effect of increasing VPP virtual output by cutting off the load, which can be used to participate in the reserve market and provide reserve capacity.

$$L_t^{curt} = \sum_{m=1}^{n_m} L_{m,t}^{curt} \tag{10.22}$$

$$0 \leq L_{m,t}^{curt} \leq k_m^{curt} R_t^{el} \tag{10.23}$$

$$0 \leq L_t^{curt} + R_t^{el} \leq \sum_{m}^{n_m} k_m^{curt} P_t^{el} \tag{10.24}$$

In the formula, k_m^{curt} represents the interruption level coefficient of the m-th load; P_t^{el} is the power at time . Time period load; R_t^{el} is the interruptible load reserve capacity for time period, which is a decision variable; L_t^{curt} is the interruptible load for time period t .

3. Energy storage system constraints:

$$0 \leq P_{v,t}^{esc} \leq P_v^{cmax} \tag{10.25}$$

$$0 \leq P_{v,t}^{esd} \leq P_v^{dmax} \tag{10.26}$$

$$S_v^{\min} \leq S_{v,t}^{es} \leq S_v^{\max} \tag{10.27}$$

$$S_{v,t}^{es} = S_{v,t-1}^{es} + \eta_v^{esc} P_{v,t}^{esc} - \frac{P_{v,t}^{esd}}{\eta_v^{esd}} \tag{10.28}$$

$$S_{v,0}^{es} = S_v^{esi} \tag{10.29}$$

$$S_{v,24}^{es} = S_v^{esf} \tag{10.30}$$

In the formula, $P_v^{c\max}$ and P_v^{dmax} are respectively the v The maximum charging and discharging power of a storage unit; $S_{v,t}^{es}$ is the charge of the v th storage unit in the t time period; $S_v^{\min}$ and $S_v^{\max}$ are the upper and lower limits of the charge of the v th storage unit; η_v^{esc} and η_v^{esd} are the charging and discharging efficiencies; S_v^{esi}, S_v^{esf} represents the charge amount at the start and end of each day.

4. Power balance constraint:

$$\sum_{i}^{n_i} P_{i,t}^{dg} + \sum_{v}^{n_v} P_{v,t}^{esd} = P_t^{el} + P_t^{em} - L_t^{curt} + \sum_{v}^{n_v} P_{v,t}^{esc} \tag{10.31}$$

5. Market capacity constraint:

$$R_t^{srm} = \sum_{i}^{n_i} R_{i,t}^{dg} + R_t^{el} \tag{10.32}$$

$$F_t^{frm} = \sum_{i}^{n_i} F_{i,t}^{dg} + \sum_{v}^{n_v} F_{v,t}^{str} \tag{10.33}$$

In the formula, $F_{v,t}^{str}$ represents the capacity of the v th energy storage unit participating in the frequency regulation market at time .

The two-stage distribution robust optimization model needs to find the probability distribution value of the worst-case scenario. In order to make the probability distribution value sought more consistent with the actual operation data, a constraint condition based on the 1-norm and ∞-norm centered on the initial probability distribution value is constructed to constrain the discrete scenario probability distribution value.

$$\Omega = \{p_k\} \left| \begin{array}{l} p_k \geq 0, k = 1, 2, \ldots, K \\ \sum_{k=1}^{K} p_k = 1 \\ \sum_{k=1}^{K} \left| p_k - p_k^0 \right| \leq \lambda_1 \\ \max_{1 \leq k \leq K} \left| p_k - p_k^0 \right| \leq \lambda_\infty \end{array} \right. \tag{10.34}$$

Where in the formula: p_k^0 is the k-th result obtained from the initial operation data k The initial probability of a discrete scenario. The following two inequalities are constraints formed by the 1-norm and ∞-norm, respectively, λ_1 and λ_∞ are the corresponding allowable probability deviations. $\{p_k\}$ satisfies the confidence level:

$$\begin{cases} \Pr\left\{ \sum_{k=1}^{K} \left| p_k - p_k^0 \right| \leq \lambda_1 \right\} \geq 1 - 2Ke^{-\frac{2M\lambda_1}{K}} = \alpha_1 \\ \Pr\left\{ \max_{1 \leq k \leq K} \left| p_k - p_k^0 \right| \leq \lambda_\infty \right\} \geq 1 - 2Ke^{-2M\lambda_\infty} = \alpha_\infty \end{cases} \tag{10.35}$$

Where, α_1 and α_∞ represent the confidence levels of the corresponding probability distributions [18].

Since both norm constraints are absolute value constraints, it is necessary to linearize them by introducing 0–1 auxiliary variables s_k^+ and s_k^-:

$$s_k^+ + s_k^- \le 1 \tag{10.36}$$

$$0 \le p_k^+ \le s_k^+ \lambda_1 \tag{10.37}$$

$$0 \le p_k^- \le s_k^- \lambda_1 \tag{10.38}$$

$$p_k = p_k^0 + p_k^+ - p_k^- \tag{10.39}$$

In the formula, p_k^+ and p_k^- are the positive and negative offsets of the probability value p_k relative to the initial probability value p_k^0.

Therefore, the original 1-norm absolute value constraint is:

$$\sum_{k=1}^{K} \left(p_k^+ + p_k^-\right) \le \lambda_1 \tag{10.40}$$

Similarly, deal with the absolute value constraint of the ∞-norm:

$$p_k^+ + p_k^- \le \lambda_\infty \tag{10.41}$$

10.3.2 Solution Algorithm

The two-stage distribution robust optimization model constructed in the previous section makes day-ahead decisions for decision variables such as the start-stop plan of distributed power sources in VPP, the arrangement plan of demand-side response resources, and the output value of energy storage in the first stage. In the second stage, it adjusts the output of power sources, demand-side response, and energy storage based on multiple scenario load values and market electricity prices.

To solve the two-stage distribution robust optimization model, a decomposition algorithm is needed to decompose the two-stage planning problem. The Column-and-Constraint Generation (CCG) algorithm is used to decompose the model problem into the main problem MP and the subproblem SP for iterative solution. The main problem is to solve the maximum profit that meets the constraint conditions after setting a known adverse probability distribution, carry out day-ahead scheduling, and provide an upper bound for formula (10.11):

$$\max_{x \in X \quad y_0 \in Y(x \quad \varphi_0) \quad y_k^{(m)} \in Y(x \quad \varphi_k) \quad Z} \quad P = a^T x + b^T y_0 + c^T \varphi_0 + Z \tag{10.42}$$

$$Z \le \sum_{k=1}^{K} p_k^{(m)} \left(b^T y_k^{(m)} + c^T \varphi_k\right) \tag{10.43}$$

Where in the formula, x for the first stage decision variable, y_k is the second stage variable under the k th scenario, y_0 is the second stage variable under the day-ahead forecast scenario, φ_0 is the forecast value of load and electricity price, φ_k is the forecast value of load and electricity price under the kth scenario, m is the number of iterations.

The inner subproblem is to optimize the worst probability distribution under the given first stage decision variable x of the outer main problem, then return to the main problem for iteration, and provide a lower bound value for Eq. (10.11):

$$L\left(x^*\right)=\min_{\{p_k\}\in\Theta}\left(\sum_{k=1}^{k}p_k\max_{y_k\in Y\left(x^*\ \varphi_k\right)}\left(b^T y_k+c^T\varphi_k\right)\right) \tag{10.44}$$

Because in the inner max problem, each scenario is relatively independent, so we can use parallel computing methods to perform calculations, so the two-stage model can be iterated continuously through formula (10.42) and formula (10.44) until the specified precision is reached. The specific process is as follows:

1. Given a set of uncertain parameters φ as the initial worst-case scenario, set the lower bound of the final optimization result's profit $LB = -\infty$, and the upper bound $UB = +\infty$, iteration times m;
2. Assume the worst-case scenario φ_m^*, solve the main problem Eq. (10.42), take the main problem objective function value P_m^* as the new upper bound, $UB = P_m^*$;
3. Substitute the solution of the main problem x_m^* into Eq. (10.44), solve the subproblem to get the objective function value $L_m^*\left(x_m^*\right)$ and the corresponding worst-case scenario uncertain parameter value φ_{m+1}^*, and update the lower bound to $LB = \max\left(P_m^*, L_m^*\left(x_m^*\right)\right)$;
4. Set the convergence error threshold, when $UB - LB \le \varepsilon$, stop iteration and return the optimal solution x_m^* and y_m^*.

The optimization model established by the above method is a mixed integer linear programming problem, and the GAMS/CPLEX solver is used to solve this problem. Figure 10.3 shows the solution flowchart of the two-stage robust optimization method.

10.3.3 Virtual Power Plant Case

The improved IEEE30 node power system model is used for VPP case analysis. The system topology is shown in Fig. 10.4. G1 and G2 are the two generator sets in the system. Node 5 is the demand response side, which can provide up to 50MWh of interruptible load. Node 6 is a battery storage station with a total capacity of 40 MW. Nodes 7, 8, 9, and 28 are distributed power generation systems. These units together form a VPP system. Table 10.2 provides the main parameters of the system. Table 10.3 provides the contracted power, load forecast values, time-divided electricity prices for various power markets, and 500 sets of simulated scenarios generated by electric load as historical data. The penalty for load shedding is 4000USD/MWh, and the dispatch time period is 24 h.

Fig. 10.3 Solution flowchart of the two-stage distributed robust optimization model

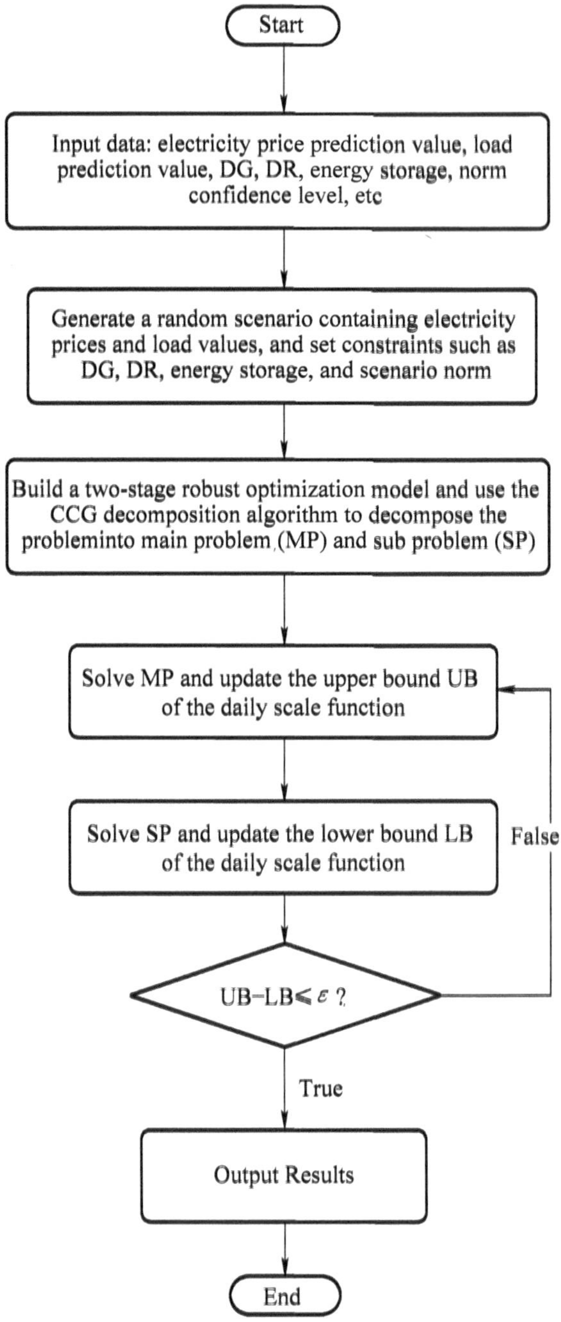

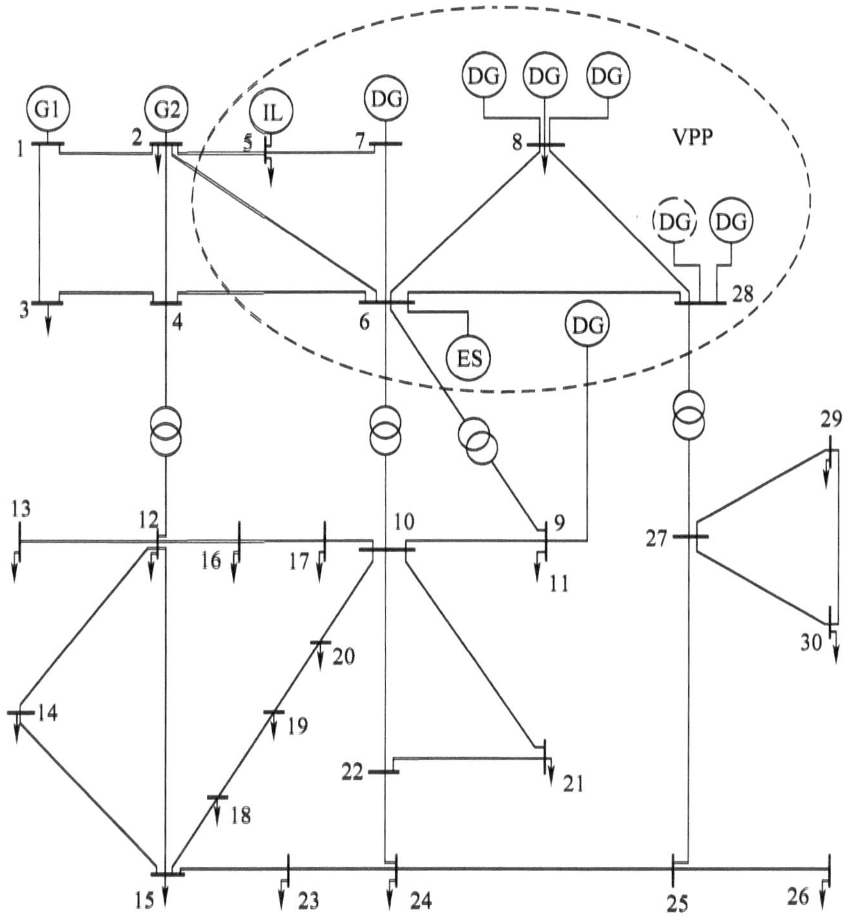

Fig. 10.4 Improved IEEE30 Node VPP System Structure Diagram

Table 10.2 VPP Power Unit Parameters

Parameter/Unit	Distributed Generator Set	Energy Storage Station
Maximum output/MW	50	±40
Minimum output/MW	10	0
Ramp rate/(MW/h)	20	/
Quadratic term/(USD/MW^2•h)	0.0175	0
Linear term/(USD/ MW•h)	2.00	0.50
Constant term (USD/h)	0	0

Table 10.3 Contracted Power and Forecast Load within Dispatch period

Time/h	Contracted Power/MWh	Contracted Reserve Power/MWh	Forecast Load/MW
1	126	12.6	84
2	128	12.9	86
3	133	13.3	90
4	135	13.5	91
5	143	13.5	91
6	160	14.3	95
7	174	15.8	105
8	182	17.4	116
9	178	18.2	125
10	178	17.1	120
11	162	16.2	108
12	149	14.9	99
13	136	13.7	90
14	147	14.7	98
15	161	16.3	108
16	170	17.2	114
17	179	18.1	120
18	185	18.5	124
19	176	17.9	118
20	173	17.5	113
21	160	16.2	106
22	147	14.8	98
23	134	13.5	89
24	125	12.6	83

The Monte Carlo method is used to simulate the four uncertain parameters of EM electricity price, SRM electricity price, FRM electricity price, and electricity load, generating 500 sets of data and establishing 20 scenarios. Based on these 20 scenarios, the optimization objective function (10.11) is solved, and the worst probability distribution of each discrete scenario is obtained by setting 1-norm and ∞-norm.

The norm constraint is determined by the confidence levels α_1 and α_∞, the size of which determines the uncertainty of the discrete scenario, directly affecting the conservatism and economy of the final decision. By setting different confidence levels, the VPP profit calculation results are analyzed, as shown in Table 10.4.

Table 10.4 VPP profit under different confidence levels

VPP Profit (USD)	$\alpha_\infty = 0.1$	$\alpha_\infty = 0.5$	$\alpha_\infty = 0.9$
$\alpha_1 = 0.1$	26,142	25,491	24,165
$\alpha_1 = 0.5$	25,293	24,623	22,489
$\alpha_1 = 0.9$	23,916	22,973	18,493

It can be seen that as the confidence level increases, the overall profit of VPP is continuously decreasing. This is because as the confidence level increases, the uncertainty of the parameters included in the model increases, the VPP system will become more conservative, and the economy will decline. And when both confidence levels are 0.9, the norm constraint is almost trivial, and the profit will drop significantly.

As shown in Fig. 10.5, under the condition of $\alpha_1 = 0.5$, compared with $\alpha_\infty = 0.1$, $\alpha_\infty = 0.5$, $\alpha_\infty = 0.9$ shows the EM bidding amount of VPP under three situations, where the positive direction of the y-axis represents VPP selling electricity to EM. It can be seen that as the confidence level increases, the decision-making scheme gradually becomes conservative. VPP needs to reduce the amount of electricity sold to EM or buy more electricity from EM for energy storage charging to ensure the balance of internal load. At the same time, the surplus electricity is maintained to provide auxiliary services to SRM and FRM, obtaining greater profits. Therefore, in the actual production process, through the setting of the confidence level, optimization results with different robustness and economy can be obtained.

In order to compare the advantages and disadvantages of the two-stage distribution robust optimization method with the traditional robust optimization method and the stochastic programming method, the VPP decision variables and profits of

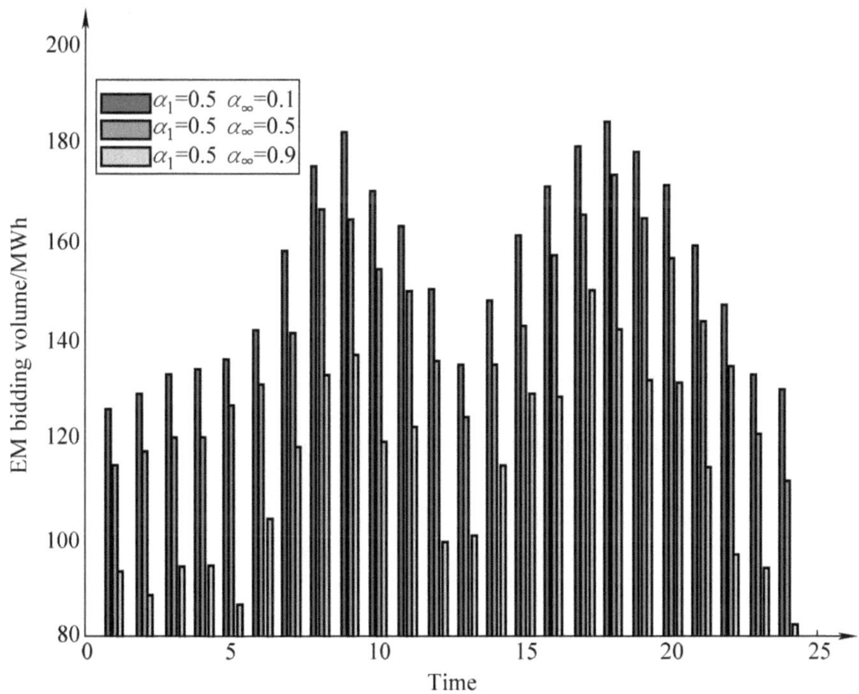

Fig. 10.5 The impact of different confidence levels on the bidding quantity of VPP participating in EM

the day-ahead scheduling are obtained through the first stage, and 1000 random scenarios are generated using the Monte Carlo method, and then the profit expectation adjustment value analysis is carried out for each scenario. The two norm confidence levels of the two-stage distribution robust optimization method are both 0.5, and the results of the three methods are shown in Table 10.5.

It can be seen that when the two-stage distribution robust optimization method solves the optimization scheduling model of VPP participating in the auxiliary market, its first-stage decision variable profit is between the stochastic programming method and the traditional robust optimization method. This shows that the two-stage distribution robust optimization method has stronger robustness than the stochastic programming method, and at the same time, it is more economical than the traditional robust optimization method that only considers the worst case. In addition, in 1000 random scenarios, whether it is the expected profit mean or the maximum profit, the two-stage distribution robust optimization method is the highest, indicating that in terms of average performance indicators of probability distribution and worst-case probability distribution performance indicators, the two-stage distribution robust optimization method is superior to the other two, and the economy is better. In summary, the two-stage distribution robust optimization method combines the advantages of traditional robust methods and stochastic programming methods, balances the disadvantages of the two, is more balanced in economy and robustness, and the final optimization scheduling result is better.

Of course, under the conditions of the spot market, the operation mechanism of VPP is quite different from traditional distributed power generation. When participating in power market dispatch, distributed power generation will directly exchange data with the dispatch center and respond to dispatch directors. However, the power resources within the VPP do not directly fall under the jurisdiction of the dispatch center when participating in dispatch, but exchange data with the dispatch center through the VPP control center. According to the operating norms of the spot market, due to factors such as grid structure and line congestion, the physical node electricity prices of each grid will vary, so the distributed power generation and energy storage units within the VPP should be calculated according to the electricity prices of their physical nodes. However, in most existing studies on VPP, usually only one electricity price curve is set for its participation in the power market, which means that all the electric energy of the VPP is exchanged with the grid through a bus node, which is different from the actual operation mechanism of the spot market

Table 10.5 Comparison of results of three algorithms

	First stage decision variable profit/USD	Expected mean profit /USD	Maximum expected profit/USD
Stochastic programming method	18,714	22,147	29,285
Two-stage robust method	15,438	24,825	33,731
Traditional robust method	10,689	20,771	32,386

and the operation status of distributed power generation. Therefore, it is necessary to consider various factors such as grid information and line congestion, and improve the operation mode and optimization dispatch strategy of VPP.

The above only introduces and analyzes the architecture, model, and dispatch strategy of VPP, and the problems involved are far more than these. For example, the effective aggregation and communication technology of resources within the VPP, improving the technical performance indicators of the VPP participating in various power markets; big data technology can improve the accuracy of VPP load forecasting, electricity price forecasting, and renewable energy output forecasting, which can improve the uncertainty parameters in the VPP optimization dispatch process; blockchain technology can control the real-name certificate of transactions between multiple entities inside and outside the VPP, making transactions more transparent and transaction costs lower; and the cooperative game problem of optimizing regulation when multiple VPPs in the same area overlap in coverage in the future.

References

1. Ferreira A C, Souza L M, Watanabe E H. Improving power quality with a variable speed synchronous condenser[C]. International Conference on Power Electronics. IET, 2002:456-461.
2. Cheng Shijie, Yu Wenhui, Wen Jinyu, et al. Energy storage technology and its application in power system stability control[J]. Power Grid Technology, 2007, 31(20):97-108.
3. Cheng Shijie, Li Gang, Sun Haishun, et al. Application and prospect of energy storage technology in the field of electrical engineering[J]. Power Grid and Clean Energy, 2009, 25(02):1-8.
4. Liu Yunren. Auxiliary service market[R]. 2019.
5. Prabha Kundur. Power system stability and control[M]. China Electric Power Press, 2002.
6. "Power System Frequency Regulation and Automatic Generation Control" Editorial Committee. Power System Frequency Regulation and Automatic Generation Control[M]. China Electric Power Press, 2006.
7. Huang Jiyuan, Li Xinran, Huang Jijun, et al. Comparative study on the effect of different types of energy storage sources participating in grid frequency regulation[J]. New Technology of Electric Power and Electric Energy, 2015(3):49-53.
8. Huang Yawei. Research on the demand scenario and control strategy of energy storage sources participating in power system frequency regulation[D]. Hunan University, 2015.
9. LIN J, DAMATO G, HAND P. Energy storage-a cheaper, faster and cleaner alternative to conventional frequency regulation[R]. Strategen, CESA, 2011: 1-15.
10. Chen Dayu, Zhang Lizhi, Wang Shu, et al. Development and enlightenment of energy storage in the US frequency regulation market[J]. Power System Automation, 2013, 37(1):9-13.
11. Li Xinran, Huang Jiyuan, Chen Yuanyang, et al. Review of research on large-scale energy storage sources participating in grid frequency regulation[J]. Power System Protection and Control, 2016, 44(7):145-153.
12. Deng Wei, Xie Yudong, Huang Jiyuan, et al. Study on the primary frequency regulation characteristics of multi-type energy storage[J]. Hunan Electric Power, 2015(1):1-4.
13. Gao Mingjie, Hui Dong, Gao Zonghe, et al. Introduction to the national wind and light storage and transmission demonstration project and analysis of its typical operation modes[J]. Power System Automation, 2013, 37(1):59-64.
14. Zhou Guopeng. Research on multi-type energy storage coordination control technology for smoothing wind power fluctuation[D]. Beijing: University of Chinese Academy of Sciences, 2014.
15. Sun Yushu, Tang Xisheng, Sun Xiaozhe, et al. Research on multi-type energy storage coordination control strategy based on MPC-HHT[J]. Transactions of China Electrotechnical Society, 2018, 38(9):2580-2588.

X. Tang et al., *Electrical Energy Storage Technologies and Applications*, https://doi.org/10.1007/978-981-96-5625-7

16. Yuan Xiaoming, Cheng Shijie, Wen Jinyu. Analysis on the application prospect of energy storage technology in solving the problem of large-scale wind power grid connection[J]. Power System Automation, 2013, 37(01):14-18.
17. North American Electric Reliability Corporation. Accommodating high levels of variable generation[R]. Washington, DC, USA: North American Electric Reliability Corporation, 2009.
18. Yuan Xiaoming. Grid connection technology of large wind farms under the condition of weak power grid with long lines[J]. Journal of Electrical Engineering Technology, 2007, 22(7):29-36.
19. Hu Juan, Yang Shuili, Hou Chaoyong, et al. Analysis and enlightenment of the current situation of typical demonstration applications of large-scale energy storage technology[J]. Power Grid Technology, 2015(04):879-885.
20. Xu Shouping, Li Xiangjun, Hui Dong. Review of the development status and demonstration application of large-scale energy storage system[J]. Grid and Clean Energy, 2013, 29(08):94-100.
21. Consultation Group of the State Grid Corporation's "Prospective Study of New Grid Technologies". Analysis of the application prospects of large-scale energy storage technology in power systems[J]. Power System Automation, 2013, 37(01):3-8.
22. Jiang Quanyuan, Gong Yuzhong. Review of the application of energy storage technology in wind power grid connection control[J]. Power Grid Technology, 2015, 39(12):3360-3368.
23. Zhang Wenliang, Qiu Ming, Lai Xiaokang. Application of energy storage technology in power systems[J]. Power Grid Technology, 2008, 32(07): 1-9.
24. Xu Shouping, Li Xiangjun, Hui Dong. Review of the development status and demonstration application of large-scale energy storage system[J]. Grid and Clean Energy, 2013, 29(08): 94-100.
25. Cheng Shijie, Li Gang, Sun Haishun, et al. Application and prospect of energy storage technology in the field of electrical engineering[J]. Grid and Clean Energy, 2009, 25(02): 1-8.
26. Energy Storage Committee of China Energy Research Society, Zhongguancun Energy Storage Industry Technology Alliance. Energy Storage Industry Research White Paper[R]. 2019.
27. Polar Star Power News Network. http://news.bjx.com.cn/html/20180309/884445.shtml .
28. Swider D J. Compressed Air Energy Storage in an Electricity System With Significant Wind Power Generation[J]. IEEE Trans on Energy Conversion, 2007, 22(1): 95-102.
29. Lee S S, Kim Y M, Park J K, et al. Compressed Air Energy Storage Units for Power Generation and DSM in Korea[J]. 2007: 1-6.
30. Teleke S, Baran M E, Huang A Q, et al. Control Strategies for Battery Energy Storage for Wind Farm Dispatching[J]. IEEE Trans on Energy Conversion, 2009, 24(3): 725-732.
31. Introduction to the scientific research progress of the Energy Storage R&D Center of the Institute of Engineering Thermophysics, Chinese Academy of Sciences [EB/OL]. http://www.etp.ac.cn /jgsz/yjxt/kybm/kybm6/201109/t20110923_3353929.html
32. Wu Yuping, Dai Xiaobing, Ma Junqi, et al. Lithium-ion Battery - Application and Practice[M]. Beijing: Chemical Industry Press, 2004.
33. Pinheiro J M S, Dornellas C R R, Schilling M T *et al* . Probing the new IEEE Reliability Test System (RTS-96): HL-II assessment[J]. IEEE Trans on Power Systems, 1998, 13(1): 171-176.
34. State Grid Corporation of China. Signing ceremony of the sodium-sulfur battery joint venture project by Shanghai Electric Power Company, Shanghai Electric (Group) Company, and Shanghai Institute of Silicate, Chinese Academy of Sciences [EB/OL]. 2011-10-14[2012-09-28]. http://www. cec.org.cn/zdlhuiyuandongtai /dianwang/2011-10-14/71394.html.
35. China Power Grid. Three sodium-sulfur battery energy storage projects of Shanghai Electric Power Company passed acceptance [EB/OL]. 2013-01-09 [2013-04-05].http://www.chinasmartgrid.com. cn/news/20130109/412489. Shtml.
36. Zhang Huamin, Zhou Hantao, Zhao Ping, et al. Current status and prospects of energy storage technology research and development[J]. Energy Engineering, 2005(3): 1-7.

37. Zhang Weiyu, Zhu Wanqiu. Key technologies and current status of flywheel energy storage[J]. Journal of Electrical Engineering Technology, 2011(07): 141-146.
38. Deng Zigang, Wang Jiasu, Wang Suyu, et al. Current status of high temperature superconducting flywheel energy storage technology[J]. Journal of Electrical Engineering Technology, 2008, 23(12): 1-10.
39. Li Dehai, Wei Haigang, Dai Xingjian. Principles, applications and research progress of flywheel energy storage technology[J]. Mechanical Engineer, 2002(4): 5-7.
40. Tang Xisheng, Liu Wenjun, Zhou Long, et al. Research on flywheel array energy storage system[J]. Energy Storage Science and Technology, 2013,2(3): 208-221.
41. Energy Storage Special Committee of China Energy Research Association, Zhongguancun Energy Storage Industry Technology Alliance, Blue Book of Energy Storage Industry Development[R]. China Petrochemical Press, 2019.1.
42. GTR Magnetic Energy Technology Co., Ltd. Application of GTR flywheel in rail transit[R]. 2019.5.
43. Tang Xisheng. Research on energy management and stability of supercapacitor energy storage in distributed power generation systems[D]. Institute of Electrical Engineering, Chinese Academy of Sciences, 2006.
44. Cao Bin, Jiang Xiaohua. Application of superconducting energy storage in improving power quality[J]. Science and Technology Review, 2008, 26(01): 47-52.
45. Yves Brunet et al., translated by Tang Xisheng et al. Energy Storage Technology and Application[M]. Mechanical Industry Press, 2018.5.
46. Energy Storage Engineering Committee of China Chemical Society. Energy Storage Technology and Application[M]. Chemical Industry Press, 2018.7.
47. Ding Ming, Zhang Zhengkai, Bi Rui. Extension of CIM for Distributed Power Generation Systems[J]. Automation of Electric Power Systems, 2008, 32(20): 83-87, 96.
48. Huang Shengli. Research on Seamless Switching Technology of Microgrid Operation Mode[D]. Institute of Electrical Engineering, Chinese Academy of Sciences. 2009.
49. Leon M Tolbert, Thomas G Habetler. Novel Multilevel Inverter Carrier-based PWM Method[J]. IEEE Trans on Industry Applications, Vol. 35, No. 5, 1999, pp. 1098-1107.
50. Sun Xiaofeng, Gu Herong, Wang Lijiao, et al. High Frequency Switching Inverter and Its Parallel Grid Connection Technology[M]. China Machine Press, 2011.2.
51. Richard Zhang, V. Himamshu Prasad, Dushan Boroyevich, Fred C. Lee. Three-Dimensional Space Vector Modulation for Four-Leg Voltage-Source Converters[J]. IEEE Trans on Power Electronics, Vol. 17, No. 3, May 2002, pp. 314-326.
52. Yang Hong. Control and Implementation of Four-Leg Three-Phase Inverter[D]. Nanjing University of Aeronautics and Astronautics, 2004.
53. Zhang Chongwei, Zhang Xing. PWM Rectifier and Its Control[M]. Beijing: China Machine Press, 2003.10.
54. Li Xun, Zhu Pengcheng, Yang Yinfu, Chen Jian. Research on Three-Phase SVPWM Inverter Based on Dual Loop Control[J]. Power Electronics, 2003, 37(5):30-32.
55. Alexis B. Rey, José M. Ruiz, Santiago de Pablo *et al*. A New Current Source Control Strategy for VSI-PWM Inverters. http://www.dte.eis.uva.es/Datos/Congresos/Ipec2000.pdf
56. P. C. Krause, O. Wasynczuk and S. D. Sudhoff. Analysis of Electric Machinery[M]. John Wiley & Sons, 2nd edition, 2002.
57. Hu Xiao. Research on Energy Storage Parallel Control Technology of Islanded Microgrid[D]. Institute of Electrical Engineering, Chinese Academy of Sciences, 2014.5
58. Chandorkar M C, Divan D M, Adapa R. Control of parallel connected inverters in standalone ac supply systems[J]. IEEE Trans on Industry Applications, 1993, 29(1): 136-143.
59. Tuladhar A, Jin K, Unger T, et al. Parallel operation of single phase inverter modules with no control interconnections[C]. Applied Power Electronics Conference and Exposition, 1997. APEC'97 Conference Proceedings 1997, Twelfth Annual. IEEE, 1997, pp.94-100.
60. Peas Lopes J A, Moreira C L, Madureira A G. Defining control strategies for microgrids islanded operation[J]. IEEE Trans on Power Systems, 2006, 21(2): 916-924.

61. Cheng Junzhao, Li Shusen, Wu Zaijun, et al. Analysis of Power Decoupling Mechanism of Virtual Reactance in Microgrid Droop Control[J]. Automation of Electric Power Systems, 2012, 36(7): 27-32.
62. Liu Wenjun. Research on Advanced Control Methods of Flywheel Energy Storage System[D]. Institute of Electrical Engineering, Chinese Academy of Sciences, 2014.5
63. Zhou Long. Design Analysis and Control Method Research of Flywheel Energy Storage System[D]. Institute of Electrical Engineering, Chinese Academy of Sciences, 2009.
64. Liu Wenjun, Zhou Long, Tang Xisheng, et al. Control Method of Flywheel Energy Storage System Based on Improved Sliding Mode Observer[J]. Transactions of China Electrotechnical Society, 2014， 34(1):71-78.
65. Gao L. Design and optimization of Fuel Cell/Battery/Supercapacitor hybrid power sources for electric vehicles[D]. College of Engineering and Information Technology, University of South Carolina, 2003.
66. Dougal R A, Liu S, White R E. Power and Life Extension of Battery-Ultracapacitor Hybrids[J]. IEEE Trans on Components and Packaging Technologies, March 2002, 25 (1): 120-131.
67. Gao L, Dougal R A. Liu S. Power Enhancement of an Actively Controlled Battery/ Ultracapacitor Hybrid[J]. IEEE Trans on Power Electronics, Jan 2005, 20(1): 236-243.
68. Zheng J P, Jow T R, Ding M S. Hybrid power sources for pulsed current applications[J]. IEEE Trans on Aerospace and Electronic Systems, Jan 2001, 37(1): 288-292.
69. Pay S, Baghzouz Y. Effectiveness of Battery-Supercapacitor Combination in Electric Vehicles[C]. IEEE Bologna PowerTech Conference 2003.
70. Schupbach R M, Balda J C, Zolot M *et al* . Design methodology of a combined battery-ultracapacitor energy storage unit for vehicle power management[C]. Power Electronics Specialist Conference 2003, 2003 IEEE 34th Annual, 15-19 June 2003, pp. 88-93.
71. Wang Xiaofeng. Development of a carbon nanotube supercapacitor composite power source for GSM mobile communication[J]. High Tech Communication, 2005, 3: 56-59.
72. Wang Xiaofeng, Wang Dazhi, Liang Ji, et al. Development of double layer capacitor and its composite power system[J]. Acta Electronica Sinica, 2002, 8: 1100-1103.
73. Zhang Yinian, Hu Shuqing, Du Zhizhong. High power density composite power source[J]. Power Supply Technologies, 2002, 26(5): 341-343.
74. Akiyama K, Nozaki Y, Kudo M et al. Ni-MH batteries and EDLCs hybrid stand-alone photovoltaic power system for digital access equipment[C]. Telecommunications Energy Conference 2000. INTELEC. Twenty-second International, 10-14 Sept 2000, pp.387-393.
75. Nozaki Y, Akiyama K, Kawaguchi H et al. An improved method for controlling an EDLC-battery hybrid stand-alone photovoltaic power system[C]. Applied Power Electronics Conference and Exposition, 2000. Fifteenth Annual IEEE, 6-10 Feb 2000, pp. 781-786.
76. Nozaki Y, Akiyama K, Kawaguchi H et al. Evaluation of an EDLC-battery hybrid stand-alone photovoltaic power system[C]. Photovoltaic Specialists Conference. 15-22 Sept 2000, pp.1634-1637.
77. Tang Xisheng, Qi Zhiping. Research on active hybrid energy storage scheme of supercapacitor battery in standalone photovoltaic system[J]. New Technologies in Electric Power & Electric Engineering, 2007, 25(3): 37-41, 67.
78. Tang Xisheng, Wu Xin, Qi Zhiping. Research on supercapacitor battery hybrid energy storage standalone photovoltaic system[J]. Journal of Solar Energy, 2007, 28(2): 178-183.
79. Tang Xisheng, Li Haidong, Qi Zhiping, et al. Research on active supercapacitor-battery hybrid energy storage system[J]. High Tech Communication, 2006, 16(12): 1273-1277.
80. Tang Xisheng, Qi Zhiping. Supercapacitor battery hybrid power source[J]. Power Supply Technologies, 2006, 30(11): 933-936.
81. Zhang Guoju, Tang Xisheng, Qi Zhiping. Application of supercapacitor and battery hybrid energy storage system in microgrid[J]. Automation of Electric Power Systems, 2010, 34(12): 1-5.

82. Zhang Guoju. Research on optimization design and control method of supercapacitor/battery hybrid energy storage system[D]. Institute of Electrical Engineering, Chinese Academy of Sciences, May 2011.
83. Zhang Guoju, Tang Xisheng, Qi Zhiping. Design of a hybrid energy storage system for smoothing intermittent power source fluctuations[J]. Automation of Electric Power Systems, 2011, 35(20): 24-28, 93.
84. Zubieta L, Bonert R. Characterization of Double-layer capacitors for power electronics applications[J]. IEEE Trans on Industry Applications, 2000, 36(1): 199-205
85. Haiping Xu, Li Kong, Xuhui Wen. Fuel cell power system and high power DC-DC converter[J]. IEEE Trans on Power Electronics, 2004, 19(5): 1250-1255.
86. Cheng Xueqi, Jin Xiaolong, Wang Yuanzhuo, et al. Overview of big data systems and analysis technologies[J]. Journal of Software, 2014, 25(09): 1889-1908.
87. Dean J, Ghemawat S. MapReduce: A flexible data processing tool[J]. Communications of the ACM, 2010, 53(1):72-77.
88. White T. Hadoop: The definitive guide[M]. O'Reilly Media, Inc., 2012.
89. Storm. http://storm.incubator.apache.org/
90. Hive. https://hive.apache.org/
91. MongoDB. http://www.mongodb.org
92. Jin Heping, Guo Chuangxin, Xu Yibin, Liao Weihan. System conception and application research of energy big data[J]. Hydropower and Pumped Storage, 2019, 5(01):1-13.
93. Torres R S, Falcao A X, Gonc M A, *et al* . A genetic programming framework for content-based image retrieval[J]. Pattern Recognition, 2009, 42(2):283-292.
94. Zhu Yongli, Yin Jinliang. Application of combined kernel related support vector machine in power transformer fault diagnosis[J]. Proceedings of the CSEE, 2013, 33(22):68-74.
95. Dominik F, Thiemo G, Bernhard S. Swiftrule: mining comprehensible classification rules for time series analysis[J]. IEEE Trans on knowledge and data engineering, 2011, 23(5): 774-787.
96. Yan Yingjie, Sheng Gehao, Chen Yufeng, et al. Abnormal detection method of transmission and transformation equipment state data based on big data analysis[J]. Proceedings of the CSEE, 2015, 35(01): 52-59.
97. Cai Yu. Research on analysis and diagnosis of thermal power unit energy consumption characteristics based on big data mining[D]. Zhejiang University, 2018.
98. Wang Yang, Yu Haitao, Zhang Xu, et al. Construction and application practice of power big data basic platform[M]. Beijing: China Electric Power Press, 2016.
99. J. Hernández, I. Gyuk and C. Christensen, DOE global energy storage database — A platform for large scale data analytics and system performance metrics[C]. 2016 IEEE International Conference on Power System Technology (POWERCON), Wollongong, NSW, 2016, pp.1-6.
100. GESDB[DB] http://www.energystorageexchange.org/
101. Wei Haiyan, Chen Xiaojie, Lv Zhiqiang, et al. Online estimation of lithium-ion battery SOH based on gray neural network model[J]. Power System Technology, 2017, 41(12):4038-4044.
102. Beijing Shuangdeng Huifeng JUNENG Product Manual[S]. Beijing Shuangdeng Huifeng JUNENG Technology Co., Ltd., 2018.
103. Lead-carbon battery technology charging and discharging strategy(LLC-1000)[S]. Shuangdeng Group Co., Ltd., 2018.
104. Hu Xinguo, Wang Dianlong, Dai Changsong. Lead Carbon Battery[M]. Chemical Industry Press, 2015.
105. Tao Zhanliang, Chen Jun. Lead Carbon Battery Energy Storage Technology[J]. Energy Storage Science and Technology, 2015, 4(06): 546-555.
106. Li Fang. Research on the Adaptive Optimization Method of K-value in K-means Algorithm[D]. Anhui University, 2015.
107. Yin Jun. Optimization of K-means Clustering Algorithm and Its Application in Image Deduplication[D]. Huazhong University of Science and Technology, 2016.

108. S. Banerjee, A. Choudhary and S. Pal. Empirical evaluation of K-Means, Bisecting K-Means, Fuzzy C-Means and Genetic K-Means clustering algorithms[C]. 2015 IEEE International WIE Conference on Electrical and Computer Engineering (WIECON-ECE), Dhaka, 2015, pp.168-172.
109. Ester Martin, Kriegel Hans-Peter, Sander, Jörg, et al. A density-based algorithm for discovering clusters in large spatial databases with noise[C]. Proceedings of the Second International Conference on Knowledge Discovery and Data Mining (KDD-96). pp.226–231
110. Li Haijun. Research on Thermal Management System of Lithium Power Battery for New Energy Vehicles[D]. Jiangsu University, 2018.
111. Deng Tao, Luo Weixing. Discussion on SOH Estimation Method of Electric Vehicle Power Battery[J]. Modern Manufacturing Engineering, 2018(05): 43-49.
112. Lasseter R H. Microgrids[C]. IEEE Power Engineering Society Winter Meeting, New York USA, 2002, 305-308.
113. Lasseter R H. Certs microgrid[C]. IEEE International Conference on System of Systems Engineering, San Antonio, Texas, USA, 2007, 1-5.
114. Lu Zongxiang, Wang Caixia, Min Yong, et al. Review of Microgrid Research[J]. Automation of Electric Power Systems, 2007, 31(19) : 100-107.
115. Katiraei F, Iravani M R, Lehn P W. Micro-grid autonomous operation during and subsequent to islanding process[J]. IEEE Trans on Power Delivery, 2005, 20 (1) : 248-257.
116. Wang Chengshan, Zheng Haifeng, Xie Yinghua, et al. Random Load Flow Calculation of Distribution System Considering Distributed Generation[J]. Automation of Electric Power Systems, 2005, 29 (24) : 39-44.
117. Faridaddin K. Dynamic analysis and control of distributed energy resources in a microgrid[D]. Department of Electrical and Computer Engineering, University of Toronto, 2005.
118. Pecas L, Moreira C L, Madureira A G. Defining control strategies for microgrids islanded operation[J]. IEEE Trans on Power Systems, 2006, 21(2) : 916-924.
119. Barton J P, Infield D G. Energy storage and its use with intermittent renewable energy[J]. IEEE Trans on Energy Conversion, 2004, 19(2) : 441-448.
120. Ribeiro P F, Johnson B K, Crow M L et al. Energy storage systems for advanced power applications[J]. Proceedings of the IEEE, 2001, 89(21): 1744-1756.
121. Cheng Hua, Xu Zheng. Energy Storage Technology in Distributed Generation[J]. High Voltage Apparatus, 2003, 39(3) : 53-56.
122. Huo Qunhai. Research on Control Strategy of Inverter Type Microsource in Microgrid[D]. Institute of Electrical Engineering, Chinese Academy of Sciences, 2011.
123. Huo Qunhai, Kong Li, Tang Xisheng. Control of Microsource Inverter Unbalanced Nonlinear Mixed Load[J]. Transactions of China Electrotechnical Society, 2010, 30(15): 10-15.
124. Hsu P, Behnke M. A three-phase synchronous frame controller for unbalanced load[C]. IEEE Power Electronics Specialists Conference, Fukuoka, Japan, 1998.
125. Ningning Li, Guoju Zhang, Xisheng Tang, et al. Research on supercapacitor energy storage system based on split capacitor[J]. Power Electronics, 2011, 45(9): 18-20, 60
126. IEEE Standards Coordinating Committee 21. IEEE 1547 for Interconnecting Distributed Resources with Electric Power Systems[S]. IEEE-SA Standards Board, 2008.
127. Xisheng Tang, Wei Deng, Ningning Li, Zhiping Qi. Operation control technology of renewable energy microgrid based on energy storage[J]. Power Automation Equipment, 2012, 32(3): 99-103, 108.
128. Hao Hao, Hong Li. Research on parallel technology of inverter power supply[J]. Power Supply Technology Application, 2009, 12(2): 16-18.
129. Prodanovic M, Green T C, Mansir H. A survey of control methods for three-phase inverters in parallel connection[J]. Eighth International Conference on Power Electronics and Variable Speed Drives, 2000.
130. Tuladhar A, Jin H, Unger T, et al. Control of parallel inverters in distributed AC power systems with consideration of line impedance effect[J]. IEEE Trans on Industry Applications, 2000, 36(1): 131-138.

131. Hu Xiao, Xisheng Tang, Ningning Li et al. Virtual Impedance Based Parallel Operation of Multi-Converters in Low Voltage Microgrids[C]. International Symposium on Instrumentation & Measurement. 2014.
132. Xisheng Tang, Xiao Hu, Ningning Li *et al* . A Novel Frequency and Voltage Control Method for Islanded Microgrid based on Hybrid Energy Storages[J]. IEEE Trans on Smart Grid, 2016, 7(1):410-419.
133. Jiang Q, Hong H. Wavelet-Based Capacity Configuration and Coordinated Control of Hybrid Energy Storage System for Smoothing Out Wind Power Fluctuations[J]. IEEE Trans on Power Systems, 2013, 28(2):1363-1372.
134. Yushu Sun , Xisheng Tang , Chunzheng Tian , et al. Comparison of different control strategies for compound energy storage based on EMD[J]. Power Construction, 2017(3): 77-84
135. Xisheng Tang, Yushu Sun, Zhiping Qi. Analysis of wind power fluctuation and its impact on low frequency oscillation of power system based on HHT[J]. Power Grid Technology, 2015, 39(8): 2115-2121.
136. Yushu Sun, Xing Li, Xisheng Tang, et al. Multi-level control strategy of multiple types of energy storage for microgrid[J]. High Voltage Engineering, 2017(01): 181-188.
137. Yushu Sun, Xisheng Tang, Xiaozhe Sun, et al. Research on energy storage capacity configuration method for smoothing wind power fluctuation[J]. China Electrical Engineering Journal, 2017(S1): 88-97.
138. Jiang Q, Wang H. Two-Time-Scale Coordination Control for a Battery Energy Storage System to Mitigate Wind Power Fluctuations[J]. IEEE Trans on Energy Conversion, 2013, 28(1): 52-61.
139. Tang Xisheng, Miao Fufeng, Qi Zhiping, et al. A review of frequency regulation technology for wind power generation. Chinese Journal of Electrical Engineering, 2014, 34(25): 4304-4314.
140. Ullah N R, Thiringer T, Karlsson D. Temporary primary frequency control support by variable speed wind turbines-potential and applications[J]. IEEE Trans on Power Systems. 2008, 23(2): 601-612.
141. Lalor G, Mullane A, O'Malley M. Frequency control and wind turbine technologies[J]. IEEE Trans on Power Systems, 2005, 20(4): 1905-19135.
142. Morren J, Haan S W H, Kling W L *et al* . Wind turbines emulating inertia and supporting primary frequency control[J]. IEEE Trans on Power Systems, 2006, 21(1): 433-434.
143. Cao Zhangjie. Research on control strategy of doubly-fed induction generator participating in system primary frequency regulation[D]. Chengdu: Southwest Jiaotong University, 2009.
144. Miller N W, Clark K, Shao M. Frequency responsive wind plant controls: Impacts on grid performance[C]. 2011 IEEE in Power and Energy Society General Meeting. San Diego, CA, USA. 2011: 1-8.
145. Zhang Zhaosui, Sun Yuanzhang, Li Guojie, et al. Frequency control of doubly-fed wind power unit coordinated with overspeed and variable pitch[J]. Automation of Electric Power Systems, 2011, 35(17): 20-25.
146. Wu Ziping. Gao Wenzhong, Wang Jianhui *et al* . A coordinated primary frequency regulation from permanent magnet synchronous wind turbine generation[C]. IEEE in Power Electronics and Machines in Wind Applications (PEMWA). Denver, Co, USA. 2012, pp.1-6.
147. Xue Yingcheng, Tai Nengling, Song Kai, et al. Research on frequency regulation reserve capacity provided by variable speed wind power generator[J]. Electric Power Automation Equipment, 2010(8): 75-80.
148. Xiang Rong, Wang Xiaoru, Tan Jin. Operation control of flywheel energy storage system with wind farm[C]. 2011 30th Chinese in Control Conference (CCC). Yantai, China, 2011, pp.6208-6212.
149. Sun Chunshun, Wang Yaonan, Li Xinran. Integrated power and frequency control of wind power system assisted by flywheel[J]. Chinese Journal of Electrical Engineering, 2008, (29): 111-116.

150. Abed N, Teleke S, Castaneda J. Planning and operation of dynamic energy storage for improved integration of wind energy[C]. 2011 IEEE in Power and Energy Society General Meeting. San Diego, CA, USA. 2011, pp.1-7.
151. Muyeen S, Hasanien H, Tamura J. Reduction of frequency fluctuation for wind farm connected power systems by an adaptive artificial neural network controlled energy capacitor system[J]. IET Renewable Power Generation, 2012, 6(4): 226-235.
152. Thatte A, Fan Z, Le X. Coordination of wind farms and flywheels for energy balancing and frequency regulation[C]. IEEE Power and Energy Society General Meeting. San Diego, CA, USA. 2011, pp.1-7.
153. Tang Xisheng, Miao Fufeng, Qi Zhiping, et al. A coordinated control method for wind-storage cluster[P]. China, 201210477712. 3. 2013.04.10.
154. Xia Yuhang, Liu Junyong. A review of research on virtual power plant based on distributed generation[J]. Electric Power Automation Equipment, 2016, 36(04): 100-106, 115.
155. Wei Zhinong, Yu Shuang, Sun Guoqiang, et al. The concept and development of virtual power plant[J]. Automation of Electric Power Systems, 2013, 37(13): 1-9.
156. Wei Zhinong, Yu Shuang, Sun Guoqiang, et al. Review of European research projects on virtual power plant[J]. Automation of Electric Power Systems, 2013, 37(21): 196-202.
157. https://www.aggreko.com/en-us/microgrid-and-storage-solutions
158. https://limejump.com/balancing-mechanism/
159. National Development and Reform Commission. Construction of the "source-grid-load-storage" friendly interactive system to enhance the power grid operation management capability. http://www.ndrc.gov.cn/fzgggz/jjyx/mtzhgl/201809/t20180927_899181.html.
160. Mashhour E, Moghaddas-Tafreshi S M. Bidding Strategy of Virtual Power Plant for Participating in Energy and Spinning Reserve Markets-Part I: Problem Formulation[J]. IEEE Trans on Power Systems, 2011, 26(2): 949-956.
161. Mashhour E, Moghaddas-Tafreshi S M. Bidding Strategy of Virtual Power Plant for Participating in Energy and Spinning Reserve Markets-Part II: Numerical Analysis[J]. IEEE Trans on Power Systems, 2011, 26(2): 957-964.
162. Liu Ran. Research on the Optimal Dispatch of Virtual Power Plant[D]. North China Electric Power University, 2018.
163. Xia Yuhang, Liu Junyong, Feng Chao, et al. Optimal Dispatch Model of Virtual Power Plant Considering Demand Response[J]. Power System Technology, 2016, 40(06): 1666-1674.
164. Zhou Yizhou, Sun Guoqiang, Huang Wenjin, et al. Bidding Model of Virtual Power Plant in Multi-class Electricity Market Considering Electric Vehicles and Demand Response[J]. Power System Technology, 2017, 41(06): 1759-1767.
165. Yuan Guili, Chen Shaoliang, Liu Ying, Fang Fang. Economic Optimal Dispatch of Virtual Power Plant Based on Time-of-use Electricity Price[J]. Power System Technology, 2016, 40(03): 826-832.
166. Ortega-Vazquez M A, Kirschen D S. Optimizing the Spinning Reserve Requirements Using a Cost/Benefit Analysis[J]. IEEE Trans on Power Systems, 2007, 22(1): 24-33.
167. Sun Guoqiang, Zhou Yizhou, Wei Zhinong, et al. Robust Optimization Model of Cogeneration Scheduling of Virtual Power Plant in Energy and Spinning Reserve Markets[J]. Proceedings of the CSEE, 2017, 37(11): 3118-3128, 3367.
168. Zhang Shang, Gu Xueping, Wang Tao. Static Security Correction Control Method for Wind Power System Based on Distributed Robust Optimization[J]. Electric Power Automation Equipment, 2019, 39(02): 58-64.
169. Zhang Liudong, Yuan Yubo, Sun Dayan, et al. Wind-storage Coordinated Operation Scheduling Model Based on Two-stage Distributed Robust Interval Optimization[J]. Electric Power Automation Equipment, 2018, 38(12): 59-66, 93.